Cold War Cities
Politics, Culture and Atomic Urbanism, 1945–1965

Edited by Richard Brook, Martin Dodge and Jonathan Hogg

LONDON AND NEW YORK

First published 2021
by Routledge
2 Park Square, Milton Park, Abingdon, Oxon OX14 4RN

and by Routledge
52 Vanderbilt Avenue, New York, NY 10017

Routledge is an imprint of the Taylor & Francis Group, an informa business

© 2021 selection and editorial matter, Richard Brook, Martin Dodge and Jonathan Hogg; individual chapters, the contributors

The right of Richard Brook, Martin Dodge and Jonathan Hogg to be identified as the authors of the editorial material, and of the authors for their individual chapters, have been asserted in accordance with sections 77 and 78 of the Copyright, Designs and Patents Act 1988.

All rights reserved. No part of this book may be reprinted or reproduced or utilised in any form or by any electronic, mechanical, or other means, now known or hereafter invented, including photocopying and recording, or in any information storage or retrieval system, without permission in writing from the publishers.

Trademark notice: Product or corporate names may be trademarks or registered trademarks, and are used only for identification and explanation without intent to infringe.

British Library Cataloguing-in-Publication Data
A catalogue record for this book is available from the British Library

Library of Congress Cataloging-in-Publication Data
A catalog record has been requested for this book

ISBN: 978-1-138-57361-1 (hbk)
ISBN: 978-0-203-70147-8 (ebk)

Typeset in Times New Roman
by codeMantra

Contents

Acknowledgements viii

Cold War cities: spatial planning, social and political processes, and cultural practices in the age of atomic urbanism, 1945–1965 1
RICHARD BROOK, MARTIN DODGE AND JONATHAN HOGG

PART I
Planning the Cold War city 35

1 **Properties of science: how industrial research and the suburbs reshaped each other in Cold War Pittsburgh** 37
   PATRICK VITALE

2 **The city of Bristol: ground zero in the making** 55
   BOB CLARKE

3 **Towards a prosperous future through Cold War planning: Stalinist urban design in the industrial towns of Sillamäe and Kohtla-Järve, Estonia** 77
   SIIM SULTSON, PHD

4 **Nuclear anxiety in postwar Japan's city of the future** 97
   SEBASTIAN SCHMIDT

   Visual essay: Urbanism of fear: a tale of two Chinese Cold War cities 115
   TONG LAM

PART II
Building the Cold War city   123

  5 **The Warsaw metro and the Warsaw pact: from deep cover to cut-and-cover**   125
     ALEX LAWREY

  6 **Competing militarisation and urban development during the Cold War: how a Soviet air base came to dominate Tartu, Estonia**   148
     DANIEL BALDWIN HESS AND TAAVI PAE

  7 **In-between the East and the West: architecture and urban planning in 'Non-Aligned' Skopje**   167
     JASNA MARIOTTI

  8 **Atomic urbanism under Greenland's ice cap: Camp Century and cold war architectural imagination**   182
     KRISTIAN H. NIELSEN

     Visual essay: Welfare or Warfare? Civil defence in Danish urban welfare architecture   201
     ROSANNA FARBØL

PART III
Culture and politics in the Cold War city   217

  9 **Urban space, public protest, and nuclear weapons in early Cold War Sydney**   219
     KYLE HARVEY

10 **In the middle of the atomic Arena: visible and invisible NATO sites in Verona during the fifties**   236
     MICHELA MORGANTE

11 **Conceiving the atomic bomb threat between West and East: mobilisation, representation and perception against the A-bomb in 1950s Red Bologna**   256
     ELOISA BETTI

12 **Making a 'free world' city: urban space and social order in Cold War Bangkok**   281
     MATTHEW PHILLIPS

**Visual essay: Cold War telecommunication and urban
vulnerability: underground exchange and microwave tower
in Manchester** 301
MARTIN DODGE AND RICHARD BROOK

Index 317

Acknowledgements

The genesis of this book was a series of conference sessions convened by Richard Brook and Martin Dodge at the International Conference of Historical Geographers in London in the summer of 2015. Jonathan Hogg was one of the speakers in the sessions and in conversations afterwards agreed to participate in the project to edit a book on the theme of Cold War cities. From the conference sessions, several speakers developed their papers into chapters in this book. Other original chapters and visual essays were solicited by the editors.

All the editors wish to extend their thanks to all the contributors to the book for their great forbearance. There have been many lengthy delays in getting their work into print, but we really hope the end result has been worth the wait! We really appreciate their contributions.

Thanks to the Routledge editorial team, particularly Faye Leerink, who have been very supportive of the project.

Richard acknowledges the support of the Manchester School of Architecture and extends his thanks to all of the archivists who make historical research possible.

Every effort has been made to contact copyright holders of historic images used in the chapters in this book. The editors will be pleased to hear from any copyright holders who were not contacted and full acknowledgement will be given in a future edition.

Cold War cities
Spatial planning, social and political processes, and cultural practices in the age of atomic urbanism, 1945–1965

Richard Brook, Martin Dodge and Jonathan Hogg

MANCHESTER SCHOOL OF ARCHITECTURE; UNIVERSITY OF MANCHESTER; UNIVERSITY OF LIVERPOOL

Introduction

The period 1945–1965 was an era of massive urban growth, with significant redevelopment and reconstruction of cities across much of the world. Its pace was driven by the need for essential inner-city renewal after the Second World War in several industrial countries. This met with demands of welfare paternalism and calls for better housing provision, the emergence of more interventionist planning underpinned by scientific rationality, the rise of suburban consumerism in the West, Soviet directed socialist renewal in the East and processes of decolonisation in the Global South. During this time, many urban spaces were conceived, designed and built in response to these conditions; however, here we are more concerned about how cities also received – or were reshaped to fit – the contours of the global Cold War.

Cold War Cities reconsiders the complicities and complexities of the Cold War read through the analytical lens of urbanism. This agenda is inspired by recent scholarship which has moved focus away from a handful of cities of symbolic importance, choosing instead to map the wider Cold War urban experience in more diverse settings. In our conceptualisation of the book, the aim is to say something novel for Cold War history by focussing the discussions at the level of individual cities rather than larger scales, typical of many existing Cold War scholarly narratives. We believe that the city scale sheds new light on existing debates that have become stale when worked through in terms of other spatial levels of analysis, such as ideologically opposed blocs, the superpowers and their closely allied nation states, continental-wide containment and hemispherical geopolitical struggles, global risks and extra-territorial strategising. Second, we hope the range of original contributions solicited for the book can say something interesting for urban studies, town planning and architecture by deliberately foregrounding the Cold War contexts in the crucial decades of urban growth and change after the Second World War, showing how existential threats of atomic warfare led to efforts to reshape cities. In contrast, many cities were not reshaped in the atomic age, despite political rhetoric and large-scale civil defence initiatives worldwide.

For us, the notion of 'atomic urbanism' is central to understanding this Cold War dynamic, and it is concerned with how the threat of nuclear attack affected perceptions of city life and informed the planning of space in the Cold War era. In *Cold War Cities*, we ask a series of questions that seek to uncover some of the physical and material effects of atomic urbanism on everyday living. These include how did the perceived threat of nuclear war through the 1950s and into the 1960s, affect planning at a range of geographic scales? What were the patterns of the built environment, architectural forms and material aesthetics of atomic urbanism in different places? Moreover, how did the 'Bomb' manifest itself in political processes and civic governance, and through cultural practices and portrayal in popular media, arts and academia?

In planning the book, we concluded that a case studies approach was a feasible way to survey diverse settings for atomic urbanism. We appreciate that there is explanatory power in detailed examination of cities as singular entities and encouraged authors to foreground the particularities of places and distinctiveness of historically embedded contingent processes. The cities under examination are diverse in population size, physical territory and political economy, which mean they can provide unique, localised understandings of how global Cold War doctrines enrolled in atomic urbanism played out in different socio-political and cultural contexts. While not explicitly requiring comparative analysis, we did also push chapter authors to draw relevant connections to wider issues in post-war urbanism and debates in Cold War historiography.

Presented here are 12 original chapters which represented a fascinating and diverse selection of cities; they came from targeted invitations to established scholars, a number of volunteered chapters (often from early career researchers) and some serendipitous contributions. There was no top-down schema from us as editors to generate a representative coverage, for instance, seeking an equal split between Communist and capitalist political economies, or having equal coverage of each continent. We acknowledge that the result is disparate in terms of coverage and there are obvious gaps – we have no case study city within the Soviet Union itself, nor on the continents of Africa or South America, and no analysis of what would often be regarded as the iconic cities of Cold War geopolitics.[1] Yet, the range of material presented in this volume provides new insights into atomic urbanism. Many of the 12 different cities analysed in this volume are situated *beyond* the global superpowers and can be thought of as representing more liminal urban territories – particularly those on the blurry edges in geopolitical terms between the eastern and western blocs. While we did not necessarily seek this disparate geographical focus, it is interesting that these previously overlooked urban places have attracted the attention of the scholars who have unpicked their stories. As a consequence, the narratives of 12 individual cities start to challenge the conventionally held ideological binaries, such as east-west, or communism-capitalism. Although there is bias to European experiences with seven out of the 12 chapters examining cities on this continent, along with two out of three of the visual essays, the book is also biased in favour of cities in capitalist economies with consideration of only three places in the communist bloc and two – Bangkok and Skopje – that might be considered 'non-aligned'.

Besides the geographical specificity and city scale agenda, the book has a defined time frame on the early Cold War decades, although some chapters do necessarily have narratives that flow beyond these temporal boundaries, looking at the situation before the Second World War and the legacies of decision-making after the 1960s. The years 1945–1965 have interpretative coherence in Cold War historiography, and these two decades were a pivotal time in urban redevelopment globally. The cut-off date in the mid-1960s has relevance marking the end of what is often regarded as a junction point in post-war history and the phases of the Cold War, with the 1970s and 1980s marked by different political economic forces, urban pressures and changing strategic nuclear relations between the superpowers (Painter, 1999).

As with any historical urban study, a change in geographic scale can change the way in which past events and long-standing processes are viewed. While the finely nuanced situations of a particular place cannot wholly explain the broader geo-political conditions of a period, they do, however, offer valuable details of the lived experience and the on-the-ground conditions of those affected by the greater narratives unfolding around them.

This publication is also an opportunity to illustrate some of the different types of the urban Cold War using a range of perspectives. The chapters presented here are from a range of social science and cognate humanities disciplines, and draw on a range of different evidential bases, archival research, methods, media hermeneutics, and personal histories and lived experiences. The inclusion of several visual essays narrates the physical manifestation and aspects of lived experience of atomic urbanism through historical photography and other visual sources.

The 12 chapters and three visual essays in *Cold War Cities* are organised into three sections based on broad thematic labels: 'Planning the Cold War City', 'Building the Cold War City' and 'Culture and Politics'. In these sections, the contributors focussed their consideration on how 'atomic urbanism' was visible in terms of (1) location planning, (2) aspects of architectural design or (3) with regard more to spatiality of politics and civic life, and their expression through popular culture in the Cold War city.

4 *Richard Brook et al.*

Figure I.1 A Cold War vision of a dispersed North-American city, with pro-active planning generating a spatial structure to be resilient to nuclear attack. This was the central illustration in an article 'How U.S. cities can prepare for atomic war' printed in *Life* (18 December 1950), a leading and widely read American news magazine of that era. The sketch was drawn by A. Leydenfrost, a well-known illustrator for space-flight and speculative fiction.

Researching urbanism in the Cold War

The renewed interest in the early Cold War and its multi-scaled histories and geographies has highlighted how much scholarship concerning this time period has traditionally focussed on international relations, continental blocs and global geopolitical struggles.

There is some historical work published on Cold War planning, politics and culture at the national level, which is often strongly urban in focus (e.g. Farish, 2010; Stupar, 2015; Zipp, 2010). However, within the broadening and decentring trend in Cold War historiography, there has been less analysis of the city as a context in which physical plans, social politics and cultural practices played out in distinctive ways. Besides this volume, there are two notable recent exceptions: first, an excellent themed issue of the journal *Urban History*, November 2015, edited by Matthew Farish and David Monteyne and entitled: 'Histories of Cold War Cities'; second an edited collection entitled *Cold War Cities: History, Culture and Memory* edited by Katia Pizzi and Marjatta Hietala (2016). At one level, these two publications make it clear that the sheer diversity of the stories of Cold War cities is still largely untold.

The *Urban History* themed issue has seven diverse papers on issues including civil defense, international diplomacy, architectural design and urban renewal. The places covered are largely non-core, second tier cities, such as Baltimore and Liverpool, along with two in the Global South (Buenos Aires and Managua, Nicaragua). In their intellectual conception of the Cold War city, Farish (a Canadian geographer) and Monteyne (a Canadian architectural historian) argue that the 'making of a Cold War city was a point of contention', and so it is necessary to connect 'the complex social, economic, political and cultural histories of the Cold War to specific urban histories' (Farish and Monteyne, 2015, p. 546). We strongly concur with this sentiment. Furthermore, they point to the political epicentres of the Cold War – Berlin, Washington DC and Moscow – as typically explored and make the case for 'papers which ground their arguments in urban spaces, policies, practices and the everyday lives of residents' (p. 544), a point also made by Pizzi and Hietala in the introduction to their edited book. The chapters here assembled contribute further to this gap in knowledge and address these conditions in a variety of ways.

Pizzi and Hietala's edited book is focussed on history, culture and memory (as signalled by its subtitle), considering, via 11 urban case studies, the 'rich fabric of memories and cultural intersections that made up the lived experience of individuals and communities during the long Cold War' (p. 14). The conception here is that 'the Cold War left indelible marks on the urban fabric of Europe and the wider word', and textual and physical sites were 'where Cold War dynamics, both overt and covert, are powerfully played out' (p. 14). Their book largely focusses on the cultural Cold War, as seen primarily from European capital cities, and says little on the material form of cities, nor their planning and changing layouts and architecture forms. Much of the analysis focusses on contemporary commemoration of the Cold War and the legacies of conflict and how this is recalled in personal memories, and collective memory.

In terms of scholarly analysis of physical planning and the built environment during the early Cold War period, perhaps too much attention has been focussed on the formal structures of state defence, for example, in the typological analysis of regional command centres, large military bunkers, bomber bases and the fascinating paraphernalia of early warning systems (e.g. Clarke, 2009; McCamley, 2013). One of the most comprehensive and scholarly efforts undertaken was for UK with Wayne Cocroft's (2003) encyclopaedic volume charting the whole range of state-funded Cold War military spaces and built infrastructures. Indeed, technical interest can lead to fetishisation for the 'secret' spaces and abandoned bunkers of the Cold War, typified by photographic projects (e.g. Catford, 2010) and unofficial Urbex-style documentation. For the United States, Michael Kubo (2009) has analysed the architecture of the RAND Corporation in relation to the technological advances of the Cold War; as he noted 'the RAND model was intimately related to the dynamics of the Cold War military-industrial complex' (p. 50) and its 'environment emerged' as 'a reflection of the delicate balance of tensions at the heart of the strategic Cold War mentality in an era of containment' (p. 24). Looking at one iconic city, Samuel Zipp has examined the social and formal impact of Cold War policy working in harness with calls for comprehensive renewal on the urban fabric of New York (Zipp, 2010). His findings show that certain political powerbrokers thought that 'A renewed Manhattan could project an image of modernization and prosperity to compete with the equally grandiose vision of progress simultaneously motivating the Soviet Union.' (p. 5) He contrasts this with the lived and built experience that had alternative outcomes: 'If proponents had envisioned urban renewal as a Cold War Bulwark, shoring up the nation's readiness for John F. Kennedy's 'long twilight struggle', those who had to live with its interventions increasingly saw it as a liability in that contest precisely because they came to associate it with their fears about the Cold War enemy' (pp. 5–6). In a similar vein Jennifer Light's (2005) *From Warfare to Welfare: Defense Intellectuals and Urban Problems in Cold War America* provides a powerful analysis of urban policy in the Cold War era in the North American context. She notes how 'In a climate of concerns about reducing urban vulnerability to atomic attack, military strategists, urban planners, atomic scientists, social welfare advocates and local government officials came together for a sustained conversation about improving the nation's physical and social infrastructure in the postwar period' (p. 3).

Research that has focussed on aspects of space planning, politics and cultural discourse typically operate at the scale of the nation state. For example, the work of historians of nuclear cultures such as Gabrielle Hecht's (2011) *Entangled Geographies: Empire and Technopolitics in the Global Cold War* that considered Westad's [2005] conclusion that 'the most important aspects of the Cold War were neither military nor strategic, nor Europe-centered, but connected to political and social development in the Third World' (p. 6) by using techno-politics as a means to investigate the persistence of colonial structures, statecraft with alternative spatial, temporal and political scales. Jonathan Hogg's (2016) *British Nuclear Culture* sought to trace the contrast between official and unofficial nuclear narratives in Britain and to begin the work of analysing local, regional and national manifestations of the nuclear state. A special issue of *Contemporary British History* co-edited by Jonathan Hogg and Kate Brown (2019) offered a series of articles that focussed on the social history of nuclear mobilisation, with specific focus on the history of emotions, the planning of nuclear infrastructure, urban history, Welsh nuclear culture and transnational links in the Cold War era. The political economy of the Cold War and relations to urban planning, governance and national politics are examined in volumes such as historian of science and technology David Edgerton's (2005) *Warfare State* in which he asks why ideas of the 'welfare state' and 'declinism' prevailed in the historiography of twentieth-century Britain and what an alternative lens of the 'warfare state' might have to offer to the 'dichotomies and other binary oppositions which have been central in understanding the British state' (p. 7). In *The Secret State: Preparing For The Worst 1945–2010*, constitutional historian Peter Hennessy's (2010) deep reading of archival records allows a dissection of the processes of decision-making at times of crisis in Britain. Matthew Grant (2009), *After the Bomb: Civil Defence and Nuclear War in Britain, 1945–68*, follows a similar path looking at elite decision-making processes, based predominantly on British government archival sources.

The role of the Cold War as an economic driver in the production of cities has also been explored, for example, by Margaret Pugh O'Mara in *Cities of Knowledge: Cold War Science and the Search for the Next Silicon Valley* (2004). Her case study around big science and massive military technological investment in America demonstrates how the 'pork-barrel politics of Cold War spending became a major driver of a massive shift of population and employment from the Northeast and Midwest – the Rustbelt – to the Sunbelt states of the South and West' (p. 54). There is also Kate Brown's (2015) *Plutopia: Nuclear Families, Atomic Cities, and the Great Soviet and American Plutonium Disasters* in which she makes a case for the fear of nuclear war turning states on their own citizens in the name of defence. Brown analyses two cities – Richland, Washington, and Ozersk, Russia – to show how preparedness for war created both welfare and warfare in each place – subsidised housing and good health care contrasted with a surveillance culture and radioactive contamination wrought of the same imperative. Existing work has been heavily focussed on the US context, but the Cold War impacted differentially around the globe. There is less research on the ripple effect of policies and technologies on cities not at the core of the conflict and an absence of a comparative discourse with which to further explore these relations.

Jeffrey Engel's (2007) edited collection, *Local Consequences of the Global Cold War*, has value in that it demonstrates the utility of researching specific local conditions and the effects of the Cold War at an alternative scale – albeit that the essays tend towards focus on local political and socio-economic situations rather than physical change or aesthetic production.

Pressures and processes influencing Cold War cities

Cities were evidentially prime target as the brutal experience of aerial bombing during the Second World War had shown in Europe and, especially, in Japan. The vulnerability of urban populations could only be exacerbated in subsequent years. In *Cold War Cities*, we seek to explore the ways in which cities were affected by preparedness for atomic conflict and how the social, economic and political pressures were reflected in the changing shape and culture of the case study cities. While the existence of nuclear bombs had been revealed in the dreadful denouement to the Second World War in August 1945, it was unclear through most of the rest of the decade whether atomic attacks were a real risk that could and should be planned for.

The stakes were dramatically escalated by the surprise testing of the first Soviet bomb in August 1949. Despite this, advocates of greater action in preparing cities for a new kind of warfare still struggled to focus decision-makers on the real possibility of atomic war. The post-war period was socially and politically turbulent in many countries, and there were myriad potential geopolitical trigger points and superpower confrontations that could have 'gone nuclear' (Table I.1). Many people feared that local disputes and proxy wars – such as the Berlin Blockade or conflict on the Korean peninsula – would inevitably escalate to the use of atomic weapons. There were also concerns about unstable leadership and aggressive militaristic impulses that could make pre-emptive 'first-strikes' with atomic weapons seem like a plausible strategy; '[t]he concentration of great power in a few fallible hands

8 *Richard Brook et al.*

Table I.1 Some of the key in events in early Cold War

Hiroshima destroyed by atom bomb	6 August 1945
Berlin Blockade starts	June 1948
Soviet Union tests its first atom bomb	August 1949
Korean war starts	1950
Test of first hydrogen bomb	1952
Sputnik launch	4 October 1957
American U2 spy plane shot down over the Soviet Union	1 May 1960
Berlin wall building commences	August 1961
Cuban missile crisis	1962

makes possible unilateral initiation of nuclear holocaust through mistake, irrationality or satanical calculation' (Clayton, 1960, p. 112). As nuclear weapons proliferated – Britain successfully tested its first atomic bomb in October 1952 – the risk of incidents that could unleash rapid response from strategic forces that were kept on high alert became more widespread. Many countries, such as UK and other states in Western Europe, would have been the likely battlefield, with their major cities being prime targets. False alerts, accidents with the handling of atomic bombs and sub-critical nuclear weapons explosions were actually worryingly common throughout the Cold War, although the incidents were usually cloaked in deep secrecy at the time so as not to undermine deterrence and unsettle public opinion about atomic weapons (cf. Schlosser, 2013).

In preparing for atomic attacks, the amount of warning time available to cities and citizens was important. At the start of the Cold War, the approach of enemy bombers could potentially be spotted by radar and such a signal could alert civil defences with several hours warning. However, better targeting and the rapid development of missile technology through the 1950s, including those with intercontinental range – heralded dramatically by USSR's launch of Sputnik in 1957 – meant a reduction in warning time from hours to minutes, or, perhaps, no warning at all of an attack in a surprise 'first strike' strategy. In this context, how would it be possible for cities to predict and prepare? There would be no possibility of evacuation of urban populations being enacted – would people even have time to get to the nearest air raid shelter?

Atomic urbanism was clearly an existential challenge, but it was unclear for city planners how to respond effectively. Framed by Cold War doctrines and an omnipresent, but often distant, threat of destruction, politicians and civic officials managed risk in different ways and offered up various ideologically informed visions of urban development in the 1950s. In North America, there were competing pressures between continuing economic concentration in large metropolitan cores and a perceived logic of dispersing strategically important industries and essential infrastructures. As the Board of Governors of the American Institute of Planners noted in 1953:

Somewhere between these two extremes is a form of urban organization that will provide an optimum combination of immunity to damage from airborne weapons and efficiency and economy in producing goods, services, and amenities of modern urban living. It is a prime responsibility of the science of city planning, working with other technologies concerned with urban development and national defense, to define that form of organization and to develop the procedures by which it may be obtained.

(American Institute of Planners, 1953, p. 3)

Understanding and describing how cities would be likely damaged by an atomic bomb was a preoccupation in the early Cold War years. The reference, in the quote above, to 'the science of city planning' is one indicator of how warfare and the threat of destruction influenced the quantification of urbanism as it moved away from the beaux-arts towards a form of systematised rationalism. Various statistical charts of destruction and pseudo-scientific imagery were deployed in public communication of results; a common approach was to envision the blast impact as concentric rings (Figure I.2). Such diagrammatic explanations were often presented on an abstracted space, or in other cases, the circular zoning was overlaid onto a map of an actual city. There was some direct learning translated from detailed US and British military surveys of Japanese cities of Hiroshima and Nagasaki (e.g. Anon., 1946), along with simulation and model scenarios informed by atomic bomb testing (cf. Kirk, 2012). In reality, the pattern of damage experienced would have been more variable on different cities depending on topography, morphology and building types. Harder to understand and draw would be the dispersal pattern of radiation from fallout with widely varying intensities and dangers to human health, influenced in part by the weather patterns at the time of the attack. Contours of contamination could not be easily predicted nor adequately planned for; this invisible and lethal force became one of the most imponderable aspects of atomic urbanism.

While there were lots of uncertainty in how to translate and technically display models of destruction upon particular urban contexts, there were many more evocative images of devastation deployed to shock the general public. For example, the use of before and after photographic montages showed how familiar sites would appear after being obliterated by an atomic bomb – one such influential example was written by Manhattan Project scientist Ralph Lapp in 1949, entitled *Must We hide*. There were also images in circulation of typical American housing being blown apart in seconds from the 'Doom Town' tests in 1953 (Figure I.3). However, arguably the most potent vision of the destructive heart of atomic urbanism was the stark unreality and alien looking mushroom clouds superimposed above the centre of well-known cities. The photographer Richard Ross's view is that, '[i]f you put the images of people's fears into a visual representation, the clear winner is the mushroom cloud – the poster child of the Cold War' (Ross, 2004, pp. 12–13). Multiple versions

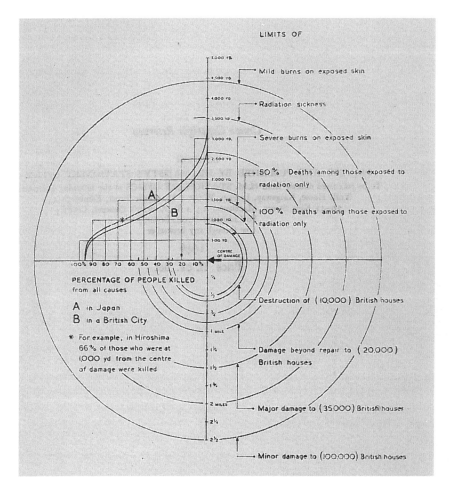

Figure I.2 The abstract concentric ring diagram of destruction and damage was common in official literature that sought to explain the impact of atomic weapons on city spaces. Here direct correspondences are being made between the evidence from Hiroshima and Nagasaki and the impact on a typical British city.

Source: author's scan from *The Effects of the Atomic Bombs at Hiroshima and Nagasaki* (HMSO, 1946, p. 21).

were presented in the media; perhaps, the most emotive and spectacular were the ones of Manhattan drawn by talented sci-fi illustrator Chesley Bonestall and published in mass market magazine *Collier's* in 1950 (cf. Jacobs and Broderick, 2012). As Page (2008, p. 108) noted, during the 1950s 'in numerous movies, apocalyptic novels, federal disaster scenarios, and planning documents,

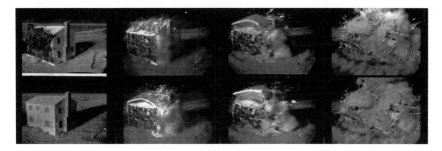

Figure I.3 The complete destruction of a typical American home in a couple of seconds caused by a nuclear test explosion in March 1953. The bomb yield was 16 kilotons.

Source: Courtesy of the National Nuclear Security Administration/Nevada Field Office; source: https://www.flickr.com/photos/nnsanevadasiteoffice/albums/72157699569032084.

Americans were reminded that New York's destruction was an inevitable parts of perhaps an unstoppable nuclear war'.

Defending cities in the Cold War

There were various strands of what might be termed active defence and overt deterrence – such as intelligence gathering, detection systems, fighter aircraft and interceptor missiles and propaganda around atmospheric nuclear weapons testing – that feed into the thinking of how to defend urban centres. However, more relevant to atomic urbanism was raft of passive activity that sought a reduction of vulnerability of cities to attack (Table I.2).

Table I.2 Passive defences in atomic urbanism

- Changing the organisation of cities to make them less target rich and more resilient: dispersal, strategic routes for motorways and the provision of wider roads
- Improving building design standards to minimise destruction
- Measures to ensure the survival of key aspects of government, industry and culture: target hardening, bypass cities, duplication and redundancy, fall-back systems and positioning key facilities underground
- Population survival strategies: evacuation and shelters, protective devices
- Civil defence preparation: distributed volunteer organisation for rescue and recovery, emergency medical personnel and facilities
- Provision of facilities and material resource to aid post-attack recovery: stockpiling of fuel, food, equipment and building materials

De-concentration of cities and dispersal planning

As strategic atomic weapons technologies improved, the capability to annihilate urban populations became central to the political-military strategy of deterrence. Major cities across the world were considered to be the primary targets of atomic weaponry in the Cold War period, and city planners, especially in North America, considered decentralised and dispersed city design as a precaution against such an attack and means to ensure better survivability.

A major feature of understanding atomic urbanism in early Cold War period is the need to document how ideas for dispersal planning of cities were promulgated. There were serious papers in specialised journals of urban planning (e.g. papers by Tracy Augur), output from military funded think-tanks like RAND and MITRE (cf. Light, 2005), analysis in substantial monographs (e.g. Lapp, 1949) and reports from major US-government funded studies (prominent was Project East River, cf. Monson, 1953). The need for new cities that were planned in different ways was also featured in mainstream news reporting. Popular publications often gave a platform to leading thinkers on Cold War strategy, such as Hermann Kahn's piece in *Fortune* magazine, 'Civil defense is possible', 1958, or the major feature in *Life* magazine in 1950. In the latter, the journalist reported academic research by Norbert Wiener at MIT – writing of the scientific underpinnings for urban dispersal and by-pass strategies[2]. The article discussed what a new kind of urbanism of deconcentration might look like, illustrated by a large double spread showing a birds-eye view of an imaginary city (Figure I.1). Wide axial roads would serve as routes for evacuation and escape, and support facilities would be located within an outer ring of highways and railroads. The result would be a kind of 'life belt' for the city, with requirements to leave open land to provide a safety zone around the existing core. (It is evident now, with hindsight, that some of this kind of dispersal replanning was happening as an elemental part of ongoing socio-economic changes in many American industrial cities: middle-class white flight from inner neighbourhoods and car-dependent ex-urban sprawl.) The *Life* article acknowledged that it would be expensive to build and tough to achieve politically; '[t]he cost of constructing the belts in terms of dollars and man-power would be huge – right-of-way for the tracks alone can run to $1 million per mile – and probably will bring the greatest opposition to the plan. ... [but] the price of not building the rail and highway rings could be even greater. Wiener thinks that price might be national disaster' (Anon., 1950, p. 81). The piece was realist in outlook, acknowledging the lack of progress in translating ideas into action in the United States thus far, but pointing to the positive benefits of deeper strategic replanning of urban-industrial society:

> [t]he blunt truth at this moment is that not one American city has so far made even a fair start toward minimum preparedness. Even more

disturbing is the possibility that local and federal plans now waiting to be put into effect are nothing more than halfway measures, ... aimed at protecting people only from the direct effects of a bomb burst. Real preparedness must reckon with what happens to the people in the days and weeks which follow the dropping of the bomb. These consequences are not only the most dangerous but are also the most preventable
(Anon., 1950, p. 77)

Similar sentiments were expressed by Tracy Augur, an influential consultant, who served as President of the American Institute of Planners and wrote widely in the late 1940s and early 1950s on the need for urban dispersal to counter atomic threats. Augur noted not only the failure to replan for defence at the start of the atomic age in America, but also the gathering need for more interventionist action in America because 'the form and size and location of our cities is a matter of national concern, to be set by the mandates of national welfare rather than the whims of individual builders' (Augur, 1948, p. 29). Augur advocated that dispersal planning could be a win–win for national defence and also economic performance. He argued that urban density, necessary for industrialism of the nineteenth century, was giving way to a new economy of services and consumerism that needed more flexible settlement patterns, supported by new forms of transport and telecommunication. '[A] sound program of dispersal for our cities would improve our position for a life of peace at the same time that it improved our chance of averting war and of surviving one if it came' (p. 31). The urban form Augur envisioned was for clusters of smaller cities (of around 50,000 people) that were spatially separated but physically linked by transport and telecommunications. These small satellite towns, well connected with each other and surrounded by productive open land, carried resonances of Garden Cities ideals from Europe. 'The cluster form of city not only disperses urban territory through the country, it also disperses the country through urban aggregations and makes it more accessible and more useful to their people' (Augur 1948, p. 34). The range of beneficial land use was described by Clayton (1960, p. 116): '[s]pace between satellites would contain forested watersheds, parks, reserves, airports, farms, institutions, large-lot cottages, and like uses. Low-density areas, freeways, drainage reservations, park strips, and so on would lace urban areas to inhibit conflagrations or firestorms'.

The emphasis on developing new towns beyond metropolitan cores in North America from late 1940s onwards was driven, in part, by the problem of increasing traffic congestion in many established cities as car ownership expanded. The search for better structures, seen as conducive to mass automobility, and the reorganisation of social and economic activity would in fact, in subsequent decades, lead to unsustainable urban forms dependent on private automobility. Such thinking was redolent across urbanism in the post-war period as many new imaginaries for socio-spatial configurations were put forward; for example, the three particular layout models for dispersed settlements illustrated in Ralph Lapp's 1949 book *Must We Hide?* – nucleated, doughnut and rod-like cities – were car-centric ones too. Another influential and visionary example was the work of Ludwig Hilberseimer[3], who asserted that 'In the world of the atomic bomb, city concentration can only be a preparation for man's suicide' (Hilberseimer, 1949, p. xvi). His model for settlement layout and regional organisation entailed a number of alternative proposals for urban and rural conditions developed around ideas of optimum effectiveness in the scale of production of food and goods. Also echoing Garden City principles, Hilberseimer imagined that,

> [i]f we were to meet these requirements, we should have not only a city built according to the needs of today, but also a city protected to a relatively high degree against aerial warfare. The city would be so dispersed as to become a decentralized instead of a centralized city. This decentralized city would combine the advantages of a small town with those of a metropolis. The metropolis can be located in the landscape
> (Hilberseimer, 1949, p. 149)

Hilberseimer also drew on pre-atomic precedent for city planning that advocated dispersal[4] and it is evident that, in the early Cold War era, pre-Second World War ideas on remaking existing cities along Modernist lines were matched to more scientific models of urban planning; 'In this era of faith in technical rationality', Lamb and Vale (2017, p. 17) argue, 'dispersal advocates combined rhetorical and graphical argumentation to make the case that their proposals were not only technically feasible but represented a logical evolution in the relationship between technology and the form of human settlements'.

These models were also distinguished from those of the interwar period by the replacement of increasing verticality in city centres with horizontality – linear cities, geometric cluster cities or concentrically dispersed cities – this was one of the imperatives at heart of atomic urbanism. High rise city centres with pronounced population density were an obvious target choice. Downtown skyscrapers were regarded by some commentators as relics of industrialism to be replaced by sprawl of consumerism. This was exemplified in the space extensive cities of the American west that had grown rapidly from the middle of the twentieth century; 'Los Angeles might well escape atomic attack because its population and industries are spread rather thinly over a large area' (Lapp, 1949, p. 147).

The scale of North America, and also the Soviet Union, meant that there was potentially space for dispersal, if there was political will, and to construct wholly new cities along the lines of Los Angeles. As Lapp, who was an influential advocate for dispersal planning for national defence, noted the United States had ample land and '[o]ne of the most decisive things which we as a country can do to reduce our vulnerability to atomic attack is to use this space effectively. This space is really our ally' (1949, pp. 8–9). As Monteyne points out though, Lapp thought that 'urban dispersal schemes should privilege the workers and infrastructure involved with essential wartime industries' and employed the term 'selective dispersion' to describe the most likely outcome of such a policy (2004, p. 191). There was in effect a spatial apartheid in action that may have mirrored the rise in mobility, but left the urban poor as targets, as the middle class left city centres for new homes and new jobs in dispersed industrial sites. Such dispersal was practically more difficult in the crowded and heavily urbanised regions in many European countries, although the need for urban renewal after the destruction of the Second World War provided opportunities for inventive new forms.

Building on the experience of total war across much of its territory and the struggle for survival, the Soviet state was much more prepared than Western countries to enact 'top-down' plans in the command economy for entirely new urban settlements and significantly adapt other existing towns in service of their Cold War agenda (Rowland, 1996; Vytuleva, 2012). However, in the United States, the town of Oakridge, Tennessee, was built according to a masterplan by Skidmore, Owings and Merrill as a 'city of 75,000 for atom bomb factory' (Jackson, 1981). It is thought that the construction of the facilities and housing accounted for around 60% of the Manhattan Project budget. For SOM it was 'one of its foundational urban design projects' and 'a catalysing agent for the firm to think about processes of urban planning' (som.com, 2019). The UK set about repurposing existing ordnance factories and chemical works for their nuclear programme and built new housing estates within areas scheduled for expansion as part of the post-war recovery programme (Garrat, 2016). Amongst all of these conditions, urban environments were altered, adjusted, expanded or had their growth inhibited by measures designed to allay fears or combat aggressive action.

The case for spending additional sums to achieve dispersal was justified in that it would contribute to making cities less congested and better places; it was a positive investment with lasting benefits, in contrast to the many billions that were sunk into making more deadly weapons (Lapp, 1949). Advocates were realistic that dispersal would be a gradual project, in the capitalist democracies, with encouragement to fit into a plan rather than draconian enforcement. The government would not build the houses in America, but significant state investment to provide infrastructure and public transport would help pump-prime new ex-urban development (the interstate highways in the 1950s are a prime exemplar; cf. Rose, 1990).

Despite persuasive arguments for dispersal, it is evident that over the first two decades of the Cold War, no significant physical alteration to the shape of cities took place. Augur and Lapp were writing passionately at the end of 1940s for the urgency of the task but a decade later Clayton – writing as kiloton atomic bombs were being replaced by thermonuclear weapons and intercontinental ballistic missiles – in his paper *Can we plan for the atomic bomb?* noted that, 'metropolitan areas have since grown in population and in vulnerability to attack. New industries have risen within potential target areas. ... Few planning reports discuss even briefly the risks posed to public safety by existing weapons' (1960, p. 111). This interpretation concurs with our knowledge of planning strategy reports published in the 1950s and 1960s by large British cities which failed signally to mention atomic threats let alone offer cogent plans to deal with the aftermath of an atomic attack. For example, the 1951 Development Plan produced for Manchester, UK, makes no mention of atomic threats or the need for the dispersal of essential infrastructure or facilities (Nicholas and Dingle, 1951).

Architectural design and possibilities of atomic-resistant building

A great deal of activity relating to atomic urbanism occurred below the unit scale of the city itself and concerned the work of architects, designers and engineers. At the start of the Cold War there was limited practical knowledge specific to atomic weapons and their impacts on different structural forms, materials and construction methods, although the Second World War conflict in Europe and Japan did provide experience and evidence of the effects of aerial bombing and firestorms on cities. The destructive capacity of atomic bombs used on Hiroshima and Nagasaki was useful in comparison, and some structures seemed to survive better than others. It was obvious that nothing could be done realistically in terms of construction technology to counter the physical forces (blast, heat, radiation) wrought immediately around ground zero of the atomic explosion. However, experts felt that with effort and thought something useful could be achieved by constructing new buildings some distance away from urban centres, designed to offer better protection to occupants and enhanced structural performance. For example, at the MIT symposium in 1952 on *Building in the Atomic Age*, the organiser Robert Hansen (academic expert in structural engineering) noted that beyond the immediate blast zone,

> the performance of buildings can be altered so that only slight damage would result. This may be accomplished by changes in design details, the provision of some lateral strength in the structure, and other methods. It is important... to bring to the attention of the designing profession that major improvements in the blast resistance of a structure are brought about with relatively small increases in building cost.
> (Hansen, 1952, pp. 2–3)

Of course many building codes already existed, these having been developed and strengthened from the late-nineteenth century onwards to meet known and relatively common hazards and it seemed logical to professionals to extend these – '[w]e strive for the greatest resistance to fire that we can afford – considering all the factors involved; and the same kind of thinking applies in connection with protection against atomic bombs' (Wilbur, 1952, p. 112).

Yet, it was evident that little that was achieved in retuning architecture and structural engineering for atomic urbanism. Architectural journals in the United States and Britain from the late 1940s carried articles about the 'Effect of atomic bombs on buildings' and 'Designing buildings to resist the atom bomb', but by the mid-1950s, the term 'atomic' in the architectural press was used to describe new research facilities, laboratories and power stations, and the discussion of hardening buildings had all but vanished from the pages. It seems as if the risks of atomic attack were simply too esoteric and difficult to comprehend in relation to daily challenges. However, this did not prevent the UK publication of 'Fire and the atomic bomb' as 'Fire Research Bulletin, No. 1' (HMSO, 1954) that outlined certain measures for fire prevention outside of the immediate blast radius and was based primarily on US research published in *The Effects of Atomic Weapons* (1950).

In some cases, there were countervailing forces in terms of building design – driven by aesthetic fashion and cost pressure, and developments of new materials technologies and construction techniques – that would make buildings *more* dangerous in the case of atomic bomb attack. The most visually obvious from the 1960s was the proliferation of glazed curtain walls in many commercial and civic buildings. 'New buildings, including schools, are more likely to feature extensive glass areas than to embody protective construction principles' (Clayton, 1960, p. 111). Daylight had been heralded as healthy and increased expanses of glazing grew as technology developed in the twentieth century (Kelley, 1996). In new commercial architecture, glass was emblematic of the democratic images commercial firms wanted to project – it allowed everyone to have a view, dissolved the distinction between inside and outside, let light into basements and reflected light into atrium spaces – the employee had the same view as the boss from their shared open plan workspace. It was this very fact – the open plan – combined with the shattering properties of glass that heightened the risk of injury in the event of an explosion.

It would be wrong to say that architecture did not respond at all in the Cold War to atomic imperatives. However, architecture was used as a retort, more than a response, it was not about designing safer buildings but did concern the construction of symbolically powerful statements for ideological purposes. Despite the exclusion of Berlin from this volume, it is of note that, '[d]uring the period of the city's division, consciously constructing an identity meant using specific architectural styles and approaches to build quite literally a "democratic" city in the west and a "socialist" one in the east' (Pugh, 2014, p. 2). As Stupar (2015, pp. 625–626) asserts, '[a]rchitecture and urban reconstruction were as important as military competition and the space race. They served as tools of propaganda, expressions of prestige, and were even a reaction to overwhelming paranoia'. Aesthetically muscular or acrobatic, prestigious public architecture and large-scale urban renewal projects, the everyday and the typical forms of military-industrial incision into urban space had their own impact and left their own legacies, as a number of the chapters here attest to.

Continuity of government challenge

Much of the thinking and planning around atomic urbanism concerned the practicalities of maintaining parts of the machinery of government after a nuclear attack. In the event of large-scale destruction, dislocation and loss of life, the capacity to rapidly repair infrastructure and to preserve essential national government functions was regarded as vital. One key aspect of achieving this target was the 'hardening' of existing facilities for the power supplies and telecommunications needed by core government sites and military command and control systems. Some effort went into planning routing of new infrastructure that could by-pass nodal city centres completely and provide duplication of routes (e.g. national telecommunications systems like the microwave backbone system built in Britain in the Cold War, cf. Cocroft 2003). Other key infrastructural facilities that were of prime importance but at high risk in certain locations could justify the expense of being positioned in deep underground facilities.[5] This was also the case with seats of government that are, for convenience of face-to-face interaction, usually highly concentrated in small districts at the centre of capital cities. Secret government bunkers were built underground to house the political elite, as well the development of 'alternative sites' for government to operate and contingencies to keep going if senior civil and military leadership was eliminated in the initial nuclear strike. The regional seats of government (RSG) in the UK, and their territorial borders, were really a product of earlier conflicts in the twentieth century. Like much military hardware, the experiences of the Second World War also informed the geography of Cold War planning (McCamley, 2013). These were often deliberately situated in obscure, non-urban locations to try to minimise the risk of attack. A great number of civic buildings across the UK – new town halls, council offices and police stations – built during the Cold War also had shelters constructed, almost as a matter of course, but rarely publicly disclosed or discussed.

Other aspects of essential infrastructure, particularly large-scale urban utilities of power, sanitation and drinking water, were much harder to reengineer for the atomic bomb era, but secure control centres and fall-back facilities were sometimes provided for key sites (e.g. stand-by diesel generators for major hospitals). There was also a need to provide facilities and material resources to aid post-attack recovery, such as the stockpiling of fuel, food, equipment and building materials. However, these were usually only amassed to support the continuity of government and essential services, they would not have met the needs of larger civilian populations that had survived atomic bombing of their cities.

The national backbones provided by transport infrastructure – large ports, strategic highways and heavy rail routes between cities, major pipelines – were considered in Cold War contingency planning. The justification of the construction of motorway systems in European countries was often premised partly on their strategic role in time of total war (Baldwin et al., 2002, p. 108). In the USA, the 1956 Interstate and Defence Highways

Act, promoted by President Eisenhower, provided billions of dollars of Federal funding for the construction of thousands of miles of fast high-capacity roads. These were typically routed widely around the circumference of major cities and served the dual purpose of supporting dispersal and encouraging suburban residential development (Ellis, 2001). At the point of crisis, they could handle large volumes of traffic and potentially provide emergency sheltering space and ad-hoc runways for aircraft (Tobin, 2002). Within lots of metropolitan areas the planning and construction of many new roads was essential in support of sprawling suburbanisation and the deconcentration of employment and education facilities.

Population survival

The logical imperative for shelters for city residents was a significant component in atomic urbanism in many countries, drawing upon the practical benefits of air raid protection measures taken in cities during the Second World War. In North America in the 1950s a considerable amount of effort went into civilian shelters and the public had the impression that they were being protected, with prominent signs in city centres to flag buildings that were adapted or allocated for public shelters (cf. Monteyne, 2011; Rose, 2001). However, the available public shelter provision was never sufficient to meet needs of the daytime working populations of cities. For affluent Americans there was also a culture of private fallout shelter building, facilitated by bespoke services provided by architects and construction companies (cf. O'Connor, 2014; Vanderbilt, 2002). Major expense was outlaid in parts of Western Europe to provide extensive civilian fallout shelters. Switzerland and the Scandinavian countries eventually provided shelters places for a near complete percentage of their populations (Ziauddin, 2017); although, even in these countries, there was often a mismatch between where people might be during the day and the situation of the majority of shelters (such as in basements of private homes and apartment blocks). In hard-line Communist Albania, the nuclear shelter was normalised – the government, under Enver Hoxha, built huge numbers of nuclear fallout shelters (c. 173,000 from the late 1950s until the early 1980s) to protect citizens and elites alike (Eaton and Roshi, 2014).

Other countries baulked at the huge expense of bunker construction needed to house millions of people and instead focussed on reimagining social facilities and propagating cultural messages to counter ideological and nuclear threats. As an extension, politics and cultural practice became a fundamental part of the Cold War battlefront. This was the case in Britain, which did not attempt to provide public shelters on a wide scale during the Cold War due to cost and strategic spending priorities on atomic weapons production. As such, civil defence could be summarised in the UK as follows:

> In the absence of a shelter for the general populace, the accent has been put on retaining the means of regeneration. That implied a need for the

survival of government. Thus the result has been on the protection of government, a policy that critics now compare with the lack of protection of the ordinary populace. 'Elitist', is the charge levelled at Britain's present Civil Defence policy.

(Evans, 1982, p.1)

The senior decision-makers articulated the difficulties of protecting citizens in cities. Their case was a hard one to argue due to the recent experiences of the Blitz during Second World War and the fact that public shelters had been provided in British towns and cities. Instead of shelter provision and effective mass population evacuation, the alternative Cold War civil defence strategy for the public in Britain was 'common-sense' advocacy to stay put and try to survive at home. For example, advice promulgated in Civil Defence Handbook No. 10, published by the British Government in 1963, directed householders to construct a crude fallout shelter within their own property made from doors and carpets; occupants were then to remain inside for up to 14 days after a blast, in order to survive the worst effects of radioactive fallout. Below the scale of the national strategy and city-wide shelter planning, there is evidence that some architects (or their clients) were alert to the needs of atomic defence through the inclusion of spaces for bomb shelters in commercial buildings (Figure I.4). However, it is unclear if the shelters shown on design drawings were manifest in the buildings when constructed and if they were actually fitted out with emergency provisions and maintained for occupation. Again, however, a handful of shelters in major buildings would not have had sufficient space for the day-time working population present in the centres of large British cities.

The architects tasked with providing civilian facilities, and advocates for mass public fallout shelters, did not typically dwell on the operational realities in the event of the Bomb being dropped. There are many practical engineering challenges in human survival when cut-off from external sources of electrical power, fresh air, clean water and so on. Genuinely bomb-proof and environmentally self-contained space would have been extremely hard to achieve, even with deeply buried bunkers and sophisticated engineering (perhaps the most obvious effective example of such spaces in the Cold War period were nuclear-powered ballistic missile submarines that could patrol submerged in the oceans for months; cf. Polmar and Moore, 2004). Beyond the practical physical needs of food, water and sanitation, what would have been the prospects for those in shelters during the hours and days after the initial atomic blast? There would have been serious psychological shock of survivors to handle, along with physical challenges. Different aspects of the engineering logics of shelters, the politics of bomb-proof protection provision, and the sociological challenges of bunker occupation and wider cultural resonances, have been dissected recently by scholars of the Cold War (cf. Bennett, 2017; Klinke, 2018).

Cold War cities 21

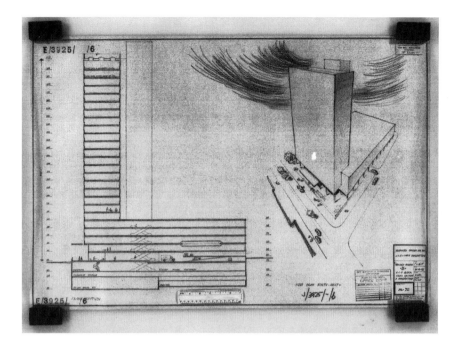

Figure I.4 A drawing by architect G.S. Hay for the most significant post-war era office block built in Manchester. Included is space on the fourth level basement labelled 'shelter'. It is unclear if the space allocated was configured as a fallout shelter and whether it was sufficient to accommodate the several hundreds of people working in the building, nor how they would have quickly accessed it in an emergency. Moreover, in the event of atomic bomb dropping on Manchester, the blast wave would have brought the 25 storeys of the office tower down upon the basement levels and likely transmuting the shelter into an inescapable tomb.
Source: Author's scan, courtesy of Manchester Archives+.

Civil defence preparations

Another important aspect of atomic urbanism was the provision of civil defence forces, typically organised as a distributed volunteer organisation that would be tasked for rescue and recovery work, along with emergency medical care (cf. Davis, 2007; Vale, 1987). The preparation had a dual purpose – initially to save lives, but the strength of a civil defence force was demonstrable of the capacity to continue fighting during an atomic war and, as such, they became an element of deterrence. As Lapp (1949, p. 149) put it '[t]he necessity for [civil defence preparation] does not arise from humanitarian motives alone – the fight for survival may require the utmost salvage of people and material from the wreckage following a surprise attack'.

22 *Richard Brook et al.*

Figure I.5 An advertisement to encourage volunteers to join the civil defence corp. in Britain in the mid-1950s. The cartoon, drawn by Leslie Illingworth, addresses directly the issue of defeatist attitudes expressed by many ordinary people in the face of the overwhelming threat of nuclear bomb attack on their cities.

Source: scanned and edited by authors. Published in Manchester Guardian newspaper, 26 Sept. 1957.

Centrally organised, the local civil defence forces were regarded as part of the national defence strategy, but should also be understood from the level of cities, that would have been the natural scale of operation. They would have been useful for responding to all manner of urban emergencies (such as natural disasters); '[a]ny organisation set up to cope with the aftermath of atomic attack will be of great value during peacetime as well' (Lapp, 1949, p. 151). However, their effectiveness depended very much on the local context in terms of people, motivation and resources; enthusiasm and capacity varied greatly in the UK for example (Grant, 2009). For a start, the civil defence organisation was never seriously funded nationally and there was much scepticism in different cities about constituting an effective corp. Through the 1950s, the organisation struggled to attract volunteers (Figure I.5). Grant (2009, p. 2) asserts that

> civil defence as it was popularly understood - dedicated to saving lives in wartime - was dead [by mid 1950s and] that the key issues now were nuclear deterrence and the machinery of government: other plans were either futile, too costly, or a combination of both.

With the greatly increased damage caused by H bombs, the scale of medical assistance needed to cope with the huge number of casualties in British cities was never seriously addressed, nor were the material needs of food and housing to keep survivors alive in the longer term. The civil defence force was formally stood-down in Britain in 1968.

Active defence

Despite clarion calls for action around atomic urbanism in the late 1940s and through the 1950s, in most countries relatively little was achieved in terms of dispersal planning, building design, shelter provision and civil defence preparations. The passive defences, while recognised as important by governments, were always constrained by budget limits, and priority spending in leading countries in NATO and the Warsaw Pact was always directed towards weapons production and military-focussed active defence.

The vast expense underpinning the production of nuclear arsenals was at the centre of Cold War doctrines for the two superpowers and was also significant for second-tier nations like Britain and France. This was first a logistically complex and highly technical industrial project to manufacture the source component of uranium and plutonium – often deliberately sited in facilities remote from cities for safety and secrecy – but over time, it gave rise to new atomic weapons-dependent forms of urbanism. In the Soviet Union, a centrally planned system of secret nuclear cities were constructed (Gentile, 2004; Vytuleva, 2012). At a lesser scale, the need to produce suitable fissile material for atomic weapons in the UK resulted in clustering of industrial facilities in Lancashire and on the coast of Cumbria (cf. Jay, 1954), which in turn gave rise to new settlement patterns, particularly to house the specialised workers in remote locales. Whilst this was not done in secrecy, some aspect of the production was 'hidden in plain sight' and the role of early civilian nuclear power station in making weapon grade plutonium as well as supplying electricity was not publicly acknowledged. The sheer complexity involved in the design, testing, production and deployment of nuclear warheads themselves required many specialised facilities and cadres of skilled technicians and scientists. The scale of administrative, geographical and technical effort for the UK's first deployed atomic bomb is well documented by Aylen (2015).

Alongside the huge investment in nuclear weapons themselves and their delivery systems, during the Cold War a great deal of technical effort and money was expended on 'active defence' which was about gaining strategic advantage by tilting the balance of information against the enemy. Many large projects, often masked by layers of secrecy and sited in remote locations, were undertaken by the superpowers. This is best illustrated in intelligence gathering, where efforts in cloak-and-dagger style spying and secret agents[6] were dwarfed by the rise of signal intelligence (e.g. Aldrich, 2010) and optical surveillance from specialised aircraft and secret satellites (Taubman, 2003). The militarisation of outer space and the strategic role of satellite platforms was highlighted in dramatic fashion by the practical demonstration of Sputnik in 1957. Satellites provided strategic vision across the globe, including over all of the enemy's territory (cf. Cloud, 2001). Cold War spending on space technology has significant legacies. Much of the geospatial technology that underpins current civilian mapping and locative services today – satnav and high-resolution imagery – had military origins; indeed the Global Positioning System is still paid for and controlled by the US Air Force (cf. Rankin, 2016). The utility of the interception of military and diplomatic radio communications had been powerfully demonstrated by the Second World War code breakers. Their highly classified work continued and diversified through the early decades of the Cold War with many tens of billions being spent on both the means to monitor enemies' communication traffic, and processing the voluminous data that was gathered.

The superpowers also developed sophisticated means of detecting strategic bombers and, later, early warning systems for ballistic missile launches. These built upon the Second World War experience with radar systems, but required much wider scale deployment of physical infrastructure (such as the U.S. Distant Early Warning (DEW) Line, which had installations built in the mid-1950s around the Arctic and sited on the Faroe Islands, Greenland and Iceland). There were also significant overlaps and direct cooperation between cutting-edge civilian space research and the secret world of strategic weapons science; for example, in relation to detection, the massive Jodrell Bank radio telescope in Cheshire, England played an important role in tracking Soviet satellites (Spinardi, 2006).

All these Cold War strategic detection systems required unprecedented investment into the development of real-time digital communication and high-performance computation machinery to generate information that would be of tactical utility to the military (cf. Aylen, 2012; Edwards, 1996). Much of this work was heavily classified in the 1950s but would lay the foundations for the new information processing industry in the next decades; military contracts helped fund start-up computer companies in Silicon Valley and the fundamental knowledge gained was a vital spur to computer science and civilian data networking into the 1960s. The lineage of the commercial internet that came into public consciousness in the early 1990s is complicated, but it does have some significant Cold War ancestry (Abbate, 2000), best illustrated in the basic design of network topology that was distributed to be resilient to attack[7].

One might also argue that the spectacle of atomic weapons testing was an integral part of active defence in the early Cold War. The imagery circulated to the media from military atmospheric testing in the 1950s can be viewed as sabre-rattling on a very dramatic scale with terribly destructive swords. Some of the most iconic images of early Cold War came from the obliteration of idyllic Pacific atolls and alien looking mushroom clouds over the deserts of the America southwest, the Australian outback and Semipalatinsk in the Kazakh Steppe in the former Soviet Union. Yet, these were deliberately remote and depopulated sites, exotic locales distant to the lived experience of most, whereas the imagery of the quotidian destruction of 'doom town' on the Nevada test site in 1953 was much more shocking, showing how feeble regular homes would be in face of atomic blast forces (Kirk, 2012; Wills, 2018). Despite their endless circulation and repetition since 1953, they are still deeply disquieting vision of urban vulnerability (see Figure I.3).

* * *

This volume demonstrates that cities were often shaped by the Cold War in subtle and not so subtle ways. *Michela Morgante's* chapter shows us the creeping urbanism of US military presence in Verona and its impact on everyday lives of both US forces and their families and the local population. *Siim Sultson* describes how Estonian towns were planned to accommodate Soviet industry and the policy led standardisation of particular aspects of urban design and architecture. *Daniel Baldwin Hess and Taavi Pae* analyse Tartu, where particular site conditions encouraged a palimpsest of aeronautical functions both civilian and military resulting in a spatial embargo that predetermined urban expansion in specific directions. *Eloisa Betti's* chapter on Bologna evidences the socio-political dimension of the Cold War city, not manifest in its form, planning or in construction but in a mass mobilisation of concerned civilians, albeit with leftist backing. The scale of protest and demonstration is viewed as significant in the lived experience of the city and the temporary use and occupation of its public spaces. In many of these chapters, the role of memory and memorialisation appears, both in commemoration and in collective loss.

More conceptually, research on the Cold War permits the exploration of the term 'city' and its power, relevance and meaning. *Kristian H. Nielsen*, in his chapter on Greenland's 'city under the ice', and *Sebastian Schmidt's* chapter on the planning of a future city, take the idea of the city and explore it in relation to the various interpretations of nation states with regard their particular needs. For Schmidt, the Cold War altered notions of urban space in post-war Japan, where the city became a confident marker for visions of not only national renewal, but a transformative future for humankind. This clearly resonates with our concept of atomic urbanism in which the overarching threat of nuclear war opened up city planning to radical alternatives, even if these were never implemented. Camp Century, the American military facility sited in Greenland, is described by Nielsen as space that embodied core American values, thus demonstrating the flexibility of dominant ideas of atomic urbanism. The ways in which it was described and classified by the United States and Danish authorities are contrasting and ask us to consider the purpose of the city as a site of resistance, a zone for technological innovation and a piece of physical propaganda.

In the global context, did the 'core values' displayed by the superpowers need representation in particular types of urban space in order to flourish?

Matthew Phillips argues that the proliferation of art galleries, tourist spectacles, handicraft showrooms, hotels and shopping centres in 1950s Bangkok provided new spaces from which to understand the Cold War. He argues that these spaces were important sites of an emerging global 'Free World' culture globally.

Secret construction plans shaped cities in subtle visual ways, as in *Bob Clarke's* chapter, marking them out as strategically important for nuclear defence, but then also vulnerable to possible nuclear attack. In contrast to the dispersal ideas present in US Cold War urbanism, UK cities had a tendency to spread sites around urban hinterlands, but to cluster certain defence technologies in particular regions. This geography was largely a result of strategies designed to serve conflict during the Second World War and the onward development of existing defence sites. These spaces often served dual purposes. *Rosanna Farbøl*, in her visual essay on Danish civil defence planning, describes how urban ideals of 'the good life' had to incorporate the prospect of atomic warfare. In Denmark, this architectural style was closely associated with ideals of internationalism, equality, solidarity, welfare, modernity and improved public health – this corresponds with Nielsen's account of how the Danish narrated the US incursion into Greenland.

The Cold War aesthetic was varied in architectural character. In this volume, we see some influenced by rules and policy and some a formal mediation between existing knowledge and new environmental conditions. Each had its own banal quality or, in this case, a generic-specificity. *Siim Sultson's* chapter addresses the repetitive architecture of the Soviet closed cities and refers to the central city as a 'stately urban ensemble' attributed to government policy designed to order and control civilian populations in service of military objectives. The Soviet secret and closed city programme ran parallel to new town planning across Western Europe – comparisons between the two are yet to be drawn. *Tong Lam's* visual essay charts the 'Third Front Construction' initiative, which involved significant financial investment in south-western China's remote and mountainous region due to the threat of nuclear attack in traditional centres of industry. Thousands of factories were strategically built, which used conventional construction technologies but with spectacular formal results due to the mountainous terrain. These residential-industrial hybrid buildings were specifically-generic and drew upon the global convention, that existed from the birth of the industrial age, of housing workers close to their site of employment. The influence of the Cold War shaped urban space, whether through order or necessity, and created buildings that only existed due to the geo-political conditions. It is also clear that the role of the state in the determination of architectural form, through legislation and statute manifested in myriad ways, with inconsistencies between politics and style. For instance, Soviet realism tended towards neo-classicism that was twisted with the systematic rationalisation of production; Yugoslavia pursued forms now classed as socialist modernism that were somehow unfettered by western architectural history or Soviet magnitude.

Alex Lawrey shows another facet to the Socialist Realist aesthetic in the reconstruction of the historic centre of Warsaw and the construction of the Palace of Culture and Science. Each had their own message attached to communist and Stalinist ideologies. Both Lawrey's essay and *Richard Brook and Martin Dodge's* visual essay consider the role of infrastructural planning and construction. Lawrey uses the drawn-out Cold War era project of the Warsaw metro system to reveal infrastructure as techno-politics and as militarised space veiled as civic enterprise. Its development, or lack of, is clearly connected to the governance of Poland and its association with Soviet control throughout the post-war and Cold War period. Similarly, certain aspects of British national technological planning were military projects with dual purpose, particularly in the realm of telecommunications. Brook and Dodge mirror Clark's observations about the clustering of certain technologies around specific centres and present the unerringly utilitarian language of these types of facility. The architecture of communications in Cold War Britain was either hidden, or hidden in plain sight. The aesthetic qualities of the new US research facilities were intended as clean, efficient and modern and aligned with their outputs of 'ideas not products' as well as with the values of white middle-class professionals who inhabited the suburbs and worked for corporations like Westinghouse in Pittsburgh, as explored by *Patrick Vitale*. This echoes work by Reinhold Martin (2003) and Michael Kubo (2009) about the 'science of architecture' and the scope offered by the military industrial complex to advance logic and systems thinking in architectural design and construction technology. Vitale also explores the socio-civic context for research and development buildings in the suburbs and connects the modern-industrial building type to ideas of the parkland picturesque. This was a space where rearmament and the American dream met to create elite enclaves, ultimately in the service of capitalism, albeit in conditions defined by the Cold War. Across these chapters, we can understand the Cold War aesthetic to be as diverse as the cities through whose contexts they are refracted.

Of course, natural disasters have the potential to shape the future of cities far more than politics. *Jasna Mariotti's* chapter on Skopje illustrates Carola Hein's observation that urban problems have the capacity to transcend political, ideological and geographical limitations (Hein, 2014). Skopje was in the non-aligned Yugoslavia where politics of east and west were filtered through Tito's agenda and ambitions. All political boundaries were put aside when a catastrophic earthquake hit the city in 1963. The international response both in rescue, rebuilding and replanning is one case of a global exchange of ideas about architecture and urbanism that was not limited by the geo-politics of the Cold War. Its narrative however shows how Yugoslavia's position in the Cold War also facilitated aid and the transfer of knowledge from a spectrum of nations.

In contrast to the sharing of knowledge, Vitale shows how the clustering of knowledge changed the shape of US suburbs, using Westinghouse's relocation in Pittsburgh as a case study. National policies met with favourable local political conditions and enabled class segregation based on industrial research, expertise and links to higher education. In an era of urban dispersal, Vitale illustrates a form of contained or clustered suburbanism that supports the idea of critical functions moving away from urban centres, whilst at the same time creating new elite suburban centres shaped by Cold War imperatives. Key stakeholders in the overall development were also residents of the same. As such, the urbanism connected to Cold War research and development bolstered already exclusive communities to entice a skilled, much-in-demand workforce (similar to Kate Brown's *Plutopia*). In the case of Westinghouse and Pittsburgh, the dynamics of communities were transformed by Cold War interests and dominant value systems. Arguments concerning quality of life and the creation of better and ideal communities were put forward and linked to ideas presented in Farbøl's visual essay on the Danish suburbs. Military imperatives and issues of national security were proffered to those who would grant licences for construction and expansion. Westinghouse sought financial gain from these conditions and thus capitalism and the Cold War created this particular urban space in a way that mirrors the communist Cold War centres of the Soviet closed cities narrated by Sultson.

Conclusion

Cold War Cities offers a focus away from the nation-state or continental-wide geopolitical struggles, where looking at individual settlements (or city regions) provides a distinctive way in which to reinterpret Cold War histories. This was also evident in Pizzi and Hietala's (2016) edited volume, where they stated that, '[t]he Cold War city, whether during or after this period, makes for a sideways, oblique and, for this very reason, all the more stimulating vantage point … [and can] problematize ossified Cold War logics and traditional historiography'. (p. 2). Rather than being a competitor, we see this book as a complementary to theirs, with a contrast in its focus on planning, architecture and politics. This book is distinctive as well – by using case study cities whose presence were not explicitly part of the grand political narratives of the period, we aim to present some alternative views of the Cold War and its geographical, political and urban cultural influence.

We also hope that when read together the range of case study cities in this volume offers a unique comparative perspective in the area of historical geography, histories of planning and architecture, urban and cultural studies and political histories. Reading across the different chapters, we think, also demonstrates the variable extent to which the development of cities was influenced by atomic urbanism, thus problematizing '*the* Cold War' as a general analytical lens.

Cold War cities 29

And yet, just how markedly were cities shaped by atomic urbanism? Despite bold claims at the start of the Cold War era – for example, Lapp writing in 1949 asserted the 'bomb is forcing a social revolution comparable in scope to the industrial revolution brought about the introduction of steam and electric power' (p. 157) – in many ways little material change in cityscapes and urban living can be directly ascribed to nuclear threats despite the discourse and correlations in socio-economic processes (e.g. lower densities through sprawl suburbanisation).

Framed by Cold War doctrines and an ever-present threat of destruction, cities managed risk in different ways and offered up various ideologically informed visions of urban development through the 1950s and well into the 1960s. However, in many respects, one of the key findings of the book's chapters is that Cold War, whilst a dominating theme and lever for huge amounts of capital spending at particular sites, did not determine the wider remaking of cities. Atomic thinking and fears of nuclear apocalypse were undoubtedly present in the mind of decision-makers in city councils – and sometimes expressed – but this seems to have surprisingly little evidential impact on the urban forms or architectural designs. This is apparent in the widespread failure to radically redesign structure in ways to mitigate damage and injuries to occupants. Clayton, an influential planner in Baltimore, USA, was writing after more than a decade of Cold War tensions and in knowledge of the power of thermonuclear weapons and the lasting toxic impact of radiative fallout on the land. He stressed the need for new thinking but acknowledged the situation had got more challenging for city planning:

> nuclear war long ago signalled the obsolescence of pre-Atomic Age urban forms, metropolitan areas have since ground in population and in vulnerability to attack. New industries have risen within potential target areas. Hospitals, food storage facilities, and essential utilities have been constructed without regard to civil defense needs.
> (Clayton, 1960, p. 111)

Moreover, as the 1950s proceeded, and the scale of exposure of populations increased, it seems there was a paradoxical 'declining attention paid to the problems of dispersal, reduction of urban vulnerability, and postattack rehabilitation' (Clayton, 1960, p. 111).

Why was so little done? In part because technological developments were so rapid – exemplified in the Space Race from the Sputnik launch in 1957 – and hard to comprehend for many. Local officials and city planners were like rabbits caught in the headlights of atomic dangers, unsure what could and should be done. Many proposals were so enormous in scope or impractical in engineering and sociological terms (e.g. underground cities) that they simply could not progress. No doubt, a good degree of apathy and fatalism pervaded the thinking of people at the time. Atomic war was too big to deal with in any practical sense. This situation in the early Cold War period is analogous to climate change today, where rhetoric about the Anthropocene is central in much discourse around urban policy, but how it is translated into concrete decision-making is much more problematic. Fatalists accepted that little useful could be done at their level of municipal control and carried on much as before. Of course, some things that had a defensive logic to justify them – like urban sprawl, low density suburban residential development – were happening in the 1950s and 1960s regardless of Cold War geopolitics. These socio-spatial changes in the cityscape were primarily driven by other large economic, sociological and technological forces.

While the analysis in this volume is inherently historical in perspective, it is evident that *Cold War Cities* has contemporary relevance to processes of urbanism. In recent years, the constantly evolving geopolitical situation in the aftermath of the post-Soviet settlement has led to the revival of East-West antagonism and territorial tensions in 'flash-point' regions, and the continued rise of new nuclear-armed states such as Pakistan, Iran and North Korea. Debates on the 'Second Cold War' are now increasingly raised in contemporary academic and public discourse, making it important to more fully understand how different cities coped in an earlier age of atomic urbanism. Additionally, the security of cities and the concern for urban resilience and citizen safety has come to the fore in recent years (Vale and Campanella, 2005). There are clear parallels and reverberations between the dangers of atomic attack in the 1950s and 1960s and the situation in cities post-9/11 where there have been significant efforts to plan and prepare for physical threats. Acts of terrorism are frequently focussed on cities, as are the threats associated with civil disorder, organised criminality, infrastructural breakdown, natural disasters and large-scale impacts of climate change. The desire to enhance safety and engender greater resilience has come to pervade everyday life, the popular media and political discourse. Understanding the age of atomic urbanism can help meet the contemporary challenges cities are facing.

Notes

1 A study of the full spatial and urban transformation of Berlin, for example, a microcosm of the global condition, though its history would warrants a book of its own, cf. Pugh, 2014.
2 For more on Wiener's contribution in Cold War strategic thinking, cf. Kargon and Molella, 2004.
3 His book *The New Regional Pattern* was published just after the Soviet's first bomb test.
4 One such source was Ralph L. Woods' book, *America Reborn. A plan for the Decentralization of Industry* (Longsman, Green and Co, 1939.).
5 Although the effectiveness of bunkers was undermined by the development of much more powerful thermonuclear weapons into the late 1950s. Increasingly the vision for survivable government was to have distributed command and mobile units (such as airborne command).

6 Spies working for the Soviet Union within the America and Britain atomic weapons laboratories did play a significant role in Russia's efforts to quickly acquire nuclear know-how.
7 Often cited in this regard is Paul Baran's report: 'On Distributed Communications: I. Introduction to Distributed Communications Network', *RAND Corporation memorandum RM3420-PR*, August 1964, http://www.rand.org/publications/RM/RM3420/RM3420.chapterl.html.

Bibliography

Abbate, Janet, *Inventing the Internet* (MIT Press, 2000).
Aldrich, Richard J, *GCHQ: The Uncensored Story of Britain's Most Secret Intelligence Agency* (HarperPress, 2010).
American Institute of Planners, (1953) "Defense considerations in city planning," *Bulletin of the Atomic Scientists*, 9, 268–269.
Anon., *The Effects of the Atomic Bombs at Hiroshima and Nagasaki* (HMSO, 1946).
Anon. (1950), "How U.S. Cities can prepare for atomic war," *Life Magazine*, 18 Dec, pp. 77–81.
Augur, Tracy B. (1948) "The dispersal of cities as a defense measure," *Journal of the American Institute of Planners*, 14(3): 29–35.
Aylen, Jonathan (2012) "Bloodhound on my trail: Building the Ferranti Argus process control computer," *International Journal for the History of Engineering & Technology*, 82(1): 1–35.
Aylen, Jonathan (2015) "First waltz: Development and deployment of Blue Danube, Britain's post-war atomic bomb," *International Journal for the History of Engineering & Technology*, 85(1): 31–59.
Baldwin, Peter, Bridle, Ron and Baldwin, Robert C.D., *The Motorway Achievement, Volume 1* (Thomas Telford, 2002).
Bennett, Luke, *In the Ruins of the Cold War Bunker: Materiality, Affect and Meaning Making* (Rowman & Littlefield, 2017).
Brown, Kate, *Plutopia: Nuclear Families, Atomic Cities, and the Great Soviet and American Plutonium Disasters* (OUP, 2015).
Catford, Nick, *Cold War Bunkers* (Folly Books, 2010).
Clarke, Bob, *Britain's Cold War* (The History Press, 2009).
Clayton, Philip (1960) "Can we plan for the atomic age?", *Journal of the American Institute of Planners*, 26(2): 111–118.
Cloud, John (2001) "Hidden in plain sight: The CORONA reconnaissance satellite programme and clandestine Cold War science," *Annals of Science*, 58(2): 203–209.
Cocroft, Wayne, *Cold War: Building for Nuclear Confrontation 1946–1989* (English Heritage, 2003).
Davis, Tracy C., *Stages of Emergency: Cold War Nuclear Civil Defense* (Duke University Press, 2007).
Eaton Jonathan, Roshi Elenita (2014) "Chiselling away at a concrete legacy: Engaging with Communist-era heritage and memory in Albania," *Journal of Field Archaeology*, 39(3): 312–319.
Edgerton, David, *Warfare State* (CUP, 2005).
Edwards, Paul N., *The Closed World: Computers and the Politics of Discourse in Cold War America* (MIT Press, 1996).

Ellis, Cliff, (2001) "Interstate highways, regional planning and the reshaping of metropolitan America," *Planning Practice and Research*, 16(3–4): 247–269.

Engel, Jeffrey A. (ed.), *Local Consequences of the Global Cold War* (Stanford University Press, 2007).

Evans, Peter (1982) "The UK front," In: The Royal United Services Institute for Defence Studies. Nuclear attack civil defence: Aspects of civil defence in the nuclear age: A symposium (Oxford).

Farish, Matthew, *The Contours of America's Cold War* (University of Minnesota Press, 2010).

Farish, Matthew and Monteyne, David (2015) "Introduction: Histories of Cold War cities," *Urban History*, 42(4): 543–546.

Garrat, J. (2016) *Atomic Spaces North West England 1945 to 1957* (MSc Thesis, University of Manchester).

Gentile, Michael (2004) "Former closed cities and urbanisation in the FSU: An exploration in Kazakhstan," *Europe-Asia Studies*, 56(2), 263–278.

Grant, Matthew, *After the Bomb: Civil Defence and Nuclear War in Britain, 1945–68* (Palgrave, 2009).

Hansen R. J., (1952) "Introduction: Purpose and need for symposium," in *Proceedings of the Conference on Building in the Atomic Age* (MIT Press), pp. 2–3.

Hecht, Gabrielle, *Entangled Geographies: Empire and Technopolitics in the Global Cold War* (MIT Press, 2011).

Hein, Carola (2014) "The exchange of planning ideas from Europe to the USA after the Second World War: Introductory thoughts and a call for further research," *Planning Perspectives*, 29, 143–151.

Hennessy, Peter, *The Secret State: Preparing for the Worst 1945–2010* (Penguin, 2010).

Hilberseimer, Ludwig, *New Regional Pattern Industries & Garden* (Paul Theobald, 1949).

Jacobs, Robert and Broderick, Mick (2012) "Nuke York, New York: Nuclear holocaust in the American imagination from Hiroshima to 9/11," *The Asia-Pacific Journal*, 10(11), https://apjjf.org/-Mick-Broderick--Robert-Jacobs/3726/article.pdf.

Jay K.E.B., *Britain's Atomic Factories* (HMSO, 1954).

Johnson, C.W. & Jackson, C.O., *City Behind a Fence: Oak Ridge, Tennessee, 1942–1946* (University of Tennessee Press, 1981).

Kahn, Hermann (1958) "Civil defense is possible," *Fortune* magazine, December.

Kargon, Robert and Molella, Arthur P. (2004), "The city as communications net: Norbert Wiener, the atomic bomb, and urban dispersal," *Technology and Culture*, 45(4): 764–777.

Kelley, Stephen J. (1996), "An image of modernity: An American history of the curtain wall," *Docomomo Journal* 15, July, 33–38.

Kirk, Andrew (2012), "Rereading the nature of atomic doom towns," *Environmental History*, 17(3): 635–647.

Klinke, Ian, *Cryptic Concrete: A Subterranean Journey into Cold War Germany* (John Wiley & Sons, 2018).

Kubo, Michael, *Constructing the Cold War Environment: The Strategic Architecture of RAND* (Michael Kubo, 2009).

Lamb, Zachary and Lawrence J. Vale (2017) "From the Cold War to the warmed globe: Planning, design-policy entrepreneurism, and the crises of nuclear weapons and climate change," *Planning Perspectives*, DOI: 10.1080/02665433.2017.1408488.

Lapp Ralph E., *Must we Hide?* (Addison-Wesley Press, 1949).
Light, Jennifer, *From Warfare to Welfare Defense Intellectuals and Urban Problems in Cold War America* (Johns Hopkins University Press, 2005).
Martin, Reinhold, *The Organizational Complex: Architecture, Media, and Corporate Space* (MIT Press, 2003).
McCamley, Nick, *Cold War Secret Nuclear Bunkers: The Passive Defence of the Western World during the Cold War* (Pen and Sword, 2013).
Monson, Donald (1953) "City planning in Project East River," *Bulletin of Atomic Scientists*, 9(7): 265–267.
Monteyne, David (2004) "Shelter from the Elements: Architecture and Civil Defense in the Early Cold War," *The Philosophical Forum*, 35: 179–199.
Monteyne, David, *Fallout Shelter: Designing for Civil Defense in the Cold War* (University of Minnesota Press, 2011).
Nicholas, R. and Dingle, P.B. *Manchester Development Plan* (City and County Borough of Manchester, 1951).
O'Connor, Lee, *Take Cover, Spokane: A History of Backyard Bunkers, Basement Hideaways, and Public Fallout Shelters of the Cold War* (CreateSpace, 2014).
Page, Max, *The City's End* (Yale University Press, 2008).
Painter, David S. *The Cold War: An Interdisciplinary History* (Routledge, 1999).
Pizzi Katia, and Hietala Marjatta (eds.) *Cold War Cities: History, Culture and Memory* (Peter Lang, 2016).
Polmar, Norman and Moore, Kenneth J., *Cold War Submarines: The Design and Construction of US and Soviet Submarines* (Potomac Books, 2004).
Pugh, Emily, *Architecture, Politics, and Identity in Divided Berlin* (University of Pittsburgh Press, 2014).
Pugh O'Mara, Margaret, *Cities of Knowledge: Cold War Science and the Search for the Next Silicon Valley* (Princeton, 2004).
Rankin, William, *After the Map: Cartography, Navigation, and the Transformation of Territory in the Twentieth Century* (University of Chicago Press, 2016).
Rose, Kenneth D., *One Nation Underground: The Fallout Shelter in American Culture* (New York University Press, 2001).
Rose, M.H., *Interstate: Express Highway Politics, 1939–1989* (University of Tennessee Press, 1990).
Ross, R. *Waiting for the End of the World* (Princeton Architectural Press, 2004).
Rowland, R.H. (1996) "Russia's secret cities," *Post-Soviet Geography and Economics*, 37(7): 426–462.
Schlosser, Eric, *Command and Control* (Allen Lane, 2013).
Spinardi, Graham (2006) "Science, technology, and the Cold War: The military uses of the Jodrell Bank radio telescope," *Cold War History*, 6(3): 279–300.
Stupar, A. (2015) "Cold War vs. architectural exchange: Belgrade beyond the confines?", *Urban History*, 42(4): 622–645.
Taubman, Philip, *Secret Empire: Eisenhower, the CIA, and the Hidden Story of America's Space Espionage* (Simon & Schuster, 2003).
Tobin, Kathleen, (2002) "The reduction of urban vulnerability: Revisiting 1950s American suburbanization as civil defence," *Cold War History*, 2(2): 1–32.
US Department of Defense and US Atomic Energy Commission, *The Effects of Atomic Weapons* (McGraw-Hill Book Co. Inc., 1950).
Vale, Lawrence J. *The Limits of Civil Defence in the USA, Switzerland, Britain and the Soviet Union* (Springer, 1987).

Vale, Lawrence J. and Campanella, Thomas J., *The Resilient City: How Modern Cities Recover from Disaster* (Oxford University Press, 2005).

Vanderbilt, Tom, *Survival City* (Princeton Architectural Press, 2002).

Vytuleva, Xenia. *ZATO: Soviet Secret Cities during the Cold War* [documentary film] (Van Alen Institute, 2012).

Wilbur, J. B., (1952) "Building in the atomic age: Closing summary," in *Proceedings of the Conference on Building in the Atomic Age* (MIT Press), pp. 112–115.

Wills, John (2018) "Exploding the 1950s consumer dream: Mannequins and mushroom clouds at Doom Town, Nevada Test Site," *Pacific Historical Review*, https://kar.kent.ac.uk/66684/.

Ziauddin, Silvia Berger (2017) "Superpower underground: Switzerland's rise to global bunker expertise in the atomic age," *Technology and Culture*, 58(4): 921–954.

Zipp, Samuel, *Manhattan Projects: The Rise and Fall of Urban Renewal in Cold War New York* (OUP, 2010).

Part I
Planning the Cold War city

1 Properties of science

How industrial research and the suburbs reshaped each other in Cold War Pittsburgh

Patrick Vitale

EASTERN CONNECTICUT STATE UNIVERSITY

The Monroeville Doctrine

In September 1969, *The Times Express*, a newspaper based in the Pittsburgh suburb of Monroeville, published a special issue devoted to the Borough's emergence as a research center. The newspaper carried its usual byline, "Serving the nation's research center." In the issue, Borough manager Carol Pickens named Monroeville "the residential research center of the nation," a claim that also appeared on its letterhead, signs, and the Chamber of Commerce's promotional materials.[1] Pickens celebrated the "quality industries" that brought "people who are interested in their community and willing to assist in the efforts to improve the community" (1969). Running alongside Pickens's article was a table showing Monroeville's growth from a rural township of 4,675 in 1940 to a satellite city of 27,701 in 1969, making it Pittsburgh's most populated and fastest growing suburb (Dieterich-Ward 2010). *The Times Express* and local political leaders attributed Monroeville's boom to its growing role as a hub of research in metals, chemicals, and nuclear power.

The remainder of the issue was devoted to advertisements from companies that called Monroeville home. Koppers, the manufacturer of chemicals and machinery, declared its embrace of "the Monroeville Doctrine," which "stands for growth, prosperity, and progress." Monroeville's government, citizens, and companies shared a commitment to "community advancement" and this was why Koppers chose the Borough for its laboratory. US Steel's advertisement described employees at its research center who by day worked with "a beam of high speed electrons" and at night were "busy with civic projects." The capstone of the special issue was a full-page advertisement showing the new Westinghouse Energy Center: the future "Nuclear Capitol of the World."[2] Koppers, US Steel, Westinghouse, and many other firms moved their laboratories to the suburbs in pursuit of the Monroeville Doctrine, the mutually beneficial relationship between middle-class suburban life and industrial research.

By 1960, the Pittsburgh region was home to at least 41 major research facilities and many smaller laboratories, nearly all of which were located in the suburbs (Lowry 1963, 2:98). Firms moved their laboratories to the suburbs because of recent highway construction, the availability of inexpensive land, federal tax incentives, and increasing federal funds for research, but they also pursued the Monroeville Doctrine because they believed that the suburbs were the ideal site for research and researchers. Executives saw building new laboratories in exclusive suburbs as key to their ability to capture federal contracts, scientific workers, and new markets.

In order to pursue the Monroeville Doctrine, firms and municipal officials had to transform the stated purpose of industrial research and the character of exclusive suburbs. Zoning in many suburbs excluded sites of employment. This chapter examines one instance, Westinghouse's lab in Churchill, a wealthy suburb directly to the west of Monroeville, where these restrictions broke down. It uses this example to explore how research laboratories and suburban communities built a mutually beneficial relationship based on social exclusion, status, and control. This mutually beneficial relationship was premised on an idealized understanding of research as an exceptional form of mental labour that was radically different from other forms of labour in general and the corporal work of manufacturing in particular. In developing this mutually beneficial relationship, exclusive suburbs and industrial firms enshrined an untenable idea of research that helped maintain racial, class, and gender exclusion in both subdivisions and laboratories. According to local politicians and residents the only suitable form of employment in the suburbs was that of highly educated scientists and engineers. In turn, industrial firms agreed to maintain the illusion that research was a radically different form of labor in order to access exclusive suburbs that were key to attracting scientists and engineers and defense contracts. During the early Cold War, in the Pittsburgh region and across the United States, industrial research and affluent suburban communities were remade in each other's exclusive image.

Research and suburbanization in Cold War Pittsburgh

Like in Boston, Los Angeles, New York, and many other metropolitan regions, research centers in Pittsburgh rapidly multiplied and suburbanized in the decades following World War II (Karafantis and Leslie 2019; Mozingo 2011; Rankin 2010). The more than 40 research laboratories in suburban Pittsburgh varied widely. Most employed less than 100, but several employed thousands. The largest labs were owned by ALCOA, Gulf Oil, US Steel, and Westinghouse. Westinghouse also operated the Bettis Atomic Laboratory, which developed nuclear reactors for submarines and aircraft carriers, on contract for the Navy and the Atomic Energy Commission. Reflecting the region's industrial past, most laboratories were devoted to research into chemicals, glass, coal, aluminum, steel, and electrical equipment. Many companies hoped to expand into defense research, but only Westinghouse, the eighteenth largest defense contractor in the United States in 1958, had much success (Markusen et al. 1991).

For corporate executives, modern laboratories were intended to present the innovative character of their firms to employees, investors, and customers. For the communities that housed laboratories, they became points of pride. For the region's elite, who hoped to transform Pittsburgh from a polluted steel town to a post-industrial metropolis, the presence of research laboratories furthered their argument that the region was transcending its industrial past (Neumann 2018; Vitale 2016). Corporate and community leaders and researchers presented pristine suburban labs as places driven by pure scientific pursuit, but the reality was very different.

The Pittsburgh region grew rapidly in the late nineteenth and early twentieth century as an industrial center devoted to manufacturing steel, aluminum, glass, chemicals, and electrical equipment (Couvares 1984; Muller 1989). It was also a leading corporate center that was home to the headquarters of US Steel, ALCOA, PPG, Heinz Foods, and Westinghouse. Despite its earlier growth, by 1945 many business and political leaders hoped to transform the region from an industrial to a post-industrial economy. Manufacturing was increasingly moving to the south and west and local leaders believed that Pittsburgh's reputation as a steel town made it difficult to attract the new industries and workers that were key to future development. In pursuit of economic growth after World War II, business and political leaders took major steps to remake the region's built environment and economy. In what was called the Pittsburgh Renaissance, they implemented pollution and flood control and urban renewal to ensure that the city's downtown remained a viable corporate center (Houston 1986; Lubove 1969; Vitale 2015). At the same time, leading industrial firms moved laboratories away from industry, into the city's more affluent suburbs, in the hope of attracting scientists and defense contracts to the region.

As part of the effort to remake the region, the Regional Industrial Development Corporation and the Pittsburgh Regional Planning Association (PRPA) obtained a grant from the Ford Foundation to fund a three-volume *Economic Study of the Pittsburgh Region* (ESPR). The 1963 study was intended to establish hard facts about Pittsburgh's declining industrial economy. The President of the PRPA described it as, "a pioneer effort… in the nation-wide attack upon the problems of large urbanized areas" (Lowry 1963, 2:v).

The ESPR highlighted how Pittsburgh's concentration of corporate headquarters and laboratories boded well "in an age when more and more of Man's work involves brain rather than brawn" (PRPA 1963a, 1:382). It set an agenda for how Pittsburgh firms must create appealing workplaces, communities, and amenities for scientists in order to advance the region's growth (PRPA 1963b, 3:117). The study emphasized that researchers were "highly selective" about where they wanted to live. In order to attract them, business and political leaders needed to create an appealing "research climate," including recreational and cultural amenities and high-quality housing and schools (PRPA 1963a, 1:415–416).

Based on interviews with corporate executives and research directors, the ESPR outlined several reasons why industrial laboratories exhibited a "virtually unanimous preference... for suburban locations." One reason was a desire to project an innovative image to potential employees and customers. In his conversations with research directors, Ira Lowry, the author of the second volume of the ESPR, observed that, "Nearly all respondents stressed the importance of visual appeal" of their research centres. To research directors, a properly designed laboratory with "a campus atmosphere helps their recruitment program" and "serve[s] as symbols of progressive management." This "visual appeal" would attract employees and "impress the company's customers and the general public." According to Lowry, laboratory design, "call[s] for low sprawling buildings, landscaped grounds, and extensive parking lots" (Lowry 1963, 2:98).

Firms did not design labs purely to boost their image. Due to Cold War defense spending, most executives shared "a vague optimism about future expansion" of research of all sorts (Lowry 1963, 2:98). Suburban land was less expensive and constrained than built-up areas and allowed for years of additions. Profit-savvy firms moved to the suburbs with an eye towards expansion, but also with awareness that they could always sell excess land later.

Many research directors told Lowry that "isolation" was "advantageous." A location removed from the factory, kept staff from becoming, "overly involved in production problems." Not only was suburban isolation important for protecting researchers' autonomy, a nearby factory also harmed a laboratory's "visual appeal." As Lowry described, "untidy production facilities tend to dominate the environment, damaging the valued image of the research center" (1963, 2:98–99). Some research also required isolation. Nuclear reactor testing could not easily take place in cities. Firms purchased large plots of land to ensure that local governments and residents did not interfere with such work.

Since the early 1900s, many managers and white-collar professionals, including scientists and engineers, preferred to live in Pittsburgh's eastern suburbs, which tended to be free of industry and occupied almost entirely by fellow middle-class white professionals. The authors of the ESPR noted that when choosing a laboratory site, "management" was "concerned about the effect of a move on the work force." As a result, each firm "first investigated the geographical distribution of the residences of the research and auxiliary staff" (Lowry 1963, 2:99). To not disrupt this workforce, firms tended to move laboratories further into the suburbs where employees already lived. Firms simultaneously followed and led their white-collar employees into the suburbs. "A move into virgin territory will soon be followed by residential subdividers eager to serve this high-income market," Lowry observed, "the development of Churchill and Monroeville... has been fostered by the cluster of research activities in that area" (Lowry 1963, 2:99). Firms in Pittsburgh and other cities often promoted the neighbourhoods that surrounded their laboratories to potential and current employees (Kaiser 2002; Kaiser 2004; Karafantis and Leslie 2019; Vitale 2016).

Labs tended to cluster in an arc to the east of Pittsburgh (see Figure 1.1). The new Penn-Lincoln Parkway gave this area easy access to downtown and universities and cultural amenities in Pittsburgh's East End. A short drive from the eastern suburbs, Carnegie Tech and the University of Pittsburgh, the region's leading research universities, had developed to serve local industry (Noble 1977; Kargon and Knowles 2002). They supplied a steady supply of freshly minted engineers and scientists and allowed the existing employees of research laboratories to pursue graduate degrees, teach, and collaborate with faculty. Local business and political leaders repeatedly

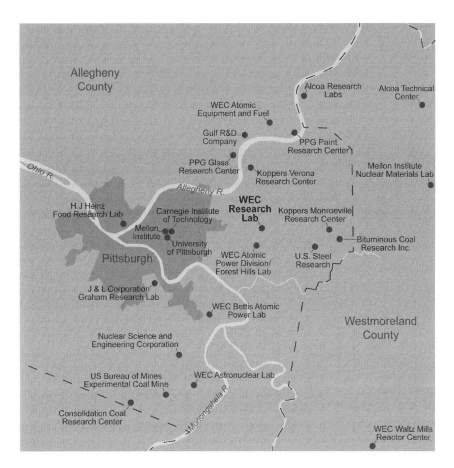

Figure 1.1 Major research facilities in Pittsburgh in 1962. WEC stands for Westinghouse Electric Corporation. (Lowry 1963, 97)

noted that proximity to universities and cultural amenities were key selling points to attract scientists and engineers to the region.

A suburban location offered an additional advantage that the ESPR did not mention: compliant local governments that rarely opposed laboratories. Municipalities embraced research centres that brought esteemed residents and taxpayers, with few of the side effects of manufacturing. Public opposition to labs was rare and usually limited to the concerns of adjacent landowners. The records of the eastern suburbs of Churchill, Monroeville, and Wilkins do not reveal a single instance of local governments seriously objecting to plans for a research lab or office complex during the 1950s or 1960s. As I describe in detail below, developers and corporate executives often served in municipal government, further limiting opposition.

Most firms developed labs in the hope of netting Federal defense and research contracts. Several other federal policies also made it advantageous to develop new facilities in the suburbs. The Armed Services Procurement Act of 1947 gave the federal government the power to use procurement procedures to encourage industrial dispersal. The federal government more effectively encouraged industrial dispersal through tax incentives (O'Mara 2005). In October 1950, the Truman administration amended the federal tax code to grant facilities given a "certificate of necessity" attesting that they did defense work full and accelerated amortization of their new facilities. Geographical dispersal was one basis by which the federal government awarded this new tax status. In 1951, the Truman administration encouraged all federal agencies to adopt measures to encourage firms to locate facilities outside of critical defense areas. The most significant federal incentive for industrial dispersal occurred in 1954, when the Department of Commerce adopted a tax policy that granted firms accelerated amortization for *all new* facilities regardless of their location. This broad tax subsidy to new construction, along with federal highway spending, encouraged firms to construct new suburban facilities in Pittsburgh and elsewhere. While generous federal tax right-offs did not assure decentralization, their combination with high taxes on corporate profits meant that new suburban facilities were often highly affordable investments for businesses (Hanchett 1996; O'Mara 2005). When coupled with subsidies and tax deductions for white suburban homeowners (Freund 2007; Jackson 1985), they accelerated the rush of laboratories and scientists to the suburbs during the Cold War.

As described above, a combination of factors contributed to firms' decisions to move labs to the suburbs during the early Cold War. Firms sought out the suburbs because they provided conditions that would allow them to attract and retain a scientific workforce, earn public renown, open new profitable areas of business, and attract government-funded research. In order to gain access to the suburbs, industrial firms needed to convince suburbanites that laboratories were a form of employment that befitted exclusive residential communities.

Building a lab to suit the suburbs

In 1956, Westinghouse, the Pittsburgh-based electrical conglomerate, moved its laboratory further into the city's suburbs from Forest Hills to Churchill (Vitale 2017). It was one of the largest laboratories in the region and at its peak employed more than 2,200, with 805 scientists and engineers engaged in research.[3] It also had the greatest diversity of researchers (ranging from metallurgists to sociologists), the strongest focus on basic research, and the closest ties to the military. Prominently located along the Penn-Lincoln Parkway, it was also the most visible laboratory in the region. In order to move to Churchill, Westinghouse executives, local residents and politicians had to adapt the laboratory and the surrounding exclusive community to fit each other's interests. Louise Mozingo (2011) and Layne Karafantis and Stuart Leslie (2018) have uncovered similar processes when research facilities moved into affluent suburbs outside of Los Angeles and New York. Digging deeply into the case of the nationally prominent Westinghouse laboratory provides a fuller sense of how industrial research and exclusive suburbs reshaped each other during the Cold War.

In 1949, Research Director John Hutcheson began to advocate within Westinghouse for a new laboratory. The reason was simple: the research needs of the company were growing, while the current laboratory could not accommodate an expanded workforce. Westinghouse was entering "new fields of endeavor," particularly electronics, and current "international tensions" were increasing military research.[4] If Westinghouse hoped to capture that work it needed to expand its workforce and the facilities that housed it.[5] Building a new lab would be highly affordable because of the growth of defense work and "the possibility of short term amortization" of the cost.[6]

Westinghouse scoured Pittsburgh's suburbs for an appropriate site. The land needed to be at least 50 acres and relatively flat. The surrounding land should be "residential rather than industrial" and preferably in a location that would allow for additional home construction. It needed to be "as close to the present site as possible."[7] Throughout the search, Hutcheson stressed the need for the new lab to be proximate to Forest Hills in order to avoid a "loss of time and personnel."[8] Given the scarcity of scientists and engineers during the early Cold War and the rapid growth of defense and research spending, similar concerns about retaining workers informed the siting of laboratories in suburbs across the United States (e.g., see: Kaiser 2004; Mozingo 2011; Karafantis and Leslie 2019).

After considering several sites, most in the eastern suburbs, Westinghouse settled on a 71-acre plot in Churchill. Churchill was no stranger to Westinghouse. The company's employees were instrumental in the Borough's secession from rural Wilkins Township in 1934. At that time, the land that became Churchill was home to the large estates of corporate executives. Owners of these estates believed that the township excessively taxed and obstructed the development of their land. In February 1934, 44 of 45 landowners voted to secede and create, "a community to which they could retire and be free of the hustle, bustle, dirt, and factories of the city" (McArdle 1934, 2). Several months after secession, the *Pittsburgh Press* observed that landowners were subdividing their estates and creating "restricted" communities. The President of Borough Council, H.C. Barton, told the *Press* that they aimed to keep Churchill, "highly restricted," in order to assure "a friendly group of neighbors; people like those residing here who hold responsible positions." Among those people were Churchill's seven new councilmen, two of which were executives at Westinghouse. In the midst of the Depression, the *Press* noted that, "the backers of Churchill Borough have built for themselves a community free from the outside world" (*Pittsburgh Press* 1934).

In 1952, when Westinghouse acquired its land, Churchill was speckled with country clubs, estates, and subdivisions. There were four African Americans among its 1,733 residents (Bureau of Census 1950). It was growing at the second fastest rate of any municipality in the region and also had the second highest median income (James 2005). Despite its growth, the Borough remained committed to exclusion and made no zoning provisions for industrial or commercial use. As the Borough secretary proudly stated in 1951, "zoning is so tight your mother-in-law can hardly move in with you" (Beachler 1951). It also remained the home of many Westinghouse executives and scientists. Westinghouse employees served extensively in local government and contributed to Churchill's development as one of Pittsburgh's most racially and socio-economically exclusive suburbs. In 1950, the wife of a former Westinghouse research director, appeared before Borough Council to support a zoning change that would prohibit smaller lot sizes because it was "for the betterment of the Borough" to "maintain zoning high."[9] Ironically, two years later, in order to finalize its land purchase, Westinghouse first had to have the exclusive community rezoned.

In January 1952, Westinghouse submitted a petition to Churchill Council requesting the re-zoning of its land. In the petition, it promised that the new lab would not produce "dust, dirt, or fumes" and "no products would be manufactured." It would be "completely landscaped" and "will enhance the appearance of the Borough."[10] Zoning Ordinance #120 created a zoning district U-7 that allowed only one additional use within the Borough: research laboratories.[11] It defined a laboratory as:

> A building or group of jointly used buildings in which research is conducted by the owner thereof to discover new scientific facts and principles for the sole and exclusive purpose of developing new and improved products to be manufactured by such owner at other locations and in which research laboratory and the land appurtenant thereto no products are manufactured or offered for sale and no services are offered for sale, and in which no operations will be conducted that will constitute a public menace.

The ordinance explicitly excluded manufacturing and sales within Churchill. It defined nuisances broadly as "odor, dust, smoke, gas, vibration, or noise" or any other uses that are "incompatible with... an essentially residential community." The ordinance restricted building height to 75 feet and only allowed build-out of 15% of land.[12] All of these restrictions were intended to ensure that only research occurred on the plot and that the building conformed with the suburban character of the surrounding neighbourhoods. The ordinance enshrined a definition of research as mental work removed from sales and manufacturing. This was inaccurate; for example, the laboratory frequently pursued outside contracts and manufactured prototypes of products, but it did create a class distinction between scientific and manufacturing labour that was essential if the lab was to fit into the exclusive community.

Prior to the public hearing of Ordinance #120, residents began to send letters in support of Westinghouse. E.M. Elkins, who worked for Westinghouse and owned 6 acres in Churchill, wrote that the proposal "will be of material benefit to the Borough."[13] Charles Williams, an employee of the University of Pittsburgh's Business School, encouraged council to "attract this highly desirable facility to our community." He described Westinghouse's "fine reputation," something, he insinuated, many council members were, "more keenly aware than I." He reminded the councilors that "the activities of a research laboratory are a far cry from those of a manufacturing plant" and that the facility would improve the tax base. As if this was not enough, he asked the councilors, "The type of person whom such a laboratory will attract to our community? Is it possible to find a better?"[14] Churchill first reviewed the proposed ordinance at an informal town meeting, where residents heaped more praise on Westinghouse and the scientists and engineers it employed. After a presentation by Westinghouse Vice President A.C. Monteith, 158 residents voted yes to the proposed ordinance, 14 no, and 9 no opinion.[15]

Support for the ordinance was similarly high when it was presented to the Churchill Council the following week. The meeting began with presentations from Monteith and Hutcheson both of whom pledged Westinghouse's appreciation of residents' "spirit... to maintain a residential atmosphere." Among Hutcheson's slides was a map of existing employees' homes, clustered in the vicinity of the new lab. The location "would mean essentially nothing to employees," he said. No employees would have to move and this "makes it very desirable for us to locate in this general area." Hutcheson emphasized that Churchill was more than a convenient location, "the people in our laboratory are scientists and engineers," and, "to these people the atmosphere in which they live is very important." Churchill offered such people a comfortable home and easy access to Pittsburgh's "cultural facilities." Lastly, he assured the audience, a laboratory is not a factory. "We are not in the business of producing apparatus," he said, "we are in the business of developing ideas." The lab required no railroad, because "ideas constitute our output" and "our production goes out... in the form of reports."[16] Once again Westinghouse maintained that there was an absolute distinction between the esteemed mental work of researchers at the lab and the labour that took place in its factories.

The vast majority of the residents in attendance were convinced. Each was invited to explain whether they supported or opposed the ordinance. Forty-six spoke in favour and five against. Of those opposed, two residents were concerned that the precedent could open the Borough to dreaded multi-family homes, "a mental institution," or "a private school... for mental deficient children." Another spoke about his fear that the Borough would be, "attacked with the atomic bomb." The most serious complaint came from the Pehna family who owned the adjoining land. Their attorney protested that, "whatever name you call it, it is still industrialization" and would, "lower the quality, characteristics and uses of the properties within the Borough."[17] This complaint took nearly a year to resolve.[18]

The Pehnas were not the last hurdle. Hutcheson became aware of the Federal Defense Production Administration's (DPA) delay in granting a Certificate of Necessity in the summer of 1952. This certificate was required for the lab to qualify for accelerated amortization. The delay stemmed from the DPA's position that the lab did not do defense manufacturing. Hutcheson found this absurd, noting that a GE laboratory in upstate New York had recently been granted a certificate. In a statement that contradicted much of what he told Churchill residents, he argued that the new lab "is as much a part of the... manufacturing facility as any laboratory... associated with any other industrial concern."[19] Westinghouse appealed to allies at the Pentagon who intervened; the lab's certificate of necessity was once again presented to the DPA for approval (Asner 2006).[20]

At this point, a different problem emerged. The DPA accepted that the lab was a defense plant, but was now concerned that it was located in a critical defense area. In correspondence with the Pittsburgh Area Industrial Dispersion Committee (PAIDC), Hutcheson explained Westinghouse's rationale for selecting Churchill. Of the six properties the company surveyed this was the only suitable site and the furthest from likely military targets. Westinghouse selected Churchill because "it is actually somewhat closer to the present homes of the laboratory employees" and it had access to utilities. Lastly, Hutcheson stressed that, "the proposed move does not... increase the potential of this area as an enemy target."[21] The DPA still did not provide a certificate and Westinghouse began construction without one.

In September 1954, Hutcheson made a final attempt to convince the DPA to grant a certificate. In a nine-page memo he explained the rationale behind the location. He began by arguing, falsely, that Westinghouse was building the lab, "due entirely to the demand [from]... the Department of Defense." There could be no dispute that the lab was a "critical" defense facility. He told the DPA that the primary criterion for choosing a site was its proximity to the homes of laboratory personnel. For scientists, he argued, this was important because they are, "individuals who have strong interests of a cultural nature," and therefore they "insist their homes be located so that it will be possible for them to find an outlet for these interests." According to Hutcheson, it was vital for national defense that Westinghouse be located in an exclusive suburb that appealed to scientists and engineers and that this outweighed concerns about defense dispersal. If Westinghouse had moved the lab to a remote location, "our scientists would have sought work elsewhere."[22] Four months after receiving Hutcheson's memo, the DPA issued a certificate for 25% of the lab's value, the portion it deemed related to defense work (Asner 2006). Like many private facilities that the federal government subsidized, Westinghouse first selected a site that met its needs and then advocated for the location's suitability for defense dispersal and corresponding tax incentives. In most cases, federal dispersal policy subsidized industries' existing preference for suburban locations.

In 1953, Westinghouse broke ground for what Hutcheson called, "the most modern Research Center in the entire electrical industry." The site, described by Hutcheson as "wooded and rolling," would become "a place of beauty as well as efficiency, surrounded by trees and shrubs on well-landscaped lawns." John Elder, President of Churchill Council, welcomed a new neighbour that would bring "beauty and desirability" and "international prestige to this area."[23]

The Westinghouse lab was "a direct copy" of the Bell Laboratories in Murray Hill, New Jersey.[24] To imitate Bell was not a surprise. As Mozingo argues, the 1941 Bell Labs with its suburban setting and flexible laboratory space, had "redefined corporate standards for research and development facilities" and "invented the fundamentals of a corporate campus" (2011, 63 & 53; Rankin 2010). Besides workshops and laboratory and office space, the new lab also included a cafeteria, lecture hall, and library. The exterior of the building was red brick selected to "recall the warmth of Colonial Williamsburg." Otherwise there was little ornamentation, "in keeping with the entire functional aspect of a great research organization."[25]

Residents vigilantly monitored the lab's construction and final appearance. At the first zoning hearing, residents quizzed A.C. Monteith about landscaping. He assured them that Westinghouse would leave most of the property "as nature has set it down" and the grounds surrounding the building would be "landscaped... the same way as you would around your home."[26] After the lab's completion the Research Director regularly invited Churchill residents and local garden clubs to tour the grounds.[27]

Expanding the lab and joining the Cold War

In 1959, Westinghouse made plans for a major expansion of the labs, now renamed a Research and Development Center. Once again it described the lab as an ideal partnership with the exclusive Churchill community. According

48 *Patrick Vitale*

to Westinghouse, by supporting the lab, Churchill would contribute to scientific advance, national security, and its tax base and desirability. In an announcement distributed to residents, Westinghouse assured them that the "new buildings and associated land areas are designed to fit into the character" of "an area of beautiful homes and natural surroundings."[28] The announcement detailed the features of the modern Skidmore, Owings, and Merrill design, which was a radical departure from the 1956 lab. The design included two long white buildings that formed a "university-type quadrangle" with a courtyard at the center. The proposed Research Center, "not only provides the proper environment for our scientists, but insures an attractive view of the center for the surrounding community."[29] This would be "harmonious and pleasing" to residents (Figure 1.2).[30]

The expansion would also draw more prized professionals to the Borough; currently "10% of the families in Churchill Borough are Westinghouse people" and "their role in the civic and cultural activities… is well-known." Not only were the lab's employees a credit to the community, but the lab also hosted community events and many gardeners had visited its "beautiful landscape." The announcement closed with the coup-de-grace: to stand in the way of the Research Center was to block the road to growth, progress,

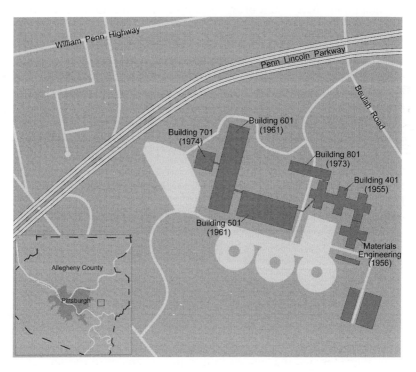

Figure 1.2 Growth of the Westinghouse Research and Development Center from 1956 to 1974.

and "national security." The "new knowledge, new concepts, and new products" created at the Center would be vital to "industrial progress" and the "economic and military power" of the United States. "World trade is partly a battle of technology," the announcement explained, and "competition in the military sphere is likewise a struggle for technical supremacy." Westinghouse made explicit the collapsing geographic scales of Cold War militarism (Cowen 2004, Farish 2010). By making space for the lab in their exclusive community, Churchill residents could contribute to American economic, military, and scientific supremacy. The announcement ended by asking residents to, "join with us in making our community one of the world's outstanding centers of research and development."[31]

S.W. Herald, Vice President of Research at Westinghouse, hit these points again when he spoke to Churchill Borough Council. He noted that, "to some degree," if the Borough agreed to the expansion then they were, "contributing to this country's research and development progress." Such progress was vital because without it "our nation's position, both as an economic and military power, is jeopardized." The Research Center would be, "a pleasant place to work" and "a place of beauty for our community." He tallied up the benefits that Westinghouse provided: it paid 13% of school taxes and 25% of Borough taxes, was a "good neighbor," "laid out the welcome mat" to local organizations, and hosted science education programs. One of the biggest benefits the lab provided was the influx of well-paid scientists and engineers into the community. "A credit to any community," he proudly noted, "Westinghouse people are the kind of people who actively support community affairs, youth programs, schools."[32] The Council approved the expansion, with five members voting yes and one no.

In 1969, Westinghouse made plans for another expansion, this time to house administrative offices. Space was desperately needed as the lab was increasingly oriented towards "division needs," "new product activity," and "cost-price improvement goals." This required more space in which to test manufacturing processes and products. More space would also allow the company to enter new fields, many of them, such as surveillance systems, modular construction, and education, opened by the federal government's efforts to develop scientific solutions to urban problems (cf. Light 2003).

The proposed addition also dealt with pressing "aesthetic considerations" and improved "the corporate research image." One of Westinghouse's goals was to block the view of the 1956 lab (renamed Building 401) and thereby, "greatly improve the outsider's impression of the R&D Center." Building 401, a symbol of modernity when it opened 13 years earlier, needed to, "be brought into conformity with contemporary Corporate Design policy." Westinghouse had painted Building 401's "warm colonial Williamsburg" brick white in 1968 to match the 1961 addition, but it nonetheless, "does not 'fit' the modern design of the complex." To be blunt, as one Borough Councilor complained, it looked like a "textile mill out of Lawrence, Massachusetts."[33] The new addition resolved this problem by blocking the view of Building 401. Improvements to the appearance of the complex were also necessary from a "community relations standpoint" because the Center's location in "Churchill Borough, an exclusive residential area, demands special attention to the architectural beauty of the site."[34]

Throughout the lab's development, Westinghouse derived local benefits from emphasizing how the character of research fit into the exclusive community. Churchill residents were highly resistant to any suggestion that it was a manufacturing facility. As a result, at nearly every meeting where Westinghouse presented plans it emphasized that the lab produced ideas, not products. This illusion was necessary to maintain the acquiescence of an exclusive community that valorized scientific labour and compliance with zoning law that explicitly stated that work at the site was limited to the search for "new scientific facts and principles." Borough residents occasionally contested whether the work at the lab was research, but usually deferred to Westinghouse, because, as one councilor observed, "once we get into the realm of talking about... what goes on at the research lab, we have lost most of us on council."[35] At one hearing, the council questioned whether a new facility to test parts for nuclear reactors was research, but ultimately accepted an affidavit from Westinghouse stating that it was. At an "exploratory meeting" concerning further rezoning of the property in 1964, Councilors also expressed skepticism about whether all of the lab's work was research. The meeting concluded with Westinghouse executives' promise that they "would not embarrass the Borough by manufacturing on the site."[36]

In fact, Westinghouse routinely violated Churchill's zoning ordinance. It was impossible to limit research to the production of ideas and the lab frequently engaged in sales, the production of prototypes, and activities related to manufacturing. Such activities only increased in the 1960s and '70s when the company moved away from basic research (Asner 2006). Nonetheless, Churchill officials and residents, who saw the lab as a perfect match to the exclusive community, ignored these activities in order to maintain the illusion that it was a place where ideas were produced, not products. Churchill residents and politicians and Westinghouse reimagined science and engineering as purely mental labour befitting an exclusive community.

Conclusion

Compared to Boston or Los Angeles, the growth of research and development in Pittsburgh's suburbs was relatively small during the Cold War. However, even in a region greatly removed from the sunbelt, the desire to attract defense contracts to the region was irresistible (Markusen et al. 1991). Industrial firms developed laboratories, in part, with the hope that they would capture a portion of the ever-growing spending of the warfare state. Even if they failed, generous tax policies assured the profitability of their new laboratories. One key component of the Monroeville Doctrine – local firms' belief that the suburbs were the best place for research – was the hope that it would allow them to secure Washington's largesse.

Attracting defense contracts was not the only component of the Monroeville Doctrine. Industrial firms build new research laboratories in order to create conditions that would allow them to open and monopolize new markets. This strategy had been the major driver of industrial research since the turn of the century (Mowery and Rosenberg 1989; Smith Jr. 1990; Hounshell 1996). There were many conditions that corporate executives believed a laboratory in the suburbs would provide. Flexible and expandable suburban workspaces could adapt to the needs of Cold-War science. Labs in the suburbs isolated researchers from the immediate demands of corporate headquarters and manufacturing divisions. A freshly built laboratory projected an image of modernity to customers, investors, and employees. The most important condition, during the early Cold War, was firms' need to attract scientists and engineers that were in scarce supply and great demand. Firms in Pittsburgh and elsewhere built new labs in exclusive suburbs because they believed this was the best way to entice scarce scientific labour (Kaiser 2004; Vitale 2016).

Firms sought out suburbs as sites for their laboratories because they believed scientists and engineers preferred exclusive communities. In order to fit their laboratories into these suburbs, firms exaggerated the separation between research and manufacturing. They claimed that industrial scientists pursued immaterial knowledge, not the manufacture of material things. Many suburban municipalities wrote this false distinction between research and manufacturing into their zoning code. Exclusive suburbs welcomed research laboratories as in keeping with their efforts to remain communities of white middle-class professionals that were free of industry, the working class, and African Americans. They saw laboratories as a way to attract prestige, tax dollars, and other outstanding residents. Cold-War era science and elite residential suburbs developed productively in concert, each reshaping the other in its exclusionary image.

Acknowledgement

A earlier version of this chapter was published in *Nuclear Suburbs: Cold War Technoscience and the Pittsburgh Renaissance* (University of Minnesota Press). Published by permission of the University of Minnesota Press.

Notes

1 Monroeville Chamber of Commerce, "Residential Research Center of the Nation," (March 1964) Monroeville folder, Carnegie Library of Pittsburgh.
2 *The Times Express* (September 25, 1969).
3 John Harlon, "Analysis of shift in research professions at Westinghouse R&D Center," (1973), Westinghouse Electric Corporation Archives (WECA), Historical Society of Western Pennsylvania (HSWP); R.D. Haun, "Westinghouse Central Corporate Research and Development," (October 7, 1971) WECA.

52 *Patrick Vitale*

4   J.A. Hutcheson, letter to L.D. Rigdon (April 19, 1951) WECA.
5   J.A. Hutcheson, letter and draft memorandum to A.C. Monteith (November 11, 1951) WECA.
6   Hutcheson letter to Rigdon (April 19, 1951).
7   J.A. Hutcheson, Letter to H.W. Reding (June 29, 1951) WECA.
8   Hutcheson, Letter to Rigdon (April 19, 1951).
9   Churchill Borough Minute Books (September 11, 1950) Churchill Borough Municipal Records (CBMR).
10 Petition of Westinghouse Electric Corporation, Westinghouse Zoning Ordinance folder (WZOF), CBMR.
11 Bell Laboratories, built in Summit, New Jersey in the early 1940s, had the first zoning ordinance that limited land use to research facilities (Mozingo 2011, 63). Similar zoning ordinances were also put in place in the exclusive Los Angeles suburb of Palos Verdes (Karafantis and Leslie 2019, 11).
12 Ordinance # 120 in WZOF, CBMR.
13 E.M. Elkin, letter to Churchill Borough Council (January 28, 1952) in WZOF, CBMR.
14 Charles A. Williams, letter to Churchill Borough Council (January 21, 1952) in WZOF, CBMR.
15 Public hearing RE proposed ordinance No. 120 (February 11, 1952), 55–58, Churchill Borough Council Minutes (1951–1953), CBMR.
16 *Ibid.*, 7–8. Tellingly, in the early 1940s, the Bell Laboratories also met with little resistance when it moved to the exclusive suburb of Summit, New Jersey. Like the Westinghouse lab, the presence of Bell employees as residents of Summit paved the way for its relocation. Frank Jewett, the Director of the Bell Labs, made a nearly identical defense to Hutcheson's. At a council meeting, he assured residents that the lab's employees were, "mostly of the scientific and engineering type who will want to live" in the communities near the lab. He added that the lab "is not a manufacturing concern: it is a laboratory." Lastly it would be landscaped by Olmsted Brothers and "will in no way harm the growth of this entire section as a residential community but rather will be of great assistance in keeping this territory on a high plain of a suburban residential community" (Mozingo 2011, 56).
17 Public hearing RE proposed ordinance No 120, 23–30 (February 11, 1952).
18 J.A. Hutcheson, letter to Earl Gulbransen (May 19, 1952) WECA; J.A. Hutcheson, letter to A.C. Monteith (September 8, 1952) WECA.
19 J.A. Hutcheson, letter to R.M. Wilson (July 18, 1952) WECA.
20 Hutcheson, letter to Monteith (September 8, 1952).
21 J.A. Hutcheson, letter to Charles R. Miller (November 28, 1952) WECA.
22 J.A. Huctheson, Memorandum Necessity Certificate – TA -21260 (September 16, 1954) WECA.
23 WEC, "Westinghouse breaks ground for most modern research center," Press Release (June 12, 1953) WECA.
24 J.A. Hutcheson, letter to Hugh Nielson and Henry Schwartz (April 25, 1952) WECA.
25 Vorhees, Walker, Foley, & Smith Architects and Engineers, "Westinghouse Research Laboratories Fundamental Design Report" (November 11, 1952) CBMR.
26 Public hearing RE proposed ordinance No. 120 (February 11, 1952) 12, CBMR.
27 Clarence Zener, "To Our Neighbors in Churchill Borough" (May 20, 1957) Westinghouse Permit #789 folder, CBMR; Westinghouse R&D Center, "Landscaping Guide" (1968) WECA.
28 "Westinghouse Research and Development in Churchill Borough," (1959) WECA.
29 WEC, press release "Westinghouse Proposes Expansion of Research Laboratories in Churchill," (June 1959) Westinghouse – Ord. #236 folder, CBMR.

30 "Westinghouse Research and Development in Churchill Borough," (1959).
31 *Ibid.*
32 "Partial Report of Proceedings before Council of the Borough of Churchill," (August 18, 1959) Westinghouse – Ord. #236 folder, CBMR.
33 Public Hearing: Westinghouse R&D Rezoning (September 15, 1964) 7 & 10, Westinghouse Expansion folder, CBMR.
34 *Ibid.*, 17–18.
35 Minutes of recessed meeting of Council (November 11, 1978) CBMR.
36 Minutes of exploratory meeting (August 31, 1964) Westinghouse Expansion R.C.H. fodler, CBMR.

Bibliography

Asner, Glen Ross. 2006. "The Cold War and American Industrial Research." Ph.D. Dissertation, Carnegie Mellon University.

Beachler, Edwin H. 1951. "Jack-and-Beanstalk Town Built Around Parkway." *Pittsburgh Press*, May 9.

Bureau of Census. 1950. *1950 United States Census of Population: Pittsburgh, PA. Census Tracts*. P-D43. Washington, D.C.

Couvares, Francis G. 1984. *The Remaking of Pittsburgh: Class and Culture in an Industrializing City, 1877–1919*. Albany: State University of New York Press.

Cowen, Deborah. 2004. "From the American Lebensraum to the American Living Room: Class, Sexuality, and the Scaled Production of 'Domestic' Intimacy." *Environment and Planning D: Society and Space* 22 (5): 755–771.

Dieterich-Ward, Allen. 2010. "From Mill Towns to 'Burbs of the Burgh': Suburban Strategies in the Postindustrial Metropolis." *Research in Urban Sociology* 10: 75–105.

Farish, Matthew. 2010. *The Contours of America's Cold War*. Minneapolis: University of Minnesota Press.

Freund, David. 2007. *Colored Property: State Policy and White Racial Politics in Suburban America*. Chicago: University of Chicago Press.

Hanchett, Thomas. 1996. "U.S. Tax Policy and the Shopping-Center Boom of the 1950s and 1960s." *The American Historical Review* 101 (4): 1082–1110.

Hounshell, David. 1996. "The Evolution of Industrial Research in the United States." In *Engines of Innovation: US Industrial Research at the End of an Era*, edited by Richard Rosenbloom and William J. Spencer, 13–85. Boston: Harvard Business School Press.

Houston, David. 1986. "A Brief History of the Process of Capital Accumulation in Pittsburgh: A Marxist Interpretation." In *Pittsburgh-Sheffield Sister Cities*, edited by Joel Tarr, 29–70. Praxis/Poetics 3. Pittsburgh: Carnegie Mellon University.

Jackson, Kenneth T. 1985. *Crabgrass Frontier: The Suburbanization of the United States*. New York: Oxford University Press.

James, Kent. 2005. "Public Policy and the Postwar Suburbanization of Pittsburgh, 1945–1990." Dissertation, Pittsburgh: Carnegie Mellon University.

Kaiser, David. 2002. "Cold War Requisitions, Scientific Manpower, and the Production of American Physicists after World War II." *Historical Studies in the Physical Sciences* 33: 131–159.

Kaiser, David. 2004. "The Postwar Suburbanization of American Physics." *American Quarterly* 56: 851–888.

Karafantis, Layne, and Stuart W. Leslie. 2019. "'Suburban Warriors': The Blue-Collar and Blue-Sky Communities of Southern California's Aerospace Industry." *Journal of Planning History* 18 (1): 3–26.

Kargon, Robert H., and Scott G. Knowles. 2002. "Knowledge for Use: Science, Higher Learning, and America's New Industrial Heartland, 1880–1915." *Annals of Science* 59 (1): 1–20.

Light, Jennifer S. 2003. *From Warfare to Welfare: Defense Intellectuals and Urban Problems in Cold War America*. Baltimore: Johns Hopkins University Press.

Lowry, Ira S. 1963. *Portrait of a Region*. Vol. 2. 3 vols. Economic Study of the Pittsburgh Region. Pittsburgh: University of Pittsburgh Press.

Lubove, Roy. 1969. *Twentieth Century Pittsburgh: Government, Business, and Environmental Change*. Vol. 1. 2 vols. Pittsburgh: University of Pittsburgh Press.

Markusen, Ann R., Peter Hall, Scott Campbell, and Sabina Deitrick. 1991. *The Rise of the Gunbelt: The Military Remapping of Industrial America*. Oxford: Oxford University Press.

McArdle, Kenneth. 1934. "'Civil War' Faced by Wilkins Township as Big Property Owners Try to Secede." *Pittsburgh Press*, February 4.

Mowery, David C., and Nathan Rosenberg. 1989. *Technology and the Pursuit of Economic Growth*. Cambridge: Cambridge University Press.

Mozingo, Louise. 2011. *Pastoral Capitalism: A History of Suburban Corporate Landscapes*. Cambridge: MIT Press.

Muller, Edward K. 1989. "Metropolis and Region: A Framework for Enquiry into Western Pennsylvania." In *City at the Point: Essays on the Social History of Pittsburgh*, edited by Samuel Hays, 181–212. Pittsburgh: University of Pittsburgh Press.

Neumann, Tracy. 2018. "Reforging the Steel City: Symbolism and Space in Postindustrial Pittsburgh." *Journal of Urban History* 44 (4): 582–602.

Noble, David. 1977. *America by Design: Science, Technology, and the Rise of Corporate Capitalism*. Oxford: Oxford University Press.

O'Mara, Margaret Pugh. 2005. *Cities of Knowledge: Cold War Science and the Search for the Next Silicon Valley*. Princeton: Princeton University Press.

Pickens, Carol. 1969. "Diversified Development Will Benefit the Borough for Many Years to Come." *The Times Express*, September 25.

Pittsburgh Press. 1934. "Council Keeps Close Eye on 'Ideal' Borough," June 10.

PRPA. 1963a. *Region in Transition*. Vol. 1. 3 vols. The Economic Study of the Pittsburgh Region. Pittsburgh: University of Pittsburgh Press.

PRPA. 1963b. *Region with a Future*. Vol. 3. 3 vols. Economic Study of the Pittsburgh Region. Pittsburgh: University of Pittsburgh Press.

Rankin, William J. 2010. "The Epistemology of the Suburbs: Knowledge, Production, and Corporate Laboratory Design." *Critical Inquiry* 36: 771–806.

Smith Jr., John Kenly. 1990. "The Scientific Tradition in American Industrial Research." *Technology and Culture* 31 (1): 121–131.

Vitale, Patrick. 2015. "Anti-Communism, the Growth Machine and the Remaking of Cold-War-Era Pittsburgh." *International Journal of Urban and Regional Research* 39: 772–787.

Vitale, Patrick. 2016. "Cradle of the Creative Class: Reinventing the Figure of the Scientist in Cold War Pittsburgh." *Annals of the American Association of Geographers* 106 (6): 1378–1396.

Vitale, Patrick. 2017. "Making Science Suburban: The Suburbanization of Industrial Research and the Invention of 'Research Man.'" *Environment and Planning A* 49 (12): 2813–2834.

2 The city of Bristol
Ground zero in the making

Bob Clarke

WESSEX ARCHAEOLOGY, UK

Introduction

Civil defence, when considered within a British context, comprises two major facets: the memory of World War II (1939–1945) and the role of the volunteer. During WWII, thousands of men and women dealt with all manner of emergencies and dangerous situations, their endeavours on the 'home Front' becoming a major propaganda weapon of the period. At the end of the conflict, the entire organisation was stood-down and the majority of its equipment, by this time mostly antiquated, disposed of. Within four years, the British Government had re-established the service and constructed a multi-layer passive network intended to ensure something of the British way of life would survive attack by atomic weapons. In 1952, the United States detonated a hydrogen yield device, and by 1955, the entire process of planning for the 'A-bomb' was rendered redundant by the realisation that the Soviet Union had not only devised a hydrogen weapon but one that was small enough to deliver by air. Moreover, Soviet intercontinental ballistic missile development meant that much of the Western Hemisphere would soon be within range of attack. This chapter focusses on the immediate post-WWII period through to 1955, and the realisation that there was no real defence against hydrogen weapons. A complex landscape of volunteer forces, their buildings and training is described. Geographically, Bristol, England and Civil Defence Region 7 are utilised here to provide context to the early Cold War period.

The landscape of Civil Defence is complex – it is also, to a greater degree, invisible. That is not to say it does not exist, more that it is heterotopic in concept, a landscape born out of deviance and crisis (Foucault 1967). The principle behind preparing for nuclear warfare is that one hopes a strong enough message will be sent to the opposing side that you are, indeed, preparing to survive. In an ideologically driven conflict such as the Cold War, a conflict where information and intelligence was the major currency, civil defence and protection for the public was just as important a symbol as was the acquisition of new and more powerful weaponry. In the late 1940s, lines of future tension such as the Berlin Wall had yet to be drawn, although with a growing realisation of Soviet intentions for Central Europe culminating in the first 'battle' of the Cold War – the Berlin Blockade – it was clear that some form of civil defence programme would be needed in the United Kingdom.

This chapter investigates the formation of the civil defence landscape, relying on both documentary and existing archaeological investigation to provide a narrative for purpose-built structures connected with the early days of the reformed Civil Defence movement in the United Kingdom. The focus of this landscape study is the City of Bristol, England where a concentration of structures was positioned in both populated areas and the surrounding countryside, linked to the defence of Britain from atomic attack. This work demonstrates that the concept of civil defence is inextricably linked with other facets of early Cold War home defence including the Royal Observer Corps and Anti-Aircraft Command. It is also almost totally dependent on the volunteer. The initial period of 'optimism' is quickly dismissed when, in 1952, the testing of hydrogen Bomb, rendered both the structures and the organisations that used them redundant.

Threats and counter threats

In early 1948, a dangerous situation, primarily over territorial gains in Germany, had arisen between the Soviet Union and its former Western Allies (Wohlstetter 1959). By September 1948, the western enclave of Berlin, which lay inside the Soviet Zone of post-war Germany, was completely isolated to surface transport (Clarke 2007). The only way to support a pro-western population of just over 1 million civilians and a substantial military garrison was by air. As tensions grew, the United States Government publically acknowledged the fact that 'atomic bombers', as the British Press called them, had been deployed to the United Kingdom (*ibid* 2007, 110). The aircraft, stationed at Royal Air Force Waddington and Scampton, Lincolnshire were nuclear capable B-29 'Silverplates' (Clarke 2005a, 153). The reality of their capability was somewhat different but regardless the move escalated the stand-off over Berlin and the British Government prepared for war.

The regeneration of a force

On 2 May 1945, the Civil Defence Service (CDS), backbone of civil involvement in home defence in World War II was officially stood-down. The process had started in mid-1944 with the relaxation of fire guard duties, although the majority of other organisations within the CDS were retained until an order, issued on 26 April 1945, indicated that the 2 May was to be the final day of the service (Woolven 2005, 35). Just four months later, the use of nuclear weapons against two Japanese cities in quick succession rekindled the debate on the value of Civil Defence in the British Isles:

> Hon. Members will remember how a few days ago we were discussing a Civil Defence Bill. We are not completely destroying our civil defences;

we are maintaining our civil defences and considering new forms of civil defence against bomb attacks. - It is quite likely that we may be the cock-pit in a future struggle between a great power in the East and a great power in the West, and we are particularly vulnerable, whether we take part in that struggle or not.

Dr. Jeger (St. Pancras, South-East)
Foreign Policy (President Truman's Declaration)
(HC Deb 07 November 1945 vol 415 cc1290–390)

The immediate post-war period was subsequently characterised by a renewed sense of government despair. The country, in the grip of an almost total economic collapse, now had to consider a new type of war, one that, if it came, had the potential to destroy on unprecedented levels. Debate throughout 1945–1947 discussed the possible dispersal of the population into the countryside and every type of industrial and infrastructure project was scrutinised for vulnerability to atomic attack (HC Deb 31 October 1946. Para 818). The dispersal proposals never left the discussion stage as no finance was available; it was also recognised as being both unachievable and pointless (Grant 2010, 13). Alongside the desire to create some form of civilian-based home defence service was a recognition that new equipment would be needed in the face of a new type of warfare. The service would also need a different type of protection and training to be effective. The issues surrounding the acquisition of equipment focussed, in 1946–1947, on what wartime stock was still fit for use; training, interestingly, had remained well provisioned for due to the government being eager to retain parts of the wartime property estate:

Mr. Alpass asked the Secretary of State for the Home Department what use is now being made of Eastwood House, Falfield, Gloucestershire.

Mr. Ede It is proposed to retain a skeleton organisation at these premises to survey and record the lessons of Civil Defence in the light of experience in this country and elsewhere during the war. The rest of the premises will be available for training purposes in connection with other Home Office activities.

Eastwood House, Falfield, Gloucestershire
(HC Deb 16 October 1945 vol 414 c979W 979W)

Falfield House was to play a pivotal role in the regeneration of civil defence throughout the 1950s and 60s, introducing a training landscape tied specifically to the Cold War.

Eastwood Park and Falfield House

Eastwood Park, Gloucestershire, is located 20 miles north of the centre of Bristol. The park and nineteenth century seat, Falfield House, was purchased

Figure 2.1 Cold war period, Civil Defence training, aimed to be as realistic as possible. Here a member of the Casualty Actors Union is 'rescued' by a member of the CDC Rescue Section during an exercise at Falfield.
Source: original copyright HMSO-Home Office Training Pack.

by the Home Office on the breakup of the estate in 1935; the following year an anti-gas school had been established on the site. 'During a debate covering the reconstruction of the civil defence network it was announced that 'It was opened in April and 150 instructors have passed through already' (HC Deb 30 July 1936. Para 1850). After WWII, the park and house were retained and turned over to police training, although, by 1948 the site was identified as the sole focus of instructor training in the Southwest of England (HC Deb 23 November 1948. Para 1101) and the first course started on 6 May 1949. The Home Office Civil Defence School, Falfield was designed to provide training for instructional staff who, once qualified, delivered the training for volunteers; the establishment also delivered specific courses such as requalifying for Warden and Rescue teams, as well as specialist scientific courses dealing with nuclear and chemical weapons (CDC School 1954, 5) (Figure 2.1).

The site was extensive, in the first year of operation a range of structures aimed at training specifically for the effects of atomic warfare were built, complementing those extant from the Anti-Gas Training School. Within a year over 1,600 potential instructors had passed through the school (CDC School 1954, 6); the majority of those from the Bristol area or Civil Defence Region 7.

The landscape of training was carefully designed to mimic the effects of an atomic attack on the centre of a large town or city. The courses were twofold; singular activities were provided with specific scenarios; these instructional areas included Fire Fighting (limited to basic techniques); Welfare (specifically emergency feeding in the open); Field Hygiene and Sanitation; Chemical and Explosive Effects Training; Rescue & First Aid and Radiation Effects and Protection. More substantial exercises were run in a large area called 'The Rescue Training Ground', an area of purpose-built semi-destroyed houses and public buildings, including later a church (Essex-Lopresti 2005, 35) (Figures 2.2a and 2.2b).

Figure 2.2a Training grounds comprised a range of scenario based structures and landscapes. This area contains a number of purpose built structures, within which 'casualties' could be places and rescued.
Source: original copyright HMSO-Home Office Training Pack.

60 Bob Clarke

Figure 2.2b Alongside buildings a range of transport modes were incorporated into training scenarios. Falfield contained road transport, a train and at least one aircraft.
Source: original copyright HMO-Home Office Training Pack.

Civil defence region 7

Alongside the regeneration of the civil defence network, the government also made plans for the continuation of a basic level of regional administration to cope with the immediate aftermath of a major air attack from the East. The Civil Defence Regions – utilised throughout WWII – were retained and each provided with a protected Regional Commissioners Office (known as a war-room) built near each region's main population centre. Geographically the United Kingdom (inc. Northern Ireland) was divided into 12 Regions with London sub-divided into four. Bristol being the largest populated area within Civil Defence Region 7 (CDR7), was the location of the war-room. Located at Flowers Hill, Brislington in a eastern suburb of the city, within a compound containing a range of Temporary Office Blocks (TOB) originally part of the national government's wartime estate, the war-room was responsible for the administration of the entire southwest of the United Kingdom had Central Government collapsed or been destroyed.

The war-room was a substantial structure. Costing around £100,000, the bunker at Flowers Hill was built to a standard design comprising a two-storey semi-sunken structure. The outer shell is constructed using re-enforced concrete 121.94 cm (48 inch) thick, the roof in excess of 200 cm (78.74 inch) creating a structure with a protection factor (pf) of 400 (the average house was between 10 and 15 pf) against radiation (Clarke 2005a, 171). Internally, a two-storey planning and map room was surrounded by offices and welfare facilities, the above-ground section has an annex containing the standby generator and an air-conditioning plant (*ibid* 2005, 173) (Figure 2.3).

Figure 2.3 The War Room at Flowers Hill, Bristol. This protected structure was the seat of Home Defence Region 7s Regional Commissioner on transition to war.
Source: Bob Clarke.

The surrounding Temporary Office Blocks contained the permanent aspects of the Civil Defence Corps, from here the day-to-day administrative tasks required to support the voluntary force were carried out. Flowers Hill acted as the southwest regional command centre in peacetime, a central permanent staff dealing with training and equipment requests as well as disseminating new orders out to the county councils in Civil Defence Region 7. On transition to war, the protected structure would be populated by key staff, everyone else would be expected to stay at their posts in the TOB. Bristol, being the largest populated area in the CD region, was naturally the focus of the largest civil defence effort. Interestingly, the Regional Commissioner did not have control of Bristol – that was devolved to the City Council who had a central control room in the Council House right in the centre of the City (McCamely 2002, 217). Bristol was further complemented by four civil defence sub-controls located at strategic positions around the main population areas, all were operational by 1954 (McCamely 2002, 198). This was in line with Government guidelines issued in 1952 under Home Office Circular 3/52 which state 'One Control Centre HQ + One Sub-Control per 100,000 head of population'.

Civil defence structure

Civil Defence, as formed in 1949, comprised a number of organisations brought under one banner. Primary, and controlled directly by the Home Office of the British Government, was the Civil Defence Corps (CDC), below that all County Authorities were responsible for the raising of volunteers,

Figure 2.4a Government poster to recruit volunteers to the auxiliary fire service. This organisation operated the Bedford self-propelled pump, otherwise known as the iconic 'Green Goddess'.
Source: original copyright HMSO-Home Office recruitment.

provision of controls and training, although, the cost of sending prospective instructors to Falfield for training was covered by Central Government. At the County level, the CDC comprised a number of divisions including Warden; Ambulance; Recue; Welfare and Headquarters sections. Allied to this was the Auxiliary Fire Service (operators of the iconic 'Green Goddess' fire pump), The National Hospital Reserve and the Special Constabulary (Cocroft *et al.* 2003, 229; Grant 2010, 64). Also considered part of the whole Civil Defence package was the St. John's and St. Andrew's Ambulance Brigade and the Women's Voluntary Service (who attended to those in the CDC while they were in pursuit of their duties) (Figures 2.4a).

In 1952, a further arm to the CDC was formed – The Industrial Civil Defence Service (ICDS). With the Civil Defence Corps being essentially a voluntary organisation, comprising very few salaried members, the service was struggling to recruit members and was way off initial targets. Just a few months into the recruitment effort the situation was less than ideal:

Mr. Osborne asked the Secretary of State for the Home Department what response he has received to his appeal for volunteers for the Civil Defence Services; how many are needed to man these Services; and how many have enrolled.

Mr. Ede From 15th November to the end of January approximately 32,000 volunteers enrolled in the Civil Defence and Allied Services in England and Wales. The ultimate requirements of the Civil Defence Services have not yet been settled, but they will certainly be many times greater than the existing number of recruits.

CIVIL DEFENCE (VOLUNTEERS)
HC Deb 09 March 1950 vol 472 c3W 3W

A year later the situation looked bleak. Many who had indicated they would be willing to join (especially those who were members during WWII) simply did not materialise:

Mr. de Freitas In England and Wales during December, 10,900 recruits joined the Civil Defence Corps, bringing the strength at the end of the month to 98,900; and 900 recruits joined the Auxiliary Fire Service, bringing the strength up to 7,900.

Recruitment
HC Deb 25 January 1951 vol 483 cc278-9

To make matters worse when the government tried to raise the profile of Civil Defence, those in the press saw the opportunity to make a story of the coming war. Indeed, these were the first rattles of a growing anti-war sentiment surrounding Britain's involvement in the Korean War – by the end of the decade many would be outwardly challenging Government Civil Defence policy through the Campaign for Nuclear Disarmament (Figure 2.4b).

Civs. Defence. 1. The Prime Minister said that he had been concerned to see, in reports of a Press Conference which the Home Secretary had held on the previous day, references to the delivery of new sirens for sounding air-raid warnings. He thought it unwise that the public should be alarmed by undue publicity about civil defence preparations at this stage.

The Home Secretary said that the purpose of his Press Conference had been to stimulate recruiting for the civil defence services, which were seriously below their peace-time establishment. It was unfortunate that the newspapers had given such prominence to his remarks on the subject of equipment. He had no intention of obtruding upon public attention the arrangements which were being made to complete the air-raid warning system.

CONCLUSIONS of a Meeting of the Cabinet held at 10 Downing Street, S.W. 1 on Tuesday, 5th February, 1952, at 11 a.m.

Figure 2.4b A bookmark from 1951. Such recruitment material was handed out at cinemas, fetes, and a range of other public gatherings.
Source: original copyright HMSO-Home Office recruitment.

Notwithstanding, the growing anti-war sentiment, it became clear that those who were joining were mostly of retirement age or women still in the home, those who the CDC would most benefit from, industrial workers over the age of 40 (younger men had National Service obligations) were simply not joining due to commitments to work and family. The answer was to target the workers directly – the Industrial Civil Defence Service was born. The organisation was intended to run parallel to the existing CDC and contain

The city of Bristol and ground zero 65

Figure 2.4c Recruitment poster c. 1957. One of the major selling points of the CDC was its Welfare Section. However, the affects of the H-bomb were radically different to that of the A-bomb, and such services would have been quickly overrun.
Source: original copyright CDC/Home Office recruitment drive, 1957.

a similar framework (Warden; Ambulance; Rescue and Headquarters – Welfare was not raised) (Clarke 2005a, 156). Requirements were simple, any company with 200 or more employees were encouraged to form their own company force – the company' management team acted as officials and formed the link between them and the Local Authority CDC. While the new organisation's title suggested heavy industry, the idea was not restricted to just that, any employer could be supported to form a Corps in any business, including offices and retail stores (Clarke 2005a, 156) (Figure 2.4c).

> It is hoped that personnel engaged in the transport undertaking will take full advantage of the training facilities available in order that they may be equipped to play their part in maintaining the services which are so vital to the life of this Nation.
> British Transport Civil Defence Training Booklet (September 1951)

The defence of Bristol

The Civil Defence plan, whilst overtly promoted in the propaganda war, was a reactionary force, the majority of the training and planning to be initiated by a transition to war, all reliant on the volunteer for the majority of its success. The protection of both Bristol and the wider population and assets fell, naturally, to the military. Interestingly those aspects linked with the protection of Bristol were also reliant on a wide range of volunteer organisations. What follows is an account of those groups and their sphere of operation.

Radar, a substantial contributing factor in early British successes such as the Battle of Britain (Clarke 2005a, 46) had been all but abandoned by 1946, the Government focussing its limited resources on social reform and industrial reconstruction instead (*ibid* 2005, 45). The increasing Soviet threat, demonstrated by the Berlin Blockade in 1948/9, and growing tensions in other parts of Europe indicated that the United Kingdom required a new air defence network, especially radar coverage, if the threat of aerial attack was to be countered. The 'Rotor' programme became one of Britain's largest post-war military developments (Cocroft *et al.* 2003, 86); aimed at providing total radar coverage across the United Kingdom, the programme struggled to keep pace with developments in technology and was never fully implemented (*ibid* 2003, 87). This is demonstrated by a range of incomplete structures at RAF Charmy Down, east of Bristol; intended to be a Readiness Ground Control Intercept station helping guide RAF Fighters onto enemy aircraft formation – the site was abandoned in 1955– the technology having been superseded (Clarke 2005a, 50). The enormity of the undertaking meant that there was no credible total radar coverage for most of the first decade after WWII, forcing the Royal Air Force to rely on more traditional forms of aircraft tracking – observation – to do so, the Government, especially the Air Ministry, turned to another voluntary wartime service – the Royal Observer Corps.

'Forewarned is forearmed'

The Royal Observer Corps (ROC) had been disbanded in May 1945 as part of the rundown of voluntary services post-WWI, however, by September 1946, the recruitment of former observers had started in earnest (Anon 1946a, 5 September, 254). The idea was to 'plug' the radar gap using the skills and lessons learned throughout the early 1940s. Observers, posted at specific locations out in the landscape, were to spot enemy aircraft, identify the type, height, range and direction and pass on the information, in real time, to a reporting centre.

Over 1,420 observation posts were reactivated across the British landscape by 1949 (Cocroft *et al.* 2003, 174), and in 1952, orders were placed by the Air Ministry with Messrs Orlit Ltd of Colnbrook, Buckinghamshire for a standard structure to replace the outdated wartime posts. Two designs, known as the 'Orlit A' and 'Orlit B' were approved and supplied. Each post was equipped with tracking instruments and communication links into the reporting network along with information from operational Rotor stations. Importantly, the Orlit Post is one of the first recognisable structures in the archaeological record linked exclusively to the Cold War (Clarke 2014) (Figure 2.5).

Figure 2.5 A promotional picture from 1952 showing a female Royal Observer Corps spare time member. The ROC actively encouraged volunteers of any background.
Source: Keystone Press.

Bristol group ROC

Within the Bristol area there were 12 ROC observation posts, all crewed by three or more volunteers at a time. A typical crew complement in the late 1940s early 1950s was 20 working a rotational series of shifts. The vast majority of those lived within sight of the post they operated, and this fostered a competitive element between sites, especially in aircraft recognition competitions. The reporting centre had a number of locations in the city during WWII, on reorganisation the Centre was located at Kings Square Avenue in the centre of Bristol until the protected structure at Lansdown became available after the disbandment of Anti-Aircraft Command (discussed below). From the ROC centre, information was passed out to a Sector Operations Centre at Rudloe Manor, northeast of Bath (Wood 1992, 254). From there it was passed to both the Fighter Command and the Anti-Aircraft Command, who, in turn, scrambled aircraft or readied the guns for contact.

Anti-aircraft command

Anti-aircraft guns, organised into a mixture of permanent and mobile batteries, were a mainstay of homeland protection throughout World War II.

Over a thousand fixed facilities were built around the United Kingdom over the period of the conflict (Cotswold Archaeology 2015, 23); by the end of the conflict sites in the Bristol area numbered at least 20 (*ibid* 2015, 7). Towards the end of 1945, just over 200 batteries were retained to form what became known as the Nucleus Force, a network of sites providing the minimum considered coverage against air attack (Cocroft *et al.* 2003, 147). If war had come guns from a series of depots around the country would quickly deploy along with mobile air defence radars to complement the existing fixed Gun Defended Areas. Interestingly, the majority of personnel in Anti-Aircraft Command were volunteers, units being raised from the recently reformed Territorial Army; the organisation was to become inextricably linked with the Cold War.

> The Secretary of State for War (Mr. Ballenger) As regards the operational role of the Territorial Army, I said a few moments ago that the Territorial soldiers will be organised in complete formations, and will also be responsible to a large extent for the anti-aircraft defences of this country.
>
> TERRITORIAL ARMY
> HC Deb 21 July 1947 vol 440 cc945-90

By 1950, this plan had already been downgraded to cover London and just a few other key industrial areas. Although, in June 1950, the onset of the Korean War, a conflict that many in the British Government considered to be a prelude to a European based war, forced a rethink and ambitious new plans were laid (Clarke 2005a, 72).

Gun defended area

The plan divided the United Kingdom into a number of groups, each in line with a Royal Air Force Sector Operations Centre (SOC). SOCs received information from the radar network and Royal Observer Corps (noted above). The data were collated and filtered at the SOC to determine the best course of action; either a response by Fighter Command or Anti-Aircraft Command (AAC). If the enemy aircraft had already penetrated the mainland airspace, instructions were passed to AAC. A purpose-built structure was at the centre of each Gun Defended Area (GDA) – the Anti-Aircraft Operations Room (AAOR) – this controlled a series of manned and automatically operated guns within a defined area (Clarke 2005a, 72). Nationally 33 GDA were formed (Cocroft *et al.* 2003, 148), each GDA protecting a number of 'high-value' industrial assets. In Bristol, these included a range of technical engineering firms connected to the defence industry – Bristol Aeroplane Company and Rolls-Royce being two key examples. Bristol was also a major port with a substantial amount of goods being imported and exported through Portbury and other docks on both the Bristol Channel and Severn Estuary.

Figure 2.6 The anti-aircraft operations room at Lansdown, west of Bristol. The square block house (centre) is the semi-sunken protected control for gun defended area two until 1956.
Source: Bob Clarke.

Units located in Bristol included 12th AA (Mixed) Signal Regiment, 2nd AA Group (West) Transport Column RASC and 2nd AA Ordnance Company. An Anti-Aircraft Operations Room stood on the ridge at Lansdown to the northeast of Bath, Avon (originally the site of a Heavy Anti-Aircraft Gun Emplacement McCamley 2002, 114). The AAOR was a standard design comprising a two storey, semi-sunken blockhouse, square in plan with a reinforced concrete outer skin 60.96 cm (24 inch) thick (*ibid* 2002, 114). The building programme appears to have begun in 1952– with the majority of structures completed by 1954– Gun Defended Area Group 2 Bristol was commissioned in 1954. The structure comprised a two-storey plotting room with a viewing gallery and offices surrounding that. The AAOR was self-contained with a plant room comprising a generator and basic filtration system; billeting for male and female operatives, along with restrooms (dependant on rank and gender) were also provided (Cocroft *et al.* 2003, 149; Clarke 2005a, 73). From here, the information provided by the Royal Air Force radar system and Royal Observer Corps network was prioritised and a number of electrically operated 5.15-inch guns were brought into action. Known sites in Bristol forming part of the GDA is restricted to Cribbs, where a 5.15inch battery was retained until at least 1955, forming part of the country-wide Nucleus Force (Cotswold Archaeology 2015, 9) (Figure 2.6).

Transition to war

The early Cold War Civil Defence Corps was structured around air raid precautions, following lessons learnt in WWII; the major difference was dealing with atomic attack, the effects of which were still little known in the late 1940s–early 1950s. How effective their brief would be in such an attack is, thankfully, pure speculation, although, detailed plans were made, the process of dealing with the effects of an attack with limited yield atomic weapons now follows.

An attack on any major population area was expected to have a similar effect to that experienced in Japan in August 1945, indeed, at the time those two events and the test site in New Mexico were the only true data sets of the effects of atomic weapons, everything else remained hypothetical. Subsequently, Point Attack, one singular detonation at the centre of a city or industrial area was expected (Clarke 2005a, 157); all the CDC force reaction was planned around this principle. Presuming a period of increased tension preceded any attack the Regional Commissioner would have already identified key staff who would operate from the war-room; interestingly, the Cemetery Superintendent was provided with an office in the bunker at Bristol (Clarke 2005a, 172). Ambulance and rescue vehicles, including heavy lifting equipment, along with a cohort of mobile communications vans and line laying equipment, all operated by CDC volunteers, then disperse to marshalling areas around the outskirts of the city. In the case of Bristol, marshalling areas have not been identified in any surviving record as yet; it is possible that this was because they were not finalised before the system was changed post-1955.

The Regional Commissioner then co-ordinated the rescue effort across the whole CD7 region. In Bristol, this meant that the immediate post-attack rescue phase fell to those at the Control Centre HQ located in the Council House in the centre of the city. If this was still functioning after an attack, the CDC team would co-ordinate information coming in from all four Civil Defence Controls located around the city. Initial information was provided by the local Wardens out in the field. Two main aspects prevailed; the level of devastation and the clearest routes into the worst affected areas, this is where training at Falfield House Rescue Training Ground came into its own. A blast wave passing down a street would cause less debris to spread out onto the road than one hitting houses at an angle, the Wardens role was to assess the level of debris in the road and then build that into an area wide map at their designated control. This was passed to the Control Centre HQ, who would update the Regional Commissioners office.

At the sub-controls

The Sub-Controls were intended to be both information hubs and welfare centres. The information flow out was intended to allow the Regional Commissioner and, eventually, Central Government understand the devastation on a national level. The immediate concern of the rest of the Sub-Control crew was to attend to the needs of those directly affected. The first aspects of Civil Defence delivered from here are known as the 'First Echelon' – described in 1950 as:

The city of Bristol and ground zero 71

First Echelon

Local Divisions of the Civil Defence Corps and auxillary services (such as the Police and Fire Services) will take the first shock of an attack. They are more or less static forces. Local Divisions will cope with the local civil defence communications, damage control, movement of refugees, rescue work, stretcher-bearing, first Aid, ambulance driving, debris clearance, decontamination of highways, escort and welfare of homeless evacuees, and a host of other civil defence matters.
Civil Defence Organisation 1950, (DS 5751/2), Paragraph 25.

Each Sub-control was located near a number of usable assets – usually a school, playing fields, parks and medical centres where close. It was here that the intended mass of casualties would be assembled and processed. As many as possible removed from the area by requisitioned public transport (Clarke 2009). If a school was nearby, as with the Sub-Control at Bedminster, then the hall and kitchens were earmarked as an emergency feeding station and reception centre.

Second Echelon

Civil Defence Mobile Columns, under the control of the Regional Commissioners, will act as the second line of defence. They will be held in reserve near target areas and are designed to reinforce Local Divisions when the scale of the attack is beyond the latter.
Civil Defence Organisation 1950, (DS 5751/2), Paragraph 25.

If the situation was bleak then the Regional Commissioner could release a mobile column into the area to assist with the rescue and evacuation effort, although, requests from local zones would be assessed against a series of criteria. As the war-room staff were responsible for a number of city-sized population areas in Civil Defence Region 7 (inc. Plymouth, Exeter, Swindon, Salisbury and Bristol), the use of such important assets was likely to be rare in the immediate aftermath.

Third Echelon

When necessary, Military Mobile Forces will, if their operational commitments permit, assist the First and Second Echelons in such emergency tasks as reconnaissance of damaged areas, elementary rescue, restoration of communications, treatment and evacuation of casualties, handling refugees and, to a limited extent, fire-fighting.
Civil Defence Organisation 1950, (DS 5751/2), Paragraph 26.

Clearly substantial planning, drawing from the lessons learnt during World War II, is in evidence here; as is consideration for the effects of an atomic device. What is not indicated but can be suggested is the notion that any attack might, in the first instance, be a horrific event but one that the nation would survive. Whether the complex structure of volunteers would hold together during a conflict escalating from conventional weapons to whole areas of destruction, does not appear to be considered in the atomic age. Concerns voice in both the House of Lords and House of Commons throughout this first decade certainly suggest anything more than a very limited period of attack would very probably become unmanageable in the short-term.

There appears to be an undercurrent of dissent in this period, driven in part by the austerity of the immediate post-war period, fuelling a growing reluctance to engage with aspects of uniformed service and warfare in general. Two situations are at play here; the inability of the Government to promote voluntary service in such organisations as anything other than preparing for war, relying initially on the recent 'Blitz Spirit' to rally people to the cause; and a population that was tired of conflict. In the 10 years between the end of WWII and the testing of the H-Bomb Britain had been involved in the Greek and Turkish Civil Wars (1946); Palestine (1945/8); Malayan Emergency (from 1948) supporting Berlin due to Soviet belligerence (1948/9) and the Korean War (1950/3). On top of this, the National Service Act (1948) required all males between the ages of 17 and 21 to do 18 months service and remain on reserve for a further four years (DEFENCE HC Deb 01 March 1948 vol 448 cc40–167). It is no wonder then that there are numerous references to the under manning of all voluntary services throughout this period. There is little evidence currently for any outward hostility to the concept of Civil Defence in the decade under scrutiny here, although the testing of the H-Bomb was to polarise both public and political opinion, bring about the rise of public disobedience and sow the seeds, by 1957, of one of the Cold War's most recognisable pressure groups – the Campaign for Nuclear Disarmament (CND) (Cocroft *et al.* 2003, 85).

The end of the age of innocence

On 1 November 1952, the United States detonated the world's first hydrogen device, the explosion, achieving 10.4 megatons, replaced the small Island of Elugalab, part of the Marshall Islands, with a crater 1.6 km wide and 60 m deep. This was swiftly followed by a Soviet device just 9 months later (Clarke 2005a, 18).

In 1954, the level of preparedness of British society in the face of this alarming development in the destructive force of new weapons were explored by a secret government commission. The findings of the study, chaired by William Strath, into the effects of a hydrogen bomb attack on the United Kingdom exposed the entire country's vulnerability to almost total shutdown (Hennessy 2003, 131; Hughes 2003, 258). What is now known as the 'Strath Report' is a pivotal document in the study of the British Cold War. Strath concluded that ten, megaton range, hydrogen weapons:

> – delivered on the western half of the UK or in the waters close in off the western seaboard, with the normal prevailing winds, would effectively

disrupt the life of the country and make normal activity completely impossible (JIC 1955).

The report had wide-ranging implications for the entire, pre-1954 Civil Defence Landscape. It was now clear that the effects of an H-Bomb attack on the United Kingdom would be an almost unmanageable catastrophe. The relatively unknown phenomenon accompanied the H-Bomb tests – radio-active fallout – now became the major concern of any future war. Prior to Strath, the residual effects of a nuclear blast had been based on the attacks on Japan; locally devastating, but with little physical effects beyond the target area. Strath recommended a new monitoring force structured on similar lines to the current ROC aircraft reporting network (Hughes 2003, 267). Indeed, a 'central fall-out plotting organisation will be required to collate the reports of the monitoring organisation and interpret them in light of current meteorological information' (*ibid* 2003, 268). In 1955, control of the ROC moved from the Air Ministry to the new organisation – the United Kingdom Warning and Monitoring Organisation (Cocroft *et al.* 2003, 176; Clarke 2005a, 142).

> Radio-Activity (Warning Organisation)
>
> Mr. Ian Harvey asked the Secretary of State for the Home Department whether he is yet able to make a statement upon the Government's plans for setting up a national monitoring organisation to give warning and to measure radio-activity in the event of air attack on the United Kingdom.
>
> Major Lloyd-George Yes. I am glad to be able to inform the House that arrangements are being made for the Royal Observer Corps, in conjunction with the Air Raid Warning Organisation, to undertake this important new function in addition to their existing duties.
>
> (HC Deb 15 June 1955 vol. 542 c18W 18W)

Other technological developments effectively removed the requirement for aircraft reporting, especially jet aircraft capable of operating at extreme height, better radar coverage and, later, rocket deliverable nuclear weapons (Clarke 2005b, 2). Other voluntary organisations were now under threat and quickly disbanded:

> Mr. Macmillan, the development of nuclear weapons and of long-range aircraft of high speed and capable of operating at great altitude has, in the Government's view, radically reduced the effectiveness of anti-aircraft gun defences. The Government have decided, after most careful consideration, that it is no longer justifiable to continue to spend money or to use manpower on the present scale for the anti-aircraft gun defences in the U.K.
>
> DEBATE ON THE ADDRESS
> (HC Deb 01 December 1954 vol 535 cc159–290)

From this point on, the Civil Defence Corps struggled to maintain the appearance of an effective defence against the horrors of the Hydrogen Bomb. Defence policy initially took a dangerous turn with the appearance of 'Mutually Assured Destruction', the stationing of American medium range Thor missiles in the eastern United Kingdom (Project Emily) from 1958 (Cocroft *et al.* 2003, 38) and the continued development of a British megaton yield weapon (successfully detonated in November 1957) (Clarke 2005a, 18) all increased public opposition to the prospect of nuclear war. The first Campaign for Nuclear Disarmament march to the Atomic Weapons Research Establishment, Berkshire at Easter in 1958 also marked the beginning of the end for the Civil Defence Corps. Any attempt to mitigate against the destructive power of a strike against the United Kingdom met with public and political scepticism; finally in April 1968, the CDC was disbanded saving an estimated £14–£20 million annually. From now on, the State would only plan for its own survival, and the citizens of Bristol – and, indeed, all citizens of the United Kingdom – were on their own.

References

Anon. (1946a, 5 September) Correspondence, Safety First: Grim Prospects and How to Avoid Them, Flight International, p. 254.

Basic Chemical Warfare. (1949) *Civil Defence Manual of Basic Training*, Vol. 11, His Majesty's Stationary Office, London.

Bingham-Hall, D. (1945) The History of Civil Defence in the No. 3 (Weston-Super-Mare) Area of Somerset County, Lawrence Bros, Weston-Super-Mare.

British Transport Commission. (1952) *British Transport Civil Defence Training Booklet*, 2nd ed., HMSO, London.

CDC. (1954) Notes for Course No. F/564: Warden Section Instructors Qualifying Course, Home Defence College Falfield, locally produced.

Clarke, B. (2005a) *Four Minute Warning: Britain's Cold War*, Tempus, Stroud.

Clarke, B. (2005b) Cold War Monuments in Wiltshire, *Wiltshire Archaeological & Natural History Magazine*, Vol. 98, pp. 1–11.

Clarke, B. (2007) *10 Tons for Tempelhof: The Berlin Airlift*, Tempus Publishing, Stroud.

Clarke, B. (2009) *The Illustrated Guide to Armageddon: Britain's Cold War*, Amberley Publishing, Stroud.

Clarke, B. (2014) An Early Cold War structure in Wiltshire, *Wiltshire Archaeological & Natural History Magazine*, Vol. 107, pp. 241–246.

Cocroft, W.D., Thomas, R.J.C., and Barnwell, P.S. (2003) *Cold War: Building for Nuclear Confrontation 1946–1989*, English Heritage, Swindon.

Cotswold Archaeology. (2015) Purdown Heavy Anti-Aircraft (HAA) Battery Stoke Park Estate, Bristol, Conservation Management Plan, CA Project: 5648, CA Report: 150762 (Unpublished Client Report).

Civil Defence. (1950) *Civil Defence Organisation: Lecture Notes*, His Majesty's Stationary Office, London.

Essex-Lopresti, T. (ed.) (2005) *A Brief History of Civil Defence, Civil Defence Association*, Higham Press ltd, Derbyshire.

Foucault, M. (1967) Des Espace Autres, Trans. Miskowiec, J. (1984) Architecture/Mouvement/Continuite.
Grant, M. (2010) *After the Bomb, Civil Defence and Nuclear War in Britain, 1945–68*, Palgrave Macmillan, London.
Hansard Millbank Extracts (https://api.parliament.uk/historic-hansard/sittings/C20).
Hansard Millbank Extracts. HC Deb 17 November 1919 vol 121 cc681–772) Government Policy – Russia.
Hansard Millbank Extracts. HC Deb 07 November 1945 vol 415 cc1290–390) Foreign Policy (President Truman's Declaration).
Hansard Millbank Extracts. HC Deb 31 October 1946 vol 428 cc793–882) Defence (Central Organisation).
Hansard Millbank Extracts. HC Deb 30 July 1936 vol 315 cc1767–853) Home Office Administration.
Hansard Millbank Extracts. HC Deb 18 April 1939 vol 346 cc170–3) Civil Defence (Regional Commissioners).
Hansard Millbank Extracts. HC Deb 16 October 1945 vol 414 c979W 979W) Eastwood House, Falfield, Gloucestershire.
Hansard Millbank Extracts. HC Deb 21 July 1947 vol 440 cc945–90) Territorial Army.
Hansard Millbank Extracts. HC Deb 01 March 1948 vol 448 cc40–167) Defence.
Hansard Millbank Extracts. HC Deb 23 November 1948 vol 458 cc1086–200) Civil Defence Bill.
Hansard Millbank Extracts. HC Deb 09 March 1950 vol 472 c3W 3W) Civil Defence (Volunteers).
Hansard Millbank Extracts. HC Deb 25 January 1951 vol 483 cc278–9) Recruitment.
Hansard Millbank Extracts. HC Deb 01 December 1954 vol 535 cc159–290) Debate on The Address.
Hansard Millbank Extracts. HC Deb 15 June 1955 vol. 542 c18W 18W) Radio-Activity (Warning Organisation).
Hennessy, P. (2003) *The Secret State: Whitehall and the Cold War*, Penguin Books, London.
Home Office, Civil Defence Circular No. 2/1957- published 19th February, 1957.
Hughes, J. (2003) The Strath Report: Britain Confronts the H-Bomb, 1954–1955, *History and Technology*, Vol. 19, No. 3, pp. 257–275.
McCamley, N.J. (2002) *Cold War Secret Nuclear Bunkers*, Leo Cooper, Pen and Sword, Barnsley.
Nuclear Weapons. (1956) Manual of Civil Defence, Vol. 1, Pamphlet No. 1, Home Office & Scottish Home Department, Her Majesty's Stationary Office. London.
Public Record Office. CONCLUSIONS of a Meeting of the Cabinet held at 1 0 Downing Street, S.W. 1 on Tuesday, 5th February, 1952, at 1 1 a.m.
Public Record Office. JIC (Joint Intelligence Committee) (1955) The 'H' Bomb Threat to the UK in the Event of a General War. January 13, 1955, NA CAB 158/20, JIC (55) 12 (Revise).
Public Record Office. WORK 12/375, Home Office Civil Defence School, Home Farm, Eastwood Park, Falfield, Gloucestershire, 1937–1957, (original reference 7A 3371/40 Part 1).
Rescue Section. (1957) *Civil Defence Instructors' Notes, Home Office & Scottish Home Department*, Her Majesty's Stationary Office, London.

Warden Section. (1957) *Civil Defence Instructors' Notes, Home Office & Scottish Home Department*, Her Majesty's Stationary Office, London.

Welfare Section. (1952) *Civil Defence Manual of Basic Training*, Vol. 1, His Majesty's Stationary Office, London.

Wohlstetter, A. (1959) The Delicate Balance of Terror, *Foreign Affairs*, Vol. 37, No. 2, pp. 211–234, Council on Foreign Relations.

Wood, D. (1992) *Attack Warning Red: The Royal Observer Corps and the Defence of Britian 1925 to 1992*, Carmichael and Sweet, England.

Woolven, R. (2005) 1945 Stand-Down, pp-32–34, in Essex-Lopresti, T. (ed.) 2005 *A Brief History of Civil Defence, Civil Defence Association*, Higham Press ltd, Derbyshire.

Wright, M.V., and Allen, A.D. (n.d.) *Industrial Civil Defence Digest*, The Society of Industrial Civil Defence Officers, private publication.

3 Towards a prosperous future through Cold War planning

Stalinist urban design in the industrial towns of Sillamäe and Kohtla-Järve, Estonia

Siim Sultson

TALLINN UNIVERSITY OF TECHNOLOGY

Introduction

This chapter focusses on the East Estonian industrial towns of Sillamäe and Kohtla-Järve and the deployment of Soviet urban design approaches from the mid-1940s through to the mid-1950s. The new industrial towns were built in order to exploit local mineral resources by the occupying Soviet regime and were examples of Stalinist planning, intended as utopias. The stately urban ensembles[1] of such towns formulated a paradigm that was unfamiliar to the existing local urban design and architectural traditions During the 1950s, both towns had a morphology that allowed them to be developed into much larger industrial centres. Sillamäe and Kohtla-Järve were designed to be model Soviet industrial towns that would demonstrate the route to a prosperous future.

Sillamäe was a Soviet 'closed city' due to the processing of uranium oxides for the nuclear industry and military needs. Kohtla-Järve was an agglomeration of six satellite settlements (Järve, Ahtme, Kukruse, Sompa, Oru, Viivikonna with Sirgala) that were collectively regarded as an oil-shale mining and processing complex in order to produce electricity in large thermal power plants. Unlike other Estonian towns, the inhabitants of Sillamäe and Kohtla-Järve were imported from other places in the Soviet Union. Both towns were designed under the guidance of state architectural branches *Lengorstroyproyekt* and *Lengiproshacht* based in Leningrad (St. Petersburg). Despite the imposition of foreign cultures, both in urban design and by the Soviet state more broadly, these towns share characteristics with certain other Estonian cities. Conventionally, research concerning Stalinist urban space in Estonia considers these types of industrial town as unfamiliar. However, there are many planning elements in Sillamäe and Kohtla-Järve that tally with those of Tallinn and Pärnu, which were planned by local architects, with Estonian training. This chapter considers integrating partially abondaned, and mostly Russian-speaking, Sillamäe and Kohtla-Järve into the entire state by the means of legacy of local Stalinist stately urban ensembles.

In the summer of 1940, independent Estonia was occupied by the Soviet Union. A year later, the German's took over Estonia and stayed until 1944 when the Soviet Union seized the country for a second time. This occupation would last until 1991, spanning the years of the post-war Soviet Stalinist period, the beginning of Cold War and Stalin's death in March 1953.

Due to the ongoing conflict of the Second World War and the relatively short duration of the occupations, neither the first Soviet takeover, nor the German occupation, had an impact on Estonian town planning or architectural design. However, Stalinist principles greatly influenced the urban landscape after 1944, and until November 1955, when the Communist Party of the Soviet Union's Central Committee decided to move away from the characteristically exaggerated architectural style (*Ob ustranenii*, 1955).

In line with official Stalinist ideology, Soviet town planning had to stand in opposition to that of the West, so as to elevate socialist principles above those of capitalism. This was one of the crucial themes during post-war Cold War years. The stately urban ensembles of the Soviet East Estonian industrial towns were influenced markedly by Stalinist utopianism and Cold War threats. They contained architectural features and political iconography designed to demonstrate a path to a prosperous future.

This chapter focusses on East Estonian industrial towns during the postwar Stalinist period (1944–1955), portraying stately urban ensemble practices that occurred during the Stalinist period of the Cold War in Sillamäe and Kohtla-Järve agglomeration (Figure 3.1). It reflects upon the architectural identity of these towns and considers the legacy as a familiar part of Estonian Stalinist urban spatial practice. On the one hand, Stalinist stately urban ensembles contrasted with local planning traditions; on the other hand, it had much in common with the Stalinist central gridlines of Tallinn and Pärnu, planned by local architects.

Due to the fact that the first phases of Sillamäe and Kohtla-Järve were not planned by local architects (indicated by surviving Russian language records, masterplans and documentation), they are different to other Estonian towns. The Stalinist appearance of these East Estonian industrial towns is still considered unfamiliar, and even exotic today, as it is quite different to the urban form elsewhere in the country. The planning and architecture of these East Estonian industrial towns is under researched and it is important to incorporate them into wider scholarship of Estonian Stalinist urban space. The centres of Sillamäe and Kohtla-Järve agglomeration have architectural and urban compositional similarities with the centres of Tallinn and Pärnu. By the mid-1940s, the Department of Architecture of the Estonian Soviet Socialist Republic (SSR), and its head Harald Arman, were familiar with the masterplans and construction plans of both industrial towns (ERA.R-1992.2.12: 44–54; ERA.R-1992.2.1; ERA.R-1992.2.31;

A prosperous future and Cold War planning 79

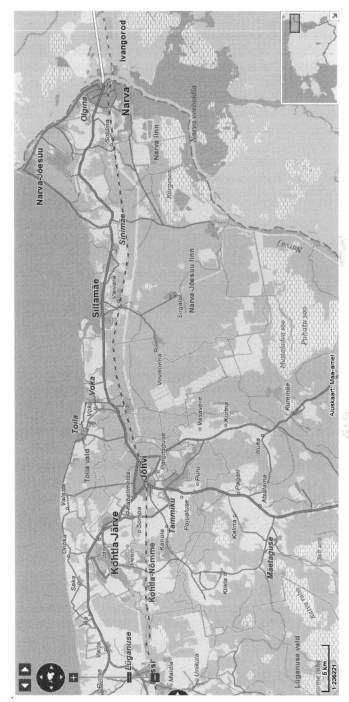

Figure 3.1 Map of the eastern coastal region of Estonia delineating the urban areas of Sillamäe and Kohtla-Järve.
Source: https://estat.stat.ee/StatistikaKaart/VKR.
Source: Eesti Statistikaamet.

ERA.R-1992.2.22; ERA.R-1992.2.41). The Stalinist period is a part of Estonian history, deeper knowledge of the period helps to define the perspectives of the state and its inhabitants.

Available literature on these topics consists of articles from the Stalinist period (mid-1940s to mid-1950s) to the post-Soviet period (early-1990s to present). Descriptive articles concerning masterplans and urban space of the East-Estonian industrial towns in the Stalinist period appeared in professional periodical publications like *ENSV Arhitektide Almanahh* (Almanac of Architects of Estonian SSR) by architect Voldemar Tippel (Tippel, 1948, pp. 54–59), the head of the Department of Architecture of the Estonian SSR Harald Arman and architect Ivan Starostin (Arman and Starostin, 1951, pp. 7–18), architects Voldemar Meigas (Meigas, 1951, pp. 19–30), Vsevolod Tihomirov (Tippel and Tihomirov, 1951, pp. 31–43). After Stalin's death, scholarship around urban space and town planning seemed much less popular. There was merely the occasional generic reference in post-Stalin Soviet period literature. For example, in the 1965 book, *Eesti arhitektuuri ajalugu* (History of Estonian Architecture), edited by Arman, the issue was discussed neutrally, more as a matter of protocol (Eesti arhitektuuri ajalugu, 1965).

In the post-Soviet period, more diverse analytical approaches emerged as a result of Estonia's political independence. The new circumstances offered the opportunity to treat the topic objectively. In 1991, architectural historian Leonid Volkov, who lived through the Stalin era, published the article, 'Eesti arhitektuurist aastail 1940–1954' (About Estonian architecture between within 1940–1954) which began to address specifically the Stalinist period (Volkov, 1991, pp. 183–213). Architectural historian Mart Kalm also reflected upon the issue in his book, 'Eesti 20. sajandi arhitektuur' (Estonian twentieth Century Architecture) (Kalm, 2001). The topic is developed further in Kalm's co-authored article, 'Perfect Representations of Soviet Planned Space. Mono-industrial towns in the Soviet Baltic Republics in the 1950s–1980s.' (Cinis, Drémaité, and Kalm, 2008, pp. 226–246). Historian David Vseviov analysed the formation and structure of the East-Estonian towns in his doctoral dissertation 'Kirde-Eesti urbaanse anaomaalia kujunemine ning struktuur pärast teist maailmasõda' (The formation and structure of the urban anomaly in northeast Estonia after the Second World War) (Vseviov, 2002). East-Estonian Stalinist urban space is considered by architectural historian Siim Sultson with a focus on the alteration in the awareness of Estonian city space. (Sultson, 2016a, pp. 49–55). More detailed overview and deeper information about the present state of research of Estonian post-war urban space one can find in Sultson's article, where the author highlights a need to incorporate East-Estonian industrial towns into research about Estonian post-war Stalinist urban space (Sultson, 2016b, pp. 283–294). Further, Sultson (2017) analyses *similarity* and *continuity* in the urban design of the post-war Stalinist period and that of the Estonia's period of independence during the 1930s (Sultson, 2017, pp. 385–409).[2]

Principles of Stalinist town planning

For ideological reasons private property was abolished in the Soviet Union, resulting in complete state ownership of the land. Consequently, it was easier to determine town structures as a means to control society via the re-planning of urban space. It was especially crucial after the Second World War – the necessity to restore wrecked towns and to establish new ones afforded the opportunity to celebrate a victorious Soviet regime and its communist ideology.

The question of urban ensemble in town planning and urban space were crucial to Stalinist-era architecture. In 1932, the previous Soviet People's Commissar of Education Anatoly Lunacharskiy stated that one task of architecture was to integrate functionality and utility to harmonise with an ideology (Kosenkova, 2009, pp. 19–20). A year later those features coalesced in the production of a new term – Socialist Realism. The Soviet Communist Party defined a new ideological method. The method was supposed to 'embody an absolute apocalyptical future where the difference between past and future abolishes significance.' (Groys, 1998, p. 859). The method had to embrace literature, music, art and architecture. In 1934, the Soviet Writer's Union of the USSR formulated Socialist Realism as a method that demanded an artist to depict reality faithfully and to link historical events in an explicit way with revolutionary development. The depiction should be closely allied to educating workers (Soviet citizens) in the socialist spirit. Socialist Realism as evolving and ambient category was intended to provide artists with vast opportunities to choose different forms, styles and genres (ENE, VII, 1975, p. 243).

Soviet architect and theoretician, Ivan Zholtovskij, emphasised the importance of composition in the urban ensemble. In 1933, he defined five principles of Soviet urban ensemble: the unity of different forms; the tectonic accuracy of architectural forms; the dynamic and organic growth of architectural forms; the natural architectural organism; and the unexpected compatibility of different elements of architectural forms (Zholtovskij, 1933).

In 1940, Zholtovskij stated that every architect should take ensemble as unity into account. According to Socialist Realism, architecture in a city space was supposed to be an ordered ensemble: every house had to be ruled by this unified order. This meant organising cities according to a certain hierarchy, in which every component of the urban landscape had to abide by the principle of unity. In Zholtovskij's (1940) words, '[t]here is no architecture outside the urban order. The architect is responsible for one's people's architecture, the town's architecture, and street architecture whilst designing a house.' Urban ensemble, e.g. stately urban ensemble, became one of the crucial principles established through the Socialist Realist method after the Second World War (in order to celebrate the victory) and during the Cold War (in order to demonstrate preferences of socialist practices compared to the capitalist ones).

During the Second World War, according to a plenary resolution of the USSR's Soviet Architect's Union, architects were expected to be ready to undertake massive post-war restoration works (Plenum Pravlenija, 1943). The instructions, given by the USSR's Council of People's Commissars and the Soviet Communist Party, compelled architects to design and restore wrecked towns in a grandiose manner, according to state ideology (*Iz istorii*, 1978, pp. 95–102, 109). Perhaps the greatest transformation of the Stalinist urban space and ensemble occurred in Stalingrad (Volgograd) during and after the Second World War, where architect Karo Alabyan developed new designs (1943) (*Iz istorii*, 1978, pp. 78, 142, 172). Stalingrad, given its association to the nation's leader and being one of the most symbolic battlefields of the war, was rebuilt as a proving ground for new ideas (Sultson, 2017, p.395). Other cities in the USSR were being re-designed to have similar shapes, composition and principles to Stalingrad. As a means to express order, harmony, and an inevitably prosperous future, planning grids of new Soviet towns had to adopt more classical motifs with axes, squares, forums, junctions, and the composition of urban ensembles more symmetrical, hierarchical (the highest houses in the centres and around squares), and dynamic with perimetral housing. While cities like Rome, and Paris were still regarded as exemplars, Leningrad, Moscow and Stalingrad became mandatory models for other Soviet towns after 1945. The squares and streets needed to be significantly scaled to enable mass processions and parades during state events and anniversaries (Barhhin, 1986, pp. 127–128). These principles, that became compulsory for every Soviet architect, tallied with the victorious message collectively propagated in the years following the Second World War.

In 1945, the head of the USSR's State Committee of Architecture, Arkadi Mordvinov, formulated seven principles of Soviet post-war town planning that were compulsory for all architects: town planning was supposed to be interrelated with the natural environment in order to expose its beauty; town plans needed a balanced compositional centre (for instance, city centre – main street – railway station square); monumental public buildings had to be erected on junctions; residential quarters had to be planned in complex ways and designed as one ensemble; buildings should only be painted in light colours (to echo the dream of a positive future); the functionality and high quality of structures and infrastructure (electricity, water supply, etc.) was a priority; and the thorough quality controls both of architects' projects and the building process was necessary (Kosenkova, 2009, p. 42). These principles, based on Soviet experience and ideology, were intended to guarantee the flourishing of post-war town planning. However, much of that remained a dream, ideals that were predominantly limited to implementation in town centres. Even at the more prestigious end, these policies were only partially effective in reshaping the bigger city centres (Moscow, Stalingrad, Leningrad, Kiev, Minsk) and closed nuclear or industrial towns. In the majority of cases, the ideas to reshape towns for socialism were effectively window dressing.

East Estonian industrial towns

Estonian architect Harald Arman was compelled to go to the Soviet Union with the Red Army in 1941. He returned to Estonia in 1944 as head of the Department of Architecture of the Estonian SSR. Earlier in that year, whilst still in the Soviet Union, he had already begun to organise the restoration of Estonian towns (Gorich, 1946). However, after the Second World War, Estonian architecture and urban planning was subject to perceived contradictory influences. Outwardly, the urban design of the Estonian post-war Stalinist period followed the Soviet doctrine in concept, forms and building materials – including in Tallinn, Pärnu, Kohtla-Järve and Sillamäe. The most radical solution, as happened in Narva, meant the replacement both of city and its inhabitants. On the other hand, however, there was also some similarity and cultural continuity inherited from urban design practices developed during the 1930s – as seen in Tallinn, Pärnu, Tartu and elsewhere. Whilst Estonian urban planning of the Stalinist period (1944–1955) is usually considered to be in conflict with that of the 1930s, in fact, Estonian architecture, town planning, and urban design were already characteristically similar to the later Stalinist period by 1940. Indeed, there seems to have been a smooth transition to Stalinist urban design practice from that of the independence period, both in the appearance and the composition of buildings. Estonian 1930s architecture and urban design became increasingly Corinthian and stately up to 1940; Estonian architecture in the mid-1940s was rather similar to that of the late 1930s. The Stalinist urban ensembles of Tallinn, Pärnu and Tartu were designed by the same architects who had worked during Estonia's period of independence. During the Stalinist era, local architects (with the notable exception of H. Arman) were not involved in the planning of Sillamäe and Kohtla-Järve.

Similarly, the oil-shale industry was not anything new in Estonia before 1940. The Republic of Estonia made its first attempts to exploit mineral resources for producing oil-shale oil and gasoline in Kohtla-Järve in the mid-1920s. However, during the seizure of the country by the Soviet Union, the new regime considered East-Estonia as a resource for oil-shale that could supply Leningrad and north-west Russia with liquid fuels, natural gas and electricity.

The East Estonian towns of Sillamäe and Kohtla-Järve, which were fairly small and widely dispersed settlements before the Soviet time, became exemplars for socialist industrialisation and the deployment of Soviet urban design between the mid-1940s and the mid-1950s. During these years Sillamäe and Kohtla-Järve (that was developed to agglomeration) were designed to be new mono-industrial towns and were supposedly examples of Stalinist utopias and future economic prosperity. Thus, compared to other Estonian towns, Sillamäe and Kohtla-Järve were newly designed settlements. (Figure 3.2) East-Estonian industrial zones became the most important Soviet territory in the country, after Tallinn. The area

Figure 3.2 Lenin Square in Kohtla-Järve, early 1950s, by *Lengorstroyproyekt*. Photograph taken in 1950s.
Source: RM f 207.5.

was rapidly urbanised. During the two war-time occupations, the area lost the majority of its population due to deaths, deportations to Russia and emigration to the West. Only about 60% of the native population remained there by 1944. Unlike other Estonian towns, new inhabitants of Sillamäe and Kohtla-Järve urban agglomeration were forcibly relocated from other parts of the Soviet Union. The formation of this type of urban space in the East-Estonian industrial zone was brought about by the economical-political policy of the Soviet authorities, and is a unique phenomenon in the history of post-war Europe (Vseviov, 2002, p. 8). As early as November 1944, the Soviet authorities imposed a 7 kilometre closed zone from the coast of Gulf of Finland in order to guarantee the secrecy of sensitive military sites. The closed zone established conditions that prevented natives from returning to their homes (ERA.R-1.5.95:48–49).

Sillamäe

Sillamäe enjoyed a special position in the East Estonian industrial region as it was thought that processing uranium oxides. Consequently, it was a classified and closed town until the end of Soviet occupation.

In May 1945, the Council of the People's Commissars of the USSR arranged for a detailed geological investigation in the East Estonian oil-shale area (ERA.R-1.5.104:74–76). According to the memoirs of the scientific director of the Institute of Geology of the USSR, M. Altgausen, the decision was reached at a secret meeting in the Kremlin concerning the raw material for atomic weapons development. The oil-shale under Sillamae seemed to contain uranium oxides – it was vital for the Soviet Union at the beginning of the Cold War (Vseviov, 2002, pp. 26–27).

At the end of 1945, the Leningrad Department of Building Factories (Lengazstroi) acting on behalf of Glavgaztopprom of the Council of the People's Commissars of the USSR sent a request to the Department of Architecture of the Estonian SSR to select a 78-hectare plot of land to establish a settlement for the workers of Viktor Kingissepp oil-shale processing and distillation factory. The local department agreed the request. The department examined and processed the masterplan of Sillamäe, designed by the architects of *Lengorstroyproyekt*, through to the autumn of 1946 (ERA.R-1992.2.2:63–66).

Meanwhile, in summer 1946, following the regulations of the Council of Ministers of the USSR both Council of Ministers of the Estonian SSR and the Central Committee of the Communist Party of the Estonian SSR issued a top secret regulation concerning uranium oxides-oriented oil-shale research in Factory No 7 and development of the classified settlement in Sillamäe (ERA.R-1.5.133). It is supposed that, following the issue of this regulation, the further development of that masterplan ceased (ERA.R-1992.2.12:44).

The Soviet nuclear industry-oriented regulation finally established Sillamäe's special position both in Estonia and the Soviet Union: as a town dedicated to enrichment of uranium oxides it received homogenous Stalinist urban design within ten years of the end of the Second World War. Presumably due to need for loyalty to the Soviet authorities, new inhabitants for the rapidly growing settlement were relocated from the inner regions of the Soviet Union itself (ERA.R-1.5.179:17).

During the Stalinist-era, Sillamäe was a settlement that emerged from practically nothing and within ten years became a town with a population of over 12,500 by 1950 (Vseviov, 2002, pp. 35, 48). According to the first masterplan of 1946 by Lengorstroyproyekt, the population of Sillamäe was originally planned to be only 2,000 (ERA.R-1992.2.12). By January 1949, it had a population of 7,000 and encompassed area of about 900 hectares (ERA.R-3.3.1071:1–4). According to the 1959 census population of Sillamäe was 8,210 (Vseviov, 2002, p. 44). On June 29th, 1957 Sillamäe became a town under the authority of Tallinn (ERA.R-3.3.3034).

Sillamäe's centre, in terms of its stately urban ensemble, consisted of three main streets and a boulevard crossing the central square and recreational area (Figure 3.3). The existing coastal topography informed the position of two major boulevards, 30 metre-wide Kalda and Kesk, in roughly a south-east-northwest orientation. They formed the boundary of a linear park and converged upon one another, also meeting at the Central Square. The square itself was flanked with a Corinthian styled Palace of Culture, a towered town hall and a grand staircase that descends to 40 m wide and 250 m long Mere Boulevard and connects the town to the sea. The boulevard in lined with trees and monumental four-storied apartment houses. The composition of the boulevard and the staircase is similar to the monumental staircase

Figure 3.3 Design for Stalinist central square of Sillamäe, 1948 by GSPI-12.
Source: SM Sillamäe keskuse perspektiivvaade.

in Stalingard that connects the Volga River and Alley of Heros (*Alleya Geroyev*). The rest of the stately urban ensemble is composed from neoclassicist apartment houses, public buildings ornamented with bas-reliefs, balustrades, pediments, columns and other decoration (Figures 3.4 and 3.5).

The current grid of the town was developed within the years of 1946/1947 – 1949. Whilst up to September 1946 the settlements masterplan was designed only by architects of Lengorstroyproyekt (Leningrad filial of all-Soviet Union Gorstroyproyekt (Planning Institute for Town Building under the Ministry of Heavy Industry of the USSR), the next versions of the masterplans were made by Lengorstroyproyekt with Leningrad filial of NII-9 (Scientific Research Institute number 9) in 1946 – 1947 and with Leningrad filial of GSPI-12 in 1948 – 1949 (Stately Specialised Planning Institute number 12). Both latter organisations belonged to the Soviet nuclear industry. (SM Generalny proekt, NII-9, 1947; SLV Detalny proekt, Tom I, GSPI-12, 1948; SLV Proekt planirovki, GSPI-12, 1949). The near complete implementation of the Stalinist masterplan in Sillamäe was a fairly rare occurrence both in Estonia and the former Soviet Union. (Figure 3.6 and 3.7) Processing of uranium oxides for the Cold War gave the new Soviet settlement Sillamäe a special position and promoted its rapid and planned development. The same conditions prevailed in other nuclear towns attached to the nuclear programme such as Ozersk, Zheleznogorsk, Seversk.

Kohtla-Järve

Kohtla-Järve was conceived as a oil-shale mining and processing centre to supply Leningrad and North-West Russia with liquid fuels, natural gas and electricity. The urban agglomeration formed in the wake of this demand consisted of six satellite towns (Järve, Ahtme, Kukruse, Sompa, Oru, Viivikonna with Sirgala) and also included Kohtla-Nõmme, Kiviõli, Jõhvi and

A prosperous future and Cold War planning 87

Figure 3.4 Stalinist centre of Sillamäe, view from staircase with sculptures along Mere Boulevard towards sea, early 1950s, by *Lengorstroyproyekt* / GSPI-12. Photograph taken in 1950s.
Source: Sillamäe Muuseum. ref. SM1F 6273.

Figure 3.5 View along Kesk street and Stalinist Central Square of Sillamäe, early 1950s, by *Lengorstroyproyekt* / GSPI-12. Photograph taken in 1950s.
Source: SM 1F 6079.

Püssi settlements during the Soviet era. The latter four are now independent towns. Regardless of the dynamics of its constitution, Järve was always the administrative and civic centre of the urban agglomeration formed in June 1946.

Figure 3.6 Stalinist centre of Sillamäe, view from staircase along Mere Boulevard, early 1950s, by *Lengorstroyproyekt* / GSPI-12. Photograph taken in 2013.
Source: Siim Sultson.

Figure 3.7 Stalinist centre of Sillamäe, view to staircase of the Central Square, early 1950s, by *Lengorstroyproyekt* / GSPI-12. Photograph taken in 2013.
Source: Siim Sultson.

Due to the changing constitution of the city region it is difficult to track population figures. However, according to the agglomerated data of the six conjoined towns, by 1950 the population of Kohtla-Järve had risen from around 7,200 in 1945 to over 17,000 in 1950 (Zabrodskaja, 2005, pp. 24, 25).

A prosperous future and Cold War planning 89

Between April 1946 and June 1947, the Department of Architecture of the Estonian SSR Estonian and Council of Ministers of the Estonian SSR examined and approved the masterplans of Kohtla-Järve. According to the masterplans, Järve was designed to accommodate 15,000 inhabitants by 1967, Sompa 15,000 inhabitants by 1966 and Ahtme 25,000 inhabitants by 1966 (Tippel, 1948, pp. 54–59). According to the 1959 census population of Kohtla-Järve was 51,200 (Vseviov, 2002, p. 44).

Whilst Kohtla-Järve's civic centre was designed by *Lengorstroyproyekt*, the rest of the agglomeration (except Oru and Sirgala) was erected mainly according to schemes prepared under *Lengiproshacht* (Leningrad branch of the State Institute for Planning Mines under the Ministry of Oil-Shale industry of the USSR). The new, Soviet era, centre of Kohtla-Järve was built about 1 kilometre to the east of the pre-war settlement. The original centre fell within a 2-kilometre sanitary zone between oil-shale mine and new town "*sotzgorod*" (socialist town) (Tippel, 1948, p. 55). The Stalinist stately ensemble of the urban space of Järve consists of two main axes that cross one another. (Figure 3.8) The southwest–northeast one, named Victory Boulevard (now Kesk Boulevard), is over 50 m wide and 700 m long. It connects the Palace of Culture with a park and the cinema (Victory) on the axis of the boulevard. (Figure 3.9) Victory Boulevard is crossed by the

Figure 3.8 Stalinist centre of Kohtla-Järve centre (Järve), designed in the early 1950s by *Lengorstroyproyekt*. Perspective view along the Victory Boulevard. Photograph taken in early 1950s.
Source: RM, ref. f 87.70.

Figure 3.9 The Palace of Culture in Kohtla-Järve centre (Järve), designed in the early 1950s by *Lengorstroyproyekt*. Photograph taken in 1952.
Source: Eesti Entsüklopeedia, 2017.

southeast-northwest axis of Rahu (Peace) Square (200 m long, 125 m wide), which was designed to concentrate local government buildings and four-storeyed apartment buildings, each assuming the giant order with columns spanning multiple floors and avant-corps projections from the facades. The rest of the houses at this intersection are two and three-storeyed apartment buildings with pitched roofs. (ERA.R-1992.2.57; ERA.R-1992.2.41) In 1956, the local architect V. Tippel from *Estongiprogorstroi* (Estonian branch of the State Institute for Town Planning and Building) made a supplementary masterplan for Järve that clearly took into account previous versions of *Lengorstroyproyekt*. Nearly three-quarters of the masterplan was implemented (EAM.3.1.281) (Figure 3.10).

Compared to Järve, the rest of the six satellites (except Oru and Sirgala as a part of Viivikonna that were started in early 1960s according to post-Stalinist modern town planning principles) were laid out using regular and simple grids, designed by *Lengiproshacht*. However, the masterplans of Sompa and a new area of Ahtme (near old, pre-war Ahtme) were not organised using orthogonal geometry and contained radial street patterns like those of Järve. Presumably, Stalinist era masterplans of Kukruse and Viivikonna were designed by *Lengiproshacht*, as well. Due to the absence of copies of the various masterplans in public archives, it is hard to estimate to what extent each of these was implemented. About one-third of the masterplan of Ahtme was implemented and in Sompa even less, approximately a quarter, was realised (EAM.3.1.248; EAM.3.1.50).

Figure 3.10 Stalinist centre of Kohtla-Järve centre (Järve), designed in the early 1950s by *Lengorstroyproyekt*. Perspective view along the Victory Boulevard. Centre: one of the main axes of Rahu Square and an apartment building in the giant order. Photograph taken in 2013.

Source: Siim Sultson.

Legacy and reflection

Though the East Estonian mono-industrial towns were mostly planned by architects from other parts of the USSR, the Stalinist central gridlines of the masterplans for Järve and Sillamäe designed by *Lengorstroyproyekt* (later on by *Estongiprogorstroi* and *Lengorstroyproyekt*/NII-9/GSPI-12 respectively) and other Soviet state entities had many compositional similarities with Stalinist central gridlines of Tallinn and Pärnu designed by Estonian architects. It was H. Arman, as head of the Department of Architecture of the Estonian SSR, who processed the masterplans and construction plans for settlements during the mid-1940s. Arman was also responsible for the masterplans of Tallinn and Pärnu through the mid-1940s and early-1950s. It would be logical to assume that the Stalinist central gridlines of Tallinn and Pärnu were inspired by the compositions designed by Lengorstroyproyekt. However, it seems to be on the contrary. Arman crossed converging axis with boulevard in the masterplan of Tallinn and its Cultural Centre composition already in 1945. He reused the solution in 1952 for the Pärnu Oblast Centre. Neither of the state designed masterplans for Sillamäe 1946 masterplan nor 1947 masterplan (Lengorstroyproyekt/ NII-9) used the same ideas. H. Arman actively processed the masterplans up to autumn 1946 and was in touch with later masterplan versions via correspondence with the state architectural firm. Meanwhile he also processed the first masterplans of Kohtla-Järve (started by Lengorstroyproyekt). (ERA R-1992.2.12: 44–54; ERA R-1992.2.1; ERA R-1992.2.31; ERA R-1992.2.22; ERA R-1992.2.41; SM

Generalny proekt, NII-9, 1947; SLV Detalny proekt, Tom I, GSPI-12, 1948; SLV Proekt planirovki, GSPI-12, 1949.

Stalinist stately urban ensembles in Estonia provided architects with an opportunity to put some of their architectural ideas from the period of independence into practice. Megalomaniac town planning and city space dreams, though expensive and unrealistic, still appealed both to the small independent Estonia and the architects. According to the global tendencies of the 1930s, local architects were interested in monumental and representative architecture. Compared to small, independent Estonia, the Soviet Union, which encompassed one-sixth of the planet, had very much larger resources to finance urban planning projects or dreams.

Meanwhile stately urban ensembles of Sillamäe and Kohtla-Järve, especially of Järve *sotzgorod* followed more orthodox Stalinist principles when compared to other Estonian towns. Responding to both Zholtovskij's and Mordvinov's principles of urban design, in the Stalinist period, the architects and designers of Sillamäe and Järve tried to embody an approach that implied a prosperous future – a kind of socialist paradise. These towns had to demonstrate the advantages of the Socialist system over its capitalist competitor by means of compositional unity and the illusion of luxurious of facades, courtyards, parks and inspiring sculptures. The prestigious appearance of these towns was inevitable due to the authorship of Leningrad-based design organisations. However, the Stalinist centres of Sillamäe and Järve were under populated, having been designed for a much larger number of inhabitants than ever settled there (Figure 3.11).

Figure 3.11 Sculpture of discus thrower (discobolus) in greenery of Victory Boulevard in Kohtla-Järve center (Järve). Photograph taken in mid-1950s.
Source: EAM, ref. Fk 2844.

The Soviet architect's handbook of 1952 suggested the need to design a 1-hectare sized central square for the town with the population of 50,000 or more (*Kratkij Spravochnik*, 1952, pp. 20–21). Using the measured guidelines contained in the handbook (that was compulsory for architects), it is possible to derive approximate numbers of the planned populations of Cold War era Sillamäe and Kohtla-Järve. The 1.68-hectare Central Square of Sillamäe implied a prospective population of 84,000 and the 2.5-hectare Rahu Square of Järve suggested a planned population of 125,000 for Kohtla-Järve – neither town reached its planned scale. An examination of population figures in Sillamäe and Kohtla-Järve agglomeration in the mid-1940s to 1950 (accordingly from planned 2,000 to 12,500 and from 7,200 to 17,000) it was sharper in Sillamäe (6.3 times against 2.4 times in Kohtla-Järve agglomeration). In the 1950s, the population of Sillamäe decreased by a factor of 1.2 (8,210 inhabitants in 1959), while in Kohtla-Järve agglomeration it increased 3 times (51,200 inhabitants in 1959). The decline was caused by the fact that in the early 1950s, Sillamäe lost its importance as one of the best uranium raw material mine in the Soviet Union, as plutonium became the preferred material for the nuclear programme. Since 1950s the Factory No 7 Sillamäe enriched uranium that was imported from socialist East European countries. However, according to Vseviov (2002, pp. 9–10), the conditions imposed on these towns by the industrial machine designed to service the Cold War had completely changed the make-up of the region. The proportion of the rural population had decreased sharply, and the East-Estonian industrial zone had become the most urbanised area in the country dominated by Russian-speaking townspeople.

The loss of industry was not enough to abate the continued trend of urbanisation and between 1959 and 1989 population growth continued in both Sillamäe and Kohtla-Järve to 20,568 and 90,828 respectively. Kohtla-Järve growth was also attributable to the addition of a further four settlements to the urban agglomeration – Kohtla-Nõmme, Kiviõli, Jõhvi and Püssi.

Even at these 1989 peaks, the population figures did not meet those obviously projected by the Soviet state for Sillamäe and Kohtla-Järve (84,000 and 125,00, respectively). In 2018, the towns are 6.3 times (Sillamäe) and 3.5 times (Kohtla-Järve) smaller than the plans of the Cold War implied. Stalinist stately ensemble urban spaces of Sillamäe and Kohtla-Järve agglomeration (based on Järve sotzgorod) following imperial over-emphasized industrialization and Cold War threats from the West were intended for a much greater number of inhabitants than ever settled in either town. Many towns amalgamated into the East-Estonian industrial legacy complex are not well preserved and others have had serious environmental problems to ameliorate. For example, Viivikonna, a part of the Kohtla-Järve agglomeration, has practically been abandoned. In 50 years of operation, until the end of Soviet occupation, Factory Number 7 in Sillamäe produced about 12 million tonnes of radioactive waste which was deposited in tailing ponds covering over 40 hectares on the Baltic coast. The project to remediate this took eight years (2000–2008) and cost upward of 21 m Euro (Tailings.info, 2017).

Conclusion

Population figures and derived statistics show that the East Estonian monoindustrial settlements played an important role in Soviet Union during the Stalinist-period of the Cold War. The large-scale immigration of non-native inhabitants to Sillamäe and Kohtla-Järve from the inner regions of Soviet Union can be viewed in two ways: firstly, as a form of colonization to import the numbers of non-native loyal people necessary to work in secretive conditions that supported the nuclear programme and in oil-shale industry, or secondly the simple displacement and assimilation of native-Estonians. Due to functional grid and planned prospective population growth the stately urban ensembles of these towns have still under used urban potential. The planning of the centres of these industrial towns had much in common with those built in Tallinn and Pärnu which were designed by Estonian architects with Estonian training. Conventionally, urban design of the Stalinist period (1944–1955) is viewed as conflicting with, or as a rupture from, the independent Estonian 1930s urban space design. Though Stalinist period Sillamäe and Kohtla-Järve were mostly planned by non-local architects, central gridlines and stately urban ensembles of these towns have much in common with the same period stately urban ensembles of Tallinn and Pärnu. Stalinist period urban space of Tallinn, Pärnu, Tartu are designed by the same architects that worked in independent Estonia. Consequently, architectural identity of Cold War 24 industrial towns Sillamäe and Kohtla-Järve agglomeration embody a legacy that is a natural and familiar part of Estonian Stalinist stately urban ensemble and space practice. Such an architectural question of familiarity via urban space research, that is social issue, as well, could contain a key to integrate mostly Russian-speaking and often unemployed north-east Estonia into the state and support the improvement of the area. The industrial Cold War urban legacy - is both stately and functional, full of potential to reuse and develop.

Note

1 The term urban ensemble marks a compositional group of buildings, as a part of the urban space, following specific design principles and consists of similar aesthetical features. The term is opened in Siim Sultson's doctoral thesis *Stalinist Urban Ensembles in East Estonian Oil-Shale Mining and Industrial Town Centres: Formation Mechanisms and Urban Space Identity as the Potential for Spatial Development*.
2 This chapter reflects author's research status on the issue as of early 2018.

References

Arman, H.; Starostin, I. (1951). Arhitektuurialastest saavutustest ja ülesannetest Nõukogude Eestis. (About Achievements and Tasks of Architecture in Soviet Estonia). *ENSV Arhitektide Almanahh IV*, pp. 7–8. (in Estonian)

Barhhin, M. G. (1986). *Gorod. Struktura i kompozitsija* (Town. Structure and Composition). Moscow: Nauka. (in Russian)
Cinis, A.; Drėmaitė, M.; Kalm, M. (2008). Perfect representations of Soviet planned space. Mono-industrial towns in the Soviet Baltic Republics in the 1950s–1980s. *Scandinavian Journal of History*, 33(3), pp. 226−246.
Eesti arhitektuuri ajalugu (History of Estonian Architecture). (1965). Arman, H. [ed.] Tallinn: Eesti Raamat. (in Estonian)
Eesti elanike arv KOV-de lõikes seisuga 01.01.2018, (2018) Siseministeerium (Ministry of the Interior). [online] Available at: https://www.siseministeerium.ee/sites/default/files/dokumendid/Rahvastiku-statistika/eesti_elanike_arv_kov_01.01.2018.pdf. (in Estonian)
Eesti Arhitektuurimuuseum (EAM) Museum of Estonian Architecture]: EAM.3.1.281 EAM.3.1.248; EAM.3.1.50.
Eesti Riigiarhiiv (ERA) [Estonian State Archives]: ERA R-1992.2.1; ERA R-1992.2.70; ERA.R-1.5.95; ERA.R-1.5.104; ERA.R-1.5.179; ERA.R-1.5.133; ERA.R-3.3.1071; ERA.R-3.3.3034; ERA R-1992.2.12; ERA R-1992.2.31; ERA R-1992.2.22; ERA R-1992.2.41 ERA.R-1992.2.57.
Gorich, I. (1946). V masterskojzodchego. (In Architect's Atelier.) *Sovetskaja Estonija*, August 10th, 1946. (in Russian)
Groys, B. (1998). Stalin-stiil. (Stalin-style), *Akadeemia* 4, pp. 855–891. (in Estonian)
Iz istorii sovetskoj arhitektury 1941–1945 gg. Dokumenty i materialy. Hronika voennyh let. Arhitekturnaja pechat (From History of Soviet Architecture 1941–1945. Chronicle of War Years. Architectural Publishing). (1978). Moscow: Izdatel'stvo "Nauka". (in Russian)
Kalm, M. (2001). *Eesti 20. sajandi arhitektuur. Estonian 20th century architecture.* Tallinn: Prisma Prindi Kirjastus. (in Estonian)
Kosenkova, J. L. (2009). *Sovetskij gorod 1940-h – pervoj poloviny 1950-h godov. Ot tvorcheskih poiskov k praktike stroitel'stva. Izd. 2-e, dop.* (Soviet Town from 1940s to the First Half of 1950s. From the Creative Searches to the Practice of Building. Ed. 2, Amended). Moscow: Knizhnyj dom. (in Russian)
Kratkij spravochnik arhitektora (Short Regulations for Architect). (1952). Moscow: Gosudarstvennoe izdatel'stvo literatury po stroitel'stvu i arhitekture. (in Russian)
Meigas, V. (1951). Eesti NSV linnaehitusest. (About urban construction of Estonian SSR). *ENSV Arhitektide Almanahh IV*, pp. 19–30. (in Estonian)
Ob ustranenii izlishestv v proektirovanii i stroitel'stve. Postanovlenie Central'nogo Komiteta KPSS i Soveta Ministrov SSSR 4 nojabrja 1955 goda (About Abandonment of Exaggerations in Planning and Building. Resolution of the Central Committee of the CPSU and the Council of Ministries of USSR in November 4th, 1955). (1955). Moscow: Gospolitizdat. (in Russian)
Plenum pravlenija Sojuza arhitektrov (Plenum of the Board of the Union of Soviet Architects.) (1943). *Pravda*, August 19th. (in Russian)
Sillamäe Muuseum (SM) [Sillamäe Museum]: SM2Ajd/0462; SM2Ajd/0463; SM2Ajd/0465
Sultson, S. (2016a). Alteration in the Awareness of Estonian City Space from Independence to Stalinism. *Periodica Polytechnica Architecture*, 47(1), pp. 49–55.
Sultson, S. (2016b). Replacement of urban space: Estonian post-war town planning principles and local Stalinist industrial towns. *Journal of Architecture and Urbanism*, 40(4), pp. 283–294.

Sultson, S. (2017). Estonian Urbanism 1935–1955: The Soviet-era implementation of pre-war ambitions. *Planning Perspectives*, latest articles. DOI: 10.1080/02665433. 2017.1348977

Tailings.info. (2017). [online] Available at: http://www.tailings.info/casestudies/sillamae.htm

Tippel, V. (1948). Eesti NSV põlevkivi-tööstuslinnadest. (About oil-shale industrial towns of Estonian SSR). *ENSV Arhitektide Almanahh 1947*, pp. 54–59. (in Estonian)

Tippel, V., and Tihomirov, V. (1951). Uusi ühiskondlikke hooneid. (Some new public buildings). *ENSV Arhitektide Almanahh IV*, pp. 31–43. (in Estonian)

Volkov, L. (1991). Eesti Arhitektuurist aastail 1940–1954. (About Estonian Architecture within 1940–1954). In *Linnaehitus ja Arhitektuur: Tallinn*, pp. 183–213. Tallinn: Ehituse Teadusliku Uurimise Instituut. (in Estonian)

Vseviov, D. (2002). *Kirde-Eesti urbaanse anaomaalia kujunemine ning struktuur pärast teist maailmasõda: Doktoritöö*. (The formation and structure of the urban anomaly in northeast Estonia after World War II. Doctoral Dissertation). Tallinn: Tallinna Pedagoogikaülikool. (in Estonian)

Zabrodskaja, A. (2005). *Vene-Eesti koodivahetus Kohtla-Järve vene emakeelega algkoolilastel* (Russian-Estonian code-switching among Russianspeaking schoolchildren in Kohtla-Järve). Tallinn: TLÜ Kirjastus. (in Estonian)

Zholtovskij (Zholtovsky), I. V. (1933). Printzip zodchestva. (Principe of Architecture). *Arhitektura SSSR* no. 5. (in Russian)

Zholtovskij (Zholtovsky), I. V. (1940). Ancambl' v arhitekture. (Ensemble in Architecture). *Stroitel'naja gazeta*, May 30th. (in Russian)

4 Nuclear anxiety in postwar Japan's city of the future

Sebastian Schmidt

HARVARD GRADUATE SCHOOL OF DESIGN

"I feel like I have come to the city of the future,"[1] says Hiroshi, protagonist in Yuasa Noriaki's 1970 monster movie *Gamera vs. Jiger* (*gamera tai daimajū jaigā*), when he is taken on a tour of the newly constructed grounds of EXPO '70 in Senrigaoka (Senri Hills), outside Osaka. The master plan for the EXPO grounds was purposely designed to be a model for a city of the future, and Hiroshi's statement suggests that the project's design mission had been accomplished. The middle school student is visiting the world's fair and is impressed with its elevated monorail and the unconventional designs of the colorful pavilion buildings. Later in the movie, Hiroshi comes to play a major part in preventing the destruction of the expo grounds by an evil quadruped named Jiger. An ancient sculpture called the 'devil's whistle' (*akuma no fue*) is moved to the expo grounds from a South Pacific island, and it produces a high-pitched sound that harms Jiger. The monster follows the sculpture to destroy it and ravages anything in its way. With the combined efforts of Hiroshi, other human actors, and Gamera—a friendly monster joining the human cause—the threat can be fought off and EXPO '70 saved.

Gamera vs. Jiger was the sixth installment in Daiei Motion Picture Company's Gamera series that had begun in 1965 as a response to Toho Studios' highly successful Godzilla franchise. The way in which EXPO '70 is integrated into the film is indicative of the importance of the event for the Japanese nation at a time of unprecedented economic growth, global business partnerships, and international political, as well as cultural, integration. Further, the very act of hosting a world's fair—an event with a distinctly Euro-American history—reflected Japan's ambition of assuming a role on the world stage of capitalist democracies. It became the first Asian country to be awarded a world's fair and it held the expo visitor record for four decades, until it was surpassed by the Shanghai Expo in 2010.[2]

The event also helped position Japan in terms of its Cold War alliances. The year 1970 saw the second renewal of the security treaty between Japan and the United States that continues to be in effect as the legal foundation for U.S. military presence in Japan today. Originally signed in 1952, the treaty's renewal and amendment in 1960 had stirred up Japan's most severe postwar political crises and mass demonstrations especially in the capital city of Tokyo. Protesters argued that supporting the treaty would be tantamount to supporting U.S. imperialist warfare in Asia. In this context, the expo's motto of Progress and Harmony for Mankind (*jinrui no shinpo to chōwa*) indicated an aspiration for universal ideals, but it was quite clear that both progress and harmony were to be realized within the governmental framework of capitalist democracy, and were to be ensured through a partnership with the United States, and through reliance on its military superiority in particular. These not-so-subtle geopolitical realities found expression in *Gamera vs. Jiger*. EXPO '70 as the emblem of a Western world order—and of its possible future form—is protected at all cost in the movie, whereas the Japanese city of Osaka sustains serious damage. Suggestively, only one building at the expo is physically attacked by Jiger before the monster is defeated. That building is the pavilion of the Soviet Union.

Japan's state-sanctioned city of the future as it was exhibited at EXPO '70 and portrayed in *Gamera vs. Jiger* was marked by a distinct hopefulness—it showed the city as something that held considerable promise for humanity, and as something to look forward to and be excited about. Further, it was recognized that Japan was well positioned for proposing an urban future that had the capacity of spreading around the world. In 1970, *Time Magazine* ran an extensive special on Japan titled "Toward the Japanese Century" that opened with a report on the world's fair and stated that "no country has a stronger franchise on the future than Japan."[3]

In this chapter, I argue that the history of the Japanese city of the future is intimately connected with World War II, and with the use of nuclear bombs on Hiroshima and Nagasaki, in particular. To this end, I will analyze the postwar development of the city of the future as a Cold War phenomenon of urbanism in Japan. EXPO'70 marks the peak moment of this development and is here positioned as the endpoint of my investigation. The beginning of the creation of the Japanese city of the future I see in the construction of a national monument to commemorate the devastation of the nuclear bomb in Hiroshima, which coincided with an important debate about the meaning and value of tradition in the conception of Japanese cultural and architectural history. From a sustained discussion of this moment in the 1950s, I follow the city of the future back to 1970 and the Osaka expo, to show that nuclear anxieties and trauma were important forces in shaping urban visions in Cold War Japan. The 'city of the future' is an unusual case study because it is not limited or confined to a single urban location, but instead is a disaggregated project consisting of both material and intellectual dimensions. Nevertheless, as an analytic theme, it gives access to evidence that helps us understand a Japanese urban imaginary in an age of nuclear trauma following World War II. However, rather than arguing for particular causalities, I demonstrate that the history of the city of the future in Japan is based in the complex relationship between domestic debates about tradition and history, as well as the experience and specter of nuclear devastation.

Hiroshima and the tragedy of modern architecture

Architect Tange Kenzō had lived in Hiroshima during his high school days, and returned to the destroyed city in 1946 with his associates, including Asada Takashi and Ōtani Sachio, to design a reconstruction master plan for the Institute for War Recovery (*sensai fukkōin*).[4] Like Tange, Ōtani and Asada also had personal ties to Hiroshima, and working on this project was a way for all of them to address and attempt to overcome a personal and national nuclear trauma.[5] The passage of the 1949 Hiroshima Peace Memorial City Building Law (*hiroshima heiwa kinen toshi kensetsuhō*) made possible the appropriation of reconstruction funds and the creation of a competition for the design of a memorial area at the center of the city.[6] Already having worked out a plan for the reconstruction, Tange entered the 1949 competition with the winning proposal. His team's plan was not fundamentally different from the consensus that had been reached at the Hiroshima municipal level, with the exception of the hypocenter of the nuclear explosion. Tange had aimed to turn this area into the city's bustling core, including the relocation of the city hall to this newly designed zone. However, while Tange's team sought to connect the memory of the bombing to people's everyday lives, the local and national governments as well as the occupying U.S. forces were leaning toward a more traditional memorial. In addition to these political issues, financial constraints made a grand redesign of the hypocenter impossible, and a symbolic peace zone was created instead.[7] Whether the inclination was to design a vital urban core or a museum-like memorial landscape, what was at stake in this project was not the past, but the future. At the site of the deadliest single attack in human history, reorganizing the urban fabric meant not only preserving the memory of that event in the past, but it carried the responsibility of showing the possibility of a future.

The central Exhibition Hall was executed according to Tange's plans and the neighboring buildings were added later, connected to the main building with enclosed elevated walkways. The cenotaph situated on the visual axis between the Exhibition Hall and the ruined Hiroshima Peace Memorial (known as the Atomic Bomb Dome, consisting of the concrete remnants of the Hiroshima Prefecture Industry Promotion Hall or *hiroshima-ken sangyō shōrei kan*) was originally designed by Japanese-American sculptor Isamu Noguchi, but ended up being created by Tange due to a controversy about Noguchi's memorial allegedly resembling human viscera, as well as uneasiness about the sculptor's US heritage.[8]

Tange's curved cenotaph constructed in thickly poured concrete was modeled after terracotta funerary objects of Japan's Kofun period (third to sixth centuries CE), especially vessels as well as the roofs of miniature houses that are called *haniwa*.[9] The Exhibition Hall that forms the centerpiece of the memorial complex consists of a cuboid concrete shape raised on pilotis, accessed from below through two staircases. The long sides of the singular volume are encased in floor-to-ceiling glass that is rhythmically broken by concrete posts, and the south-facing part has an added lattice layer that serves as a brise-soleil and echoes the structural pattern of *shōji* (Japanese paper screens set in sliding doors). The design is commonly seen as paying homage to the raised-floor tradition of the Yayoi period (300 BCE to 300 CE), and to the Ise Grand Shrine as the epitome of Yayoi architecture. Architectural critic and historian Yatsuka Hajime has described it as a "Corbusean spin on this traditional raised-floor structural type."[10]

This notion of a hybridization of Western modernist architectural expression with traditional Japanese form belies the complexity with which this exact issue had been discussed in Japan—both esthetically and historiographically. The postwar interest in Japanese tradition and history started with archeology. The 1943 discovery of the Toro archeological site in Shizuoka, dating to the late Yayoi period (100–300 CE), and its extensive excavation beginning in 1947 played an important role in the postwar refashioning of Japanese history.[11] Until Japan's capitulation in 1945, "the view of history held by most was subsumed under the ideology of *tennōsei*, the set of beliefs enveloping the emperor system."[12] This included the belief that the members of the imperial family were direct descendants of a divine lineage going back to the sun goddess Amaterasu. Thus, Japanese history until 1945 was less shaped by scientific inquiries into genealogy, and more by the mytho-historical records of the *Kojiki* and the *Nihon Shoki*, Japan's oldest works of classical history, completed in the eighth century CE. The emperor's forced renunciation of his family's divinity in 1946 suddenly opened national history to investigation in unprecedented ways, and the Toro site became a focal point of that interest.[13] The excavations at Toro were significant because they delivered proof of long-standing sophistication in rice cultivation in Japan, which was seen as a bedrock of cultural identity, and thus added to the perception of Yayoi period culture and technology as the source of an unsullied and true origin of the Japanese nation.

In 1952, Okamoto Tarō published his article "Dialogue with the Fourth Dimension: Theory of Jōmon Earthenware" (*yojigen to no taiwa. jōmon doki ron*) in the art journal *Mizue*. It was triggered by Okamoto's encounter with the Jōmon period (14,000–300 BCE) earthenware at the Tokyo National Museum in Ueno Park the year before, and Okamoto argued for the central importance of the Jōmon period in the history of Japanese art. The revived scientific interests in Yayoi culture stemming from the Toro archeological site, and Okamoto's lifting of the Jōmon out of a pre-historic and pre-Japanese past onto the map of Japanese art history constitute the context for the central debate on the role of tradition in architecture in Japan since WWII.[14]

This debate unfolded in the influential journal *shinkenchiku* ("New Architecture," better known as its English edition *The Japan Architect*), in which Tange published his project for the Hiroshima Peace Memorial Park in 1954. The following year, Tange's friend and collaborator Kawazoe Noboru, who at the time was the *shinkenchiku* editor-in-chief, invited contributions to what later came to be known as the "tradition debate" (*dentō ronsō*). The goal was to gather different viewpoints on questions of history and tradition in Japanese architecture.[15] The series opened with Tange's article "How to understand modern architecture?" ("*kindai kenchiku wo ikani rikai suru ka*"), in which he argued that beauty and function are not mutually exclusive, and that traditional architectural form should not be imported into contemporary production in a straightforward manner. The debate occurred at a moment when the war had been over for ten years, and the US occupation for three, and there was an active attempt to repair and reclaim Japanese history and tradition from a contemporary position without the problematic recent legacy of the emperor system.

As mentioned above, the question of the role of Jōmon and Yayoi culture in Japanese history was of central importance, and Okamoto and Tange expressed key positions on that question. Both saw the Jōmon as an important part of Japanese cultural development, but for different reasons. For Okamoto, that period's art was wild, asymmetrical, and open, thus enabling him to think of 'tradition' as a dynamic concept, rather than a static one; the Yayoi, on the other hand, he saw as symmetrical, hierarchical, and closed.[16]

In Tange's understanding of Japanese architectural tradition, the Jōmon and Yayoi were the two origins of Japanese architecture and there was a dialectical relationship between them. The Jōmon pit house developed into peasant dwellings, and the Yayoi raised-floor granary developed into shrines and aristocratic dwellings. The latter was deified at Ise, and also took form in the palatial structures of Heian-kyō (now Kyoto) and Edo (now Tokyo), thus demonstrating the form's openness. The pit house—and subsequent farms and merchant dwellings—was seen as closed. The Katsura Imperial Villa in Kyoto combines these trajectories by bringing the country-inspired *sukiya* tea house style together with palatial architecture, thus making it a space that is of the emperor system, but also includes the sensibilities of the Japanese people.[17]

In other words, the Jōmon were an important part of Japanese architectural history for both Okamoto and Tange, but while Okamoto saw their nonconforming 'wildness' as the key for understanding and invigorating the contemporary moment, Tange positioned them as a form of expression to be refined and elevated—metaphorically, as well as literally through the use of pilotis.

Tange's dialectical thinking about the formation of Japanese culture and tradition thus contrasted with a more integrated approach. The latter was also expressed by architect and writer Shirai Sei'ichi who in the August 1956 issue of *shinkenchiku* argued that it was a losing game to think of the formation of culture as driven by different discrete traditions. Instead, Shirai positioned tradition as a matter of infinite diversity, not isolation. He saw the positivist and isolationist understanding of tradition as the result of how the a posteriori belief in the superiority of the symmetry and hierarchy of the Yayoi period shaped the historiographical interpretation of all history.

In other words, if Yayoi culture is already assumed to be the purest form of Japanese esthetic and artistic expression then it becomes easy to see all other expressions as a stepping stone on the teleological path toward the Yayoi period as the pinnacle of Japanese culture. In his article, "The Tradition of the Jōmon Culture: on the Nirayama Mansion of the Egawa Family" ("*jōmon teki naru mono: egawa-shi kyū nirayama kan ni tsuite*"), Shirai argued that a hard look at architecture reveals the impossibility of completely separating or disentangling different cultural traditions.[18] The fluidity and diversity at the core of Shirai's thinking about tradition is also important for thinking about Japanese cultural identity along ethnic identity, and it offers a productive avenue for thinking about cultural formations within a global, non-national framework.[19]

It would be inaccurate to say that Tange favored Yayoi culture, and Shirai preferred Jōmon culture.[20] Instead, the two architects pursued different historiographical approaches in characterizing the relationship between Jōmon and Yayoi. Tange saw the Jōmon as part of an ancient past, whereas for Shirai, Jōmon culture continued as an active cultural force in the formation of Japanese tradition and contemporary esthetics. Tange's friend and *shinkenchiku* editor Kawazoe Noboru saw the contemporary moment of 1955 as most accurately defined by Tange's architecture and use of tradition. Using the pseudonym of Iwata Kazuo, he contributed to the January 1955 issue that started the "tradition debate." In his article "The Japanese Character of Tange Kenzō," he wrote that Tange "must have felt devastated" at the news of Hiroshima's destruction, alluding to the event's traumatic impact. Strikingly, he continued

> Thanks to the conceptual genius of Tange Kenzō, what ended up being only a castle in the sand—his urban planning for the Greater East Asia Co-Prosperity Sphere—has materialized in the Hiroshima Peace Memorial Park [...]
>
> The contradictions inherent in Hiroshima came to epitomize the many contradictions of postwar Japan, and it was through the design of this Hiroshima project that Tange would distinguish himself in the field of postwar architecture. For Tange, modern Japan's foremost architect, the tragedy came to exemplify all the conflicts of the world.[21]

Kawazoe thus likened the Peace Memorial to Tange's 1942 proposal for what was to become the unrealized architectural symbol of Japan's wartime colonial empire in East Asia, situated at the foot of Mount Fuji. However, he also stated that unlike in his design for the Greater East Asia Co-Prosperity Sphere, Tange specifically eschewed Ise-style architecture in Hiroshima, lest it be seen as connected to the emperor system. However, in light of a public's desire for tradition and memory as epitomized in Ise and its connection to the all-important Yayoi period, there was a contradiction in the use of the Grand Shrine's symbolism that seemed unsolvable. Similarly, Kawazoe identified a contradiction in the city of Hiroshima itself, which—ten years after the end of the war—not everyone wanted to see rebuilt as a 'peace city.'

Conservative forces were challenging article 9 of the US-imposed postwar constitution that reduced the Japanese military to self-defense, and protests against the Japanese-American military Treaty of Mutual Cooperation and Security (Anpo) that had been signed at the end of the US occupation of Japan in 1952 were ongoing.[22]

Conveniently, this now-discovered contradiction subsumes and conceals the prior imperialist connection of Tange's 'genius.' This situation itself can be seen as one of 'all the conflicts of the world' that were exemplified in the tragedy of Hiroshima. In fact, Kawazoe later broadened this by talking about the 'tragedy of Japan,' which he saw in the country's marred and, ultimately, unachieved modernization. He wrote that

> Modern architecture is achieved only in conjunction with the modernization of the peoples' daily lives, the modernization of industrial production, and the modernization of government. When modern architecture is attempted in the absence of these conditions, it starts by seeking material from foreign sources, which, when imported, undergo a startling distortion under Japan's feudalistic institutions.[23]

What is missing in Japan, according to Kawazoe, is modernization as the synthetic process and modern architecture as the synthetic expression. He argues that while Tange comes closer to achieving such synthesis than any other architect, he is "unable to escape the limitations imposed by his times," and "has come to exemplify the tragedy of Japan."[24] To take this thought one step further, we can say that the tragedy of Japan resides in the fact that the country is trapped in the unachievability of its modernization, and in the impossibility of settling questions about its history, especially as the permanence of that history had been exposed as deeply fragile by the nuclear bomb.

Tange's Exhibition Hall epitomizes the challenges of postwar Japan because it is much more than the hybridization of Le Corbusier's modernism and Yayoi raised-floor architecture. The building complicates hybridization as between domestic and foreign influences, given that at the time of its creation there was little agreement on the boundaries and historical limits of what constituted a domestic style. It is also unclear whether the pilotis are a reference to either Ise's *shinmei-zukuri* architecture or Le Corbusier, or whether they establish a safe distance from the scorched earth of Hiroshima's ground zero. The cuboid main volume, while appearing open and transparent, is simultaneously closed off with the entrance and exit hidden underneath. Tange's design, modernist at first glance, raises a critical question about the place of modernity in Japan. From the vantage point of this potentially modernist, potentially traditional building, death and destruction come into view. The visual axis connecting to the *haniwa*-inspired concrete cenotaph and ruined A-Bomb Dome in the distance is a reminder that nuclear weapons are also an outcome of that modernization that Japan can neither shake off nor, according to Kawazoe, fully achieve.

Again, in making reference to Yayoi architecture, Yatsuka described the design of the Exhibition Hall in Hiroshima as a "Corbusean spin on this traditional raised-floor structural type."[25] In his 2011 book on Metabolism, he addressed this issue in a more nuanced fashion when discussing the pilotis that the building rests on, arguing that they were inspired by Ise and the Katsura Imperial Villa, as well as Le Corbusier. However, he made the point that the design, rather than representing any particular ideology, dealt with the question of human existence.[26] Therefore, the pilotis were not straightforward imitations of either traditional Japanese architecture or European modernism, but instead played a symbolic role of demonstrating the complexity of various influences.[27]

The designs for the Hiroshima Peace Memorial Park and its buildings operate at different scales. First, they provided an avenue for Tange and his associates to confront a traumatic event that was all the more palpable through their personal connections to Hiroshima. Second, the project became an important expression of contemporary debates about what constituted Japanese architecture and architectural tradition, and it materialized a vision of what a post-traumatic city of the future could be like. Third, more than any other rebuilding project following World War II, it addressed a condition that Susan Sontag has described as

> the trauma suffered by everyone in the middle of the twentieth century when it became clear that from now on to the end of human history, every person would spend his individual life not only under the threat of individual death, which is certain, but of something almost unsupportable psychologically—collective incineration and extinction which could come any time, virtually without warning.[28]

The Peace Memorial Park does this under the pressure of presenting a viable urban future for Japan and for humanity in an age of atomic urbanism.[29] In other words, the reconstruction of Hiroshima, and the design of the Peace Memorial Park in particular, was the first articulation of a Japanese city of the future at a moment when relationships to both the past, through debates over historical origin, and to the future, through the threat of complete annihilation, had become deeply unsettled. The burden of dealing with Hiroshima exceeded the task of reconstruction as it had been previously understood, and as planner/architect and Tange's 'right hand' Asada Takashi expressed when he said that "a city annihilated by a nuclear bomb is not revived simply by rebuilding its architecture."[30] Asada's approach to urban plans operated on the level of the environment, that is, on a scale that could include concerns about a larger urban system. For him, the scale of creative and generative intervention in the city matched that of destructive intervention as had been demonstrated by the force of the atomic bomb. In other words, when the past of an entire city could be wiped out or, polemically speaking, 'redesigned' in a flash, the notion that its future could be reimagined and planned in one sweeping gesture suddenly seemed a lot more feasible. Needless to say, this development cannot be dissociated from the emergence of megastructural ideas in Japan that culminated in the formation of the Metabolist architectural movement in 1960 and the work that its members produced throughout the 1960s and into the 1970s.

The Metabolist group was started with the publication of its manifesto by Kawazoe Noboru at the 1960 World Design Conference (May 11–16 in the Sankei Kaikan Kokusai Hall). Tange served as secretary general, but he had entrusted Asada with his responsibilities of preparing the event while he was away for a semester teaching at MIT. Given Asada's central role, architect Isozaki Arata has suggested that he was the driving force behind forming the Metabolist group, and that it was in fact Asada who moved Kawazoe to publish a manifesto even though he neither joined the group or appeared in the manifesto himself.[31]

The architectural cornerstone of the World Design Conference was the presentation of Tange's well-known urban Plan for Tokyo Bay. It was first developed as a plan for Boston Harbor while Tange was at MIT. However, he said that it only "'happened to be designed on the ocean at Boston,'" saying that "'its location is not of primary importance.'"[32] Location was not seen to be of primary importance because the project was not designed for Boston or Tokyo, but for a space that, first and foremost, was conceptual. The plan, whether in Boston, Tokyo, or elsewhere, focused on providing features that became known as the tenets of Japanese Metabolist architecture—with ample infrastructure and flexible opportunities for expansion or reconfiguration; these were designed to address universally applicable challenges of urban growth and congestion, positioning the city as a globally shared common denominator amongst all nations and societies.

As much as this urban design by Tange and his associates has been discussed in relevant literature, however, a closer look at its architectural components is surprisingly rare. A detailed view of the plan's model shows its great architectural uniformity, reliant on tent-like concrete structures whose elegantly curved roofs, at their edge, are almost parallel to the water surface, as if the buildings themselves were solidified waves coming up from the ocean, randomly scattered across a gridded network of bridges. After having witnessed the reconstruction of the Ise Shrine in 1953—an event normally closed to observers—Tange and Kawazoe published the book *Ise: Prototype of Japanese Architecture* in 1962 (*ise—nihon kenchiku no genkei*, English edition 1965) with photographs by Watanabe Yoshio. One of the things that Tange emphasized in his writing was the Shrine's seamless integration with nature, and one of the vantage points he described is that from a helicopter:

> During my visits to Ise, I had an opportunity to view the Shrine from a helicopter. This made the tremendous impact of the Geku stand out even more clearly. When the helicopter approached closer, one was led

to imagine the brute strength of some primeval animal crouching on the ground, or the presence of some living, breathing earth spirit, the image of the deity enshrined in the Geku.[33]

The similarity between the photographs of the model of the 1960 Tokyo plan and the Ise shrine speaks of the role a newly discovered Japanese architecture and urbanism would play in both going back to the "brute strength" of days before the trauma of war and occupation, while moving forward into the atomic age. There is thus recognizable in Tange's mega-urban proposal for Tokyo the sensitivity and conflict that resided in Hiroshima's Peace Memorial Park, resulting in a negotiation of modern architecture as trapped between a past that was hard to recuperate and the knowledge of possible future destruction. The city of the future, beginning in Hiroshima, and continuing through 1960 as a pivotal moment in Japanese urban design, is made of this conflict. It was much less pronounced in the work of the Metabolist group members who were of a younger generation than Tange and possibly more focused on the promises of the future beyond nuclear bombings, rather than on how the future was growing out of its own nuclear undoing. Although he never became an official member of the Metabolist group, Isozaki Arata was an outlier for his generation (born 1931). He had graduated from the University of Tokyo architecture department in 1954 and was working for Tange's research lab when the Tokyo bay plan was developed. According to Yatsuka, it was Tange's work in Hiroshima that had attracted Isozaki to his mentor, but he soon developed a more pessimistic position vis-à-vis plans for the future, which became increasingly obvious toward the end of the 1960s.[34]

The city of the future is invisible/ruined

In 1968, the 'Urban Design Research Group' (*toshi dezain kenkyū tai*) published a book titled *Japanese Urban Space* (*nihon no toshi kūkan*).[35] In addition to Isozaki, other members of the group included the architectural historian, engineer, and novelist Itō Teiji, the urban planner Tsuchida Akira, the planner and community developer (*machi zukuri*) Hayashi Yasuyoshi, the architect Tomita Reiko, and the urban designer and architect Ōmura Ken'ichi. The authors offer a passionate Japanese perspective on modernist urban planning and functionalism, and use case studies of Japanese urban space to propose systematic solutions to what they argue constitute the two (connected) principal challenges facing cities: the separation between architecture and urban planning, and lack of awareness of the importance of urban structure.[36]

Rather than continuing with an esthetic focus of architectural and urban modernism that they saw in Le Corbusier's work and that appears to ignore the living conditions in cities, they argued that the goal for urban planning as a method should be to unify the quantitative analysis of urban space with the materiality of the built form.[37] Therefore, Isozaki and his co-authors offer an outspoken critique of functionalism as an urban model that they saw as having two main shortcomings: it offered an abstraction of space insufficient for understanding the conditions of actual urban space, and it lacked a deep structural understanding of the city to which the methods and mechanisms of functionalism would be applied.[38] The argument was that in functionalism, the city was not one organic body, but an aggregation of different functions because the overall structure was more arbitrary than deliberate; that is, the whole was a result of the parts, rather than the other way around.[39] Tange's 1960 Plan for Tokyo Bay is praised as an example that provides a large overall structure within which different functions are able to take up space.[40]

However, it is important not to confuse this desire for structure with a desire for a fixed shape to be filled in with parts and functions. Instead, the shape of the planned city is here 'vague, and ought to waft like fog' ("*mōrō to shite, kiri no yō ni yure ugoku mono to naru hazu de aru*").[41] In other words, there is no inherently good, or universally suitable urban form in this mode of thinking, and the realization of every urban plan depends on the thoughtful and attentive integration of actual conditions in the city with the capabilities of the built form. The authors of *nihon no toshi kūkan* argue that "there is no firm shape of the future, and the shape of the city is becoming increasingly invisible."[42] This approach—a structurally mediated compromise between the city as an object both fixed and flexible—does not preclude innovation or even radical departures from urban traditions (such as Tange's Tokyo Bay Plan), but it is strongly opposed to the notion that the city of the future could be an object or idea first in the imagination, then on paper, and finally in reality. Instead, the city of the future plays a purely symbolic role as an unknown that will come into existence, but whose emergence can never be sped up.

> That is why the future city has to be something that is always swaying and moving inside your imagination. [...]
> This moving and unfixed image can also be called 'invisible city.' It is the invisible city that really is the city of the future.[43]

In the same year that *Japanese Urban Space* was published, Isozaki had an opportunity to experiment with the invisibility of the city in his installation *Electric Labyrinth* for the XIV Triennale di Milano. With support from photographer Tōmatsu Shōmei, graphic designer Sugiura Kōhei, and composer Ichiyanagi Toshi, Isozaki had put together an installation that was to throw a critical light on narratives of reconstruction, recovery from atrocity, and ideas of overcoming the past.[44] Pictures of ghosts and demons from Japanese woodblock prints and Tōmatsu's photographs of postwar ruins in Hiroshima were printed onto aluminum sheets that would rotate in frames in response to visitor movement, variably exposing the ghosts, or

108 Sebastian Schmidt

dead bodies and nuclear devastation. On the walls was printed a panoramic view of destroyed Hiroshima, with two megastructures that looked like they were made of modular parts inserted into the ruins. However, the simulated large-scale edifices were falling into ruin themselves. The curator and art critic Hans-Ulrich Obrist quotes Isozaki's description of this part of the project. Isozaki said that he

> made a kind of collage about the ruins of Hiroshima and the megastructure it would later become, which itself was in a state of ruin: a ruined structure on the ruins, which I titled *The City of the Future is the Ruins.* I was very much obsessed by these ruins of the future. I projected many images of the future city onto the wall. We tried to show how the future city would itself constantly fall into ruin. This was on the moving panels, which, whenever they turned, would be accompanied by Toshi Ichiyanagi's strange sounds. I called the installation *Electric Labyrinth*.[45]

The only thing predictable about any rendition the city of the future as demonstrated in Isozaki's *Electric Labyrinth* is its entropic nature; that it can never escape time and the constraints of its material existence. The fact that the futuristic megastructure in the panoramic mural is itself decaying shows that the visualization of any future makes it first a matter of the present, and then, immediately, a matter of the past. The future, then, can only be seen in the form of its own past, stripped of the hopefulness of a perfect image and instead portrayed within the certainty of its own impermanence.[46]

At the same time as Isozaki was developing his *Electric Labyrinth* for Milan, he was also involved in the design of the central festival plaza (*omatsuri hiroba*) at EXPO '70 in Osaka. In terms of the development of a city of the future in postwar Japan—from Hiroshima to EXPO '70—there could have barely been a bigger ideological divide than existed between the ruins of the future in Isozaki's installation in Italy and the boundless optimism of the world's fair.

More clearly than in publicity around EXPO '70 itself, Yuasa Noriaki's 1970 monster movie *Gamera vs. Jiger* that opened this chapter articulates that optimism. The movie is able to do this because its narrative is obviously fantastical. In 1965, Susan Sontag wrote that fantasy can do two things; one is

> to lift us out of the unbearably humdrum and to distract us from terrors, real or anticipated—by an escape into exotic dangerous situations which have last-minute happy endings. But another one of the things that fantasy can do is to normalize what is psychologically unbearable, thereby inuring us to it. In the one case, fantasy beautifies the world. In the other, it neutralizes it.[47]

Sontag argues that science fiction films, especially those involving 'exotic,' that is, thoroughly 'othered' monsters and destructive disasters, do both of these things because they represent the intersection of naive narratives that downplay real or potential crises, and society's inadequate responses to the challenges and terrors that infect its consciousness.[48]

Gamera vs. Jiger renders a devastating monstrous threat quite harmless, which is only amplified by the fact that a significant part of the task of fending it off is left in the hands of a child; at the same time, the movie provides an opportunity to transfer real fear of attack onto the safe space of the filmic narrative. Similarly, the expo's aspirations for "Progress and Harmony for Mankind" beautified the esthetics of the everyday, and clearly represented an escape into a city of the future that came with a happy ending. Thereby, EXPO'70 normalized threats and fears of annihilation that were psychologically unbearable, and neutralized them by showing a path toward a world of global peace.

By contrast, In Hiroshima's 'city of the future'—be it in the historically complex design of Tange's Peace Park Memorial or in Isozaki's skeptical *Electric Labyrinth*—what we see is a representation that knows about the impossibility of representation, or, as film scholar Akira Mizuta Lippit has written, "a signifier that indicates the inability of language to absorb and stabilize the atopicality of atomic destruction."[49] This is addressed in the first line of Alain Resnais' well-known 1959 movie *Hiroshima mon amour*, when the Japanese male protagonist says to his French female counterpart *"tu n'as rien vu à Hiroshima. Rien."* ("You have seen nothing in Hiroshima. Nothing.") She protests: *"J'ai tout vu. Tout. Ainsi, l'hôpital, je l'ai vu. J'en suis sûre. L'hôpital existe à Hiroshima. Comment aurais-je pu éviter de le voir?"* ("I have seen everything. Everything. Thus, the hospital, I've seen it. I'm sure of it. The hospital exists in Hiroshima. How could I have avoided seeing it?"). The man expresses what Maurice Blanchot has written about the disaster—that it escapes our grasp. "The disaster, unexperienced. It is what escapes the very possibility of experience—it is the limit of writing."[50]

Tange's design for the Hiroshima Peace Memorial Park is the most significant memorial to WWII in Japan.[51] In its complex engagement of questions about Japanese tradition in architecture, and about the possibility of nuclear devastation, it exhibits a vision for the future, while challenging any such vision to be critical of images that appear unencumbered by the past. Throughout the 1960s, and culminating in the spectacle of EXPO '70, the rift between the future as based in its own nuclear undoing and as an improved version of the present only widened.

Hiroshi, the juvenile monster slayer in Yuasa's *Gamera vs. Jiger*, exclaimed with great confidence that 'he felt like he had come to the city of the future' when he was shown around the world's fair and was looking at its spectacular pavilions. Had Isozaki's 1968 *Electric Labyrinth* for the Triennale in Milan not been vandalized by protestors occupying the event, a visitor looking at its images of nuclear devastation and decaying architecture of the sort exhibited at EXPO '70 could have likewise said "I feel like I have come to the city of the future." What has become clear is that these environments were representative of dramatically different responses to the same set of circumstances. Following 1945, neither the past nor the future of Japanese urbanism provided any sense of certainty, from the debates over the origins of Japanese tradition to the possibility of a repetition of the nuclear trauma. Each iteration of a possible city of the future comes with its own history and characteristics, but none of them can be understood without the anxieties that World War II had left behind.

Notes

1 *"Mirai no toshi he kita yōna ki ga suru yo."*
2 Originally, Tokyo had been selected to host a world's fair in 1940, which was canceled due to the Second Sino-Japanese War. In addition to presenting Japanese industrial and cultural achievement to the world, the event was to commemorate the 2,600th anniversary of the ascension of the mythical first emperor Jimmu, thereby deepening the nationalist construction of a long and continuous historical identity.

    64 million people attended EXPO '70, equivalent to almost two thirds of Japan's total population of just over 100 million in 1970. A *New York Times* article from 2010 documented the Chinese government's active involvement in increasing the visitor count, with the specific goal of breaking the Japanese record. This indicates the importance given to the event by China, and the government's desire to perform better than Japan at an event designed to showcase Chinese economic and industrial prowess. See David Barboza, "Shanghai Expo Sets Record with 73 Million Visitors," *The New York Times*, November 2, 2010.
3 "Toward the Japanese Century," *Time* 95, no. 9, March 2 (1970): 20.
4 See Rem Koolhaas et al., *Project Japan : Metabolism Talks* (Köln, London: TASCHEN GmbH, 2011), 80–93.
5 See Hajime Yatsuka, *Metaborizumu Nekusasu [Metabolism Nexus]*, Dai 1-han. ed. (Tokyo: Ōmusha [Ohmsha], 2011), 78.
6 Mori Bijutsukan (Mori Art Museum), *Metaborizumu No Mirai Toshi Ten : Sengo Nihon - Ima Yomigaeru Fukkō No Yume to Bijon (Metabolism, City of the Future : Dreams and Visions of Reconstruction in Postwar and Present-Day Japan)* (Tokyo: Shinkenchikusha, 2011), 30.
7 See Yatsuka, *Metaborizumu Nekusasu [Metabolism Nexus]*, 78.
8 See Koolhaas et al., *Project Japan : Metabolism Talks*, 106–115.
9 Seng Kuan, Yukio Lippit, and Harvard University Graduate School of Design., *Kenzo Tange : Architecture for the World* (Baden, London: Lars Müller; distributed by Springer, 2012), 29.
10 Hajime Yatsuka, "The Social Ambition of the Architect and the Rising Nation," in *Kenzo Tange : Architecture for the World*, ed. Seng Kuan, Yukio Lippit, and Harvard University Graduate School of Design (Baden, London: Lars Müller; distributed by Springer, 2012), 47.
11 See Walter Edwards, "Buried Discourse: The Toro Archaeological Site and Japanese National Identity in the Early Postwar Period," *The Journal of Japanese Studies* 17, no. 1 (1991).
12 Ibid., 13.
13 Giving the so-called "declaration of humanity" (*ningen sengen*) on New Year's Day 1946 was part of Allied-imposed measures for the emperor to be exonerated from all potential war crime charges. See John W. Dower, *Embracing Defeat: Japan in the Wake of World War Ii*, 1st ed. (New York: W.W. Norton & Co./New Press, 1999), 308–318.

14 See Tarō Okamoto, "Yojigen to No Taiwa. Jōmon Doki Ron [Dialogue with the Fourth Dimension: Theory of Jōmon Earthenware]," *Mizue*, no. 1 (1952). Okamoto's argument, while impactful within art and architectural discourse, had little influence on archeologists. A full embrace of the Jōmon as the foundation of Japanese culture did not occur within that field until the booming interests in Japan's best known Jōmon site at Sannai Maruyama in Aomori Prefecture in the 1990s. See Junko Habu and Clare Fawcett, "Jomon Archaeology and the Representation of Japanese Origins," *Antiquity* 73, no. 281 (1999). For more recent perspectives, see also the special issue of *Asian Perspectives* on "New and Emergent Trends in Japanese Paleolithic Research" (Vol. 49, No. 2, Fall 2010)
15 For a concise summary of the debate, see architectural critic and historian Igarashi Tarō's online series on Japanese architectural theory (*nihon kenchikuron*). Tarō Igarashi, "'Minshū' No Hakken to Jōmonteki Naru Mono : Dai 5 Shō (2) [the Discovery of the Public and the Jomonization of Things: Chapter 5 (2)]," (2015), https://cakes.mu/posts/9443. "Okamoto Tarō Kara Kangaeru / Kōchaku Shita Dentōron Wo Kaitai Suru Shikō : Dai 6 Shō (1) [Thinking with Okamoto Tarō/Thoughts on Dismantling the Stiffened Tradition Discourse : Chapter 6 (1)]," (2015), https://cakes.mu/posts/9614.
16 See "Okamoto Tarō Kara Kangaeru/Kōchaku Shita Dentōron Wo Kaitai Suru Shikō : Dai 6 Shō (1) [Thinking with Okamoto Tarō/Thoughts on Dismantling the Stiffened Tradition Discourse : Chapter 6 (1)]".
17 See "Minshū No Hakken to Jōmonteki Naru Mono : Dai 5 Shō (2) [the Discovery of the Public and the Jomonization of Things : Chapter 5 (2)]".
18 For an English translation of the article, as well as a more detailed description of Shirai's positions on questions of tradition, see Toru Terakawa, "History and Tradition in Modern Japan: Translation and Commentary Upon the Texts of Sei'ichi Shirai" (McGill University, 2001).
19 Shirai's position resonates with views that employ the Jōmon for re-examining Japanese ethnicity, especially vis-à-vis Japanese minority populations in Hokkaido and Okinawa. "Broad public acceptance of a prehistoric cultural and biological link between the modern ethnic Japanese and the Ainu and Ryukyuans could result in a new paradigm of the archaeological past which would focus on prehistoric diversity rather than homogeneity and would provide space for a modern-day acceptance of ethnic diversity within Japan." Habu and Fawcett, "Jomon Archaeology and the Representation of Japanese Origins," 592. For an analysis of Shirai's thought as a form of universal or cosmopolitan architecture, see Torben Berns, "The Trouble with "Boku" - a Meditation on Cosmopolitan Architecture," *Review of Japanese Culture and Society* 13, Architecture: Re-building the Future (2001).
20 In fact, Tange saw direct continuities between Jōmon pit house construction and traditional Japanese farmhouses, and even suggested that the nature worship associated with the Yayoi period and the formation of the Shintō religion had its origins in the Jōmon period. See Kenzo Tange, Noboru Kawazoe, and Yoshio Watanabe, *Ise : Prototype of Japanese Architecture* (Cambridge, MA: MIT Press, 1965), 14–26.
21 Kazuo Iwata, "The Japanese Character of Tange Kenzō [1955]," in *From Postwar to Postmodern : Art in Japan 1945–1989: Primary Documents*, ed. Doryun Chong, et al. (New York, NY; Durham, NC: Museum of Modern Art ; Distributed by Duke University Press, 2012), 69.
22 On the changing nature of protest in postwar Japanese society, Amemiya Shōichi describes significant transformations in labor, education, and community organization, and argues that it was not in the Anpo protest that a new society was formed, but that it was a new society that made Anpo so visible. Shōichi Amemiya, *Senryō to Kaikaku [Occupation and Reform]* (Tokyo: Iwanami Shinsho,

2008), 181–188. For an analysis of issues between Japan and the US in the postwar era, see Shunya Yoshimi, *Shinbei to Hanbei : Sengo Nihon No Seijiteki Muishiki [Pro-America and Anti-America : The Political Unconscious of Postwar Japan]* (Tokyo: Iwanami Shinsho, 2007).
23 Iwata, "The Japanese Character of Tange Kenzō [1955]," 71–72.
24 Ibid., 72–73.
25 Yatsuka, "The Social Ambition of the Architect and the Rising Nation," 47.
26 *Metaborizumu Nekusasu [Metabolism Nexus]*, 90.
27 Ibid., 92.
28 Susan Sontag, "The Imagination of Disaster," *Commentary* October 1965 (1965): 48.
29 In a similar vein, Yatsuka argues that Hiroshima gives shape to the peculiar situation of the immediate postwar period. Yatsuka, *Metaborizumu Nekusasu [Metabolism Nexus]*, 77.
30 *"Genshi bakudan ni yotte kaimetsu shita toshi ha, soko ni futatabi kenchiku wo saiken suru dake de ha saisei shinai."* In Noi Sawaragi, *Sensō to Banpaku [World Wars and World Fairs]* (Tokyo: Bijutsu Shuppansha, 2005), 29, my translation.
31 Koolhaas et al., *Project Japan : Metabolism Talks*, 35.
  Kawazoe did in fact confirm that it was due to Asada that a group was formed. See Sawaragi, *Sensō to Banpaku [World Wars and World Fairs]*, 29.
32 Tange quoted in Ken Tadashi Oshima, "Rereading Urban Space in Japan at the Crossroads of World Design," in *Kenzō Tange: Architecture for the World*, ed. Seng Kuan and Yukio Lippit (Zürich: Lars Müller, 2012), 179.
33 Tange, Kawazoe, and Watanabe, *Ise : Prototype of Japanese Architecture*, 51.
34 See Yatsuka, *Metaborizumu Nekusasu [Metabolism Nexus]*, 77.
35 The book was based on the collective's 1963 article "City Invisible" that addressed many of the same core issues. Arata Isozaki et al., "'Mienai Toshi' E No Apurōchi/City Invisible [Approach to the 'Invisible City'/City Invisible]," *Kenchiku Bunka* 1963, no. December (1963). See also Oshima, "Rereading Urban Space in Japan at the Crossroads of World Design."
36 One of the core arguments that is made about Japanese urban space is that it is highly adaptable and open to different uses. For example, the absence of central plazas or town squares is more than compensated for by different parts of the streets turning into different types of public or semi-public space, and providing space for parades and processions. For example, see Henry D. Smith, "Tokyo as an Idea : An Exploration of Japanese Urban Thought until 1945," *Journal of Japanese Studies* 4, no. 1 (1978). See also Kishō Kurokawa, *Metabolism in Architecture* (London: Studio Vista, 1977), 23–40.
37 Arata Isozaki et al., *Nihon No Toshi Kūkan [Japanese Urban Space]* (Tokyo: Shōkokusha, 1968), 9–10.
38 Ibid., 16.
39 Ibid., 17.
40 Ibid., 18.
41 Ibid., 26.
42 Ibid., 26, my translation.
43 *"sore yue mirai toshi ha, itsumo anata no imeiji no naibu de yurayura to yureugoiteiru mono de nakereba naranai. […] yureugoite kotei shinai imeiji wo 'mienai toshi' to yondemo ii. Mienai toshi koso jitsu ha mirai no toshi da."* Ibid., 27, my translation.
44 Koolhaas et al., *Project Japan : Metabolism Talks*, 42.
45 http://www.castellodirivoli.org/en/mostra/arata-isozaki-electric-labyrinth/, accessed April 14, 2016.
46 See also Yatsuka, *Metaborizumu Nekusasu [Metabolism Nexus]*, 77.
47 Sontag, "The Imagination of Disaster," 42.

48 Ibid., 48.
49 Akira Mizuta Lippit, "Antigraphy: Notes on Atomic Writing and Postwar Japanese Cinema," *Review of Japanese Culture and Society* 10, Japanese Film and History, as History (1998): 58.
50 Maurice Blanchot, *The Writing of the Disaster* (Lincoln: University of Nebraska Press, 1986), 7.
51 I use the term memorial somewhat narrowly to mean that the Hiroshima Peace Memorial Park was constructed for the purpose of giving a material landscape to the memory of the nuclear bombings. Otherwise, it could be argued that the Yasukuni Shrine in Tokyo has emerged as a more significant site in a political sense, given the deep impact it has had on Japanese foreign policy. The shrine is dedicated to war dead including the casualties of World War II, with those convicted of war crimes amongst them. The act of worshipping at the shrine by sitting Prime Ministers, including Koizumi Jun'ichirō and Abe Shinzō, has attracted harsh criticism from their political opposition in Japan, as well as foreign governments—China and Korea in particular. See Tetsuya Takahashi, "The National Politics of the Yasukuni Shrine," *The Asia-Pacific Journal Japan Focus* 4, no. 11 (2006), http://www.japanfocus.org/data/takahashi_1642.pdf.

References

Amemiya, Shōichi. *Senryō to Kaikaku [Occupation and Reform]*. Tokyo: Iwanami Shinsho, 2008.
Barboza, David. "Shanghai Expo Sets Record with 73 Million Visitors." *The New York Times*, November 2, 2010.
Berns, Torben. "The Trouble with 'Boku' - a Meditation on Cosmopolitan Architecture." *Review of Japanese Culture and Society* 13, Architecture: Re-building the Future (2001): 22–30.
Blanchot, Maurice. *The Writing of the Disaster*. Lincoln: University of Nebraska Press, 1986.
Dower, John W. *Embracing Defeat : Japan in the Wake of World War Ii*. 1st ed. New York: W.W. Norton & Co./New Press, 1999.
Edwards, Walter. "Buried Discourse: The Toro Archaeological Site and Japanese National Identity in the Early Postwar Period." *The Journal of Japanese Studies* 17, no. 1 (1991): 1–23.
Habu, Junko and Clare Fawcett. "Jomon Archaeology and the Representation of Japanese Origins." *Antiquity* 73, no. 281 (1999): 587–593.
Igarashi, Tarō. "'Minshū' No Hakken to Jōmonteki Naru Mono : Dai 5 Shō (2) [the Discovery of the Public and the Jomonization of Things : Chapter 5 (2)]." (2015). Published electronically May 22, 2015. https://cakes.mu/posts/9443.
———. "Okamoto Tarō Kara Kangaeru/Kōchaku Shita Dentōron Wo Kaitai Suru Shikō : Dai 6 Shō (1) [Thinking with Okamoto Tarō/Thoughts on Dismantling the Stiffened Tradition Discourse : Chapter 6 (1)]." (2015). Published electronically June 5, 2015. https://cakes.mu/posts/9614.
Isozaki, Arata, Teiji Itō, Akira Tsuchida, Yasuyoshi Hayashi, Reiko Tomita, Ken'ichi Ōmura, Jirō Watanabe, et al. "Mienai Toshi' E No Apurōchi/City Invisible [Approach to the 'Invisible City'/City Invisible]." *Kenchiku Bunka* 1963, no. December (1963): 51–58.
———. *Nihon No Toshi Kūkan [Japanese Urban Space]*. Tokyo: Shōkokusha, 1968.

Iwata, Kazuo. "The Japanese Character of Tange Kenzō [1955]." In *From Postwar to Postmodern: Art in Japan 1945–1989: Primary Documents*, edited by Doryun Chong, Michio Hayashi, Kenji Kajiya and Fumihiko Sumitomo, 69–73. New York, NY; Durham, NC: Museum of Modern Art; Distributed by Duke University Press, 2012.

Koolhaas, Rem, Hans-Ulrich Obrist, Kayoko Ota, James Westcott, and Office for Metropolitan Architecture. AMO. *Project Japan : Metabolism Talks*. Köln; London: TASCHEN GmbH, 2011.

Kuan, Seng, Yukio Lippit, and Harvard University Graduate School of Design. *Kenzo Tange : Architecture for the World*. Baden, London: Lars Müller; distributed by Springer, 2012.

Kurokawa, Kishō. *Metabolism in Architecture*. London: Studio Vista, 1977.

Lippit, Akira Mizuta. "Antigraphy: Notes on Atomic Writing and Postwar Japanese Cinema." *Review of Japanese Culture and Society* 10, Japanese Film and History, as History (1998): 56–65.

Mori Bijutsukan (Mori Art Museum). *Metaborizumu No Mirai Toshi Ten : Sengo Nihon - Ima Yomigaeru Fukkō No Yume to Bijon (Metabolism, City of the Future : Dreams and Visions of Reconstruction in Postwar and Present-Day Japan)*. Tokyo: Shinkenchikusha, 2011.

Okamoto, Tarō. "Yojigen to No Taiwa. Jōmon Doki Ron [Dialogue with the Fourth Dimension: Theory of Jōmon Earthenware]." *Mizue*, no. 1 (1952).

Oshima, Ken Tadashi. "Rereading Urban Space in Japan at the Crossroads of World Design." In *Kenzō Tange: Architecture for the World*, edited by Seng Kuan and Yukio Lippit, 177–187. Zürich: Lars Müller, 2012.

Sawaragi, Noi. *Sensō to Banpaku [World Wars and World Fairs]*. Tokyo: Bijutsu Shuppansha, 2005.

Smith, Henry D. "Tokyo as an Idea : An Exploration of Japanese Urban Thought until 1945." *Journal of Japanese Studies* 4, no. 1 (1978): 45–80.

Sontag, Susan. "The Imagination of Disaster." *Commentary* October 1965 (1965): 42–48.

Takahashi, Tetsuya. "The National Politics of the Yasukuni Shrine." *The Asia-Pacific Journal Japan Focus* 4, no. 11 (2006): 155–180. http://www.japanfocus.org/data/takahashi_1642.pdf.

Tange, Kenzo, Noboru Kawazoe, and Yoshio Watanabe. *Ise : Prototype of Japanese Architecture*. Cambridge, MA: MIT Press, 1965.

Terakawa, Toru. "History and Tradition in Modern Japan: Translation and Commentary Upon the Texts of Sei'ichi Shirai." McGill University, 2001.

"Toward the Japanese Century." *Time* 95, no. 9, March 2 (1970): 20–38.

Yatsuka, Hajime. *Metaborizumu Nekusasu [Metabolism Nexus]*. [in In Japanese.] Dai 1-han. ed. Tokyo: Ōmusha [Ohmsha], 2011.

———. "The Social Ambition of the Architect and the Rising Nation." Translated by Nathan Elchert. In *Kenzo Tange : Architecture for the World*, edited by Seng Kuan, Yukio Lippit and Harvard University Graduate School of Design., 47–60. Baden, London: Lars Müller; distributed by Springer, 2012.

Yoshimi, Shunya. *Shinbei to Hanbei : Sengo Nihon No Seijiteki Muishiki [Pro-America and Anti-America: The Political Unconscious of Postwar Japan]*. Tokyo: Iwanami Shinsho, 2007.

Urbanism of fear
A tale of two Chinese Cold War cities

Tong Lam

UNIVERSITY OF TORONTO

Figure VE 1.1 A 1970s propaganda poster urging young men and women to "contribute their youth to the Third Front Construction."
Source: Author collection.

The geopolitical tension known as the Cold War carried different meanings in different parts of the world. For the United States, the Cold War was primarily a confrontation between the Western democracies and the Communist Bloc led by the Soviet Union. The latter, however, was hardly monolithic. Although the Soviet Union did assist China's industrialisation by offering expertise and equipment, the relation between the two nations deteriorated quickly in the late 1950s due to ideological, cultural, and leadership clashes. The rhetoric of Communist solidarity aside, the two nations eventually engaged in a brief border war in 1969, resulting in a massive mobilisation of troops.

However, even before China's dispute with the Soviet Union, some American politicians had suggested that the United States should use nuclear weapons against North Korea and China during the Korean War

Figure VE 1.2 Traversing the mountains high above the city of Panzhihua is one of the many rail bridges from the Chengdu-Kunming Railway, which linked major Cold War cities in south-western China. Completed in 1970, the 1134 km railway was regarded as a major Third Front achievement.

Source: Author photograph.

(1950–1953). The worsening relationship with the Soviet Union, therefore, simply heightened the Chinese sense of fear as the Soviet Union became an even more menacing adversary than the United States (Figure VE 1.1).

In 1964, facing looming existential threats, the Chinese leader Mao Zedong called for a large-scale defence mobilisation known as the *Third Front Construction*. This secretive geo-military project involved the relocation and duplication of the country's vital industries that had, up to that point, been located in the First Front – the north-eastern region near to the Soviet border and along the Pacific coast. These areas were perceived as being vulnerable to military attack. Meanwhile, the area between the First and the Third Fronts, known as the Second Front, was a heavily populated plateau, and was thus also difficult to defend militarily.

In the decade immediately after 1964, hundreds of factories were strategically built "in mountains, in caves, and in dispersion" in south-western China's remote and mountainous region (Wang et al. 2004). While some of these Third Front facilities were constructed in existing urban areas, others were built from scratch in uninhabited or thinly populated regions. The cities of Liupanshui in Guizhou province and Panzhihua in Sichuan province illustrated in this visual essay are prime examples of these Chinese "instant cities" of the Cold War (Figure VE 1.2).

The development of critical industries and infrastructures such as mining, steel making, petroleum refining, electricity generation, as well as conventional and nuclear weaponry was central to the massive scale of investment needed for the Third Front Construction. Although the idea of dispersing and concealing these facilities in the vast Third Front region is often attributed to Mao's leadership, such spatial strategy was also directly informed by China's World War Two experience (Xia 2017). During the bitterly fought war between China and Japan from 1937 to 1945, the challenging terrain of the south-west was a major reason why the Chinese Nationalist resistance forces were able to halt the advance of the Imperial Japanese Army. It is therefore unsurprising that the geography of the Third Front region introduced in the Cold War overlapped significantly with the unoccupied area of the previous war.

In architectural and urban design terms, the Third Front Construction was equally indebted to the earlier wars. Specifically, even before Nazi Germany unleashed its bombing campaign against London and other British cities in 1940, many Chinese cities were already under intense Japanese air assaults (Baumler 2016). Yet, for Chinese architects and city planners, the spectre of air attack was an opportunity to envision a different kind of urban future that emphasised the ideas of dispersion and concealment, which would become commonplace in Third Front factory towns. Among them, the most prolific thinker of air raid precautions was Lu Yujun, who witnessed the devastation of aerial bombing first hand as a student of architecture in Paris in the aftermath of World War1 in 1919 and then in China's wartime capital Chongqing in Sichuan province. Lu was particularly committed to integrate air raid precautions with his fascination with Le Corbusier's idea of the future city. Specifically, rather than simply promoting modernism based on logistic and aesthetic grounds, Lu saw symmetry and identical grid patterns as a way to confuse enemy pilots, and to facilitate the forming of wind tunnels that might reduce casualties in the event of a chemical attack. He also drew on prevailing notions about linear urban forms and emphasised their utility in the decentralizing of industrial and military facilities through the development of cities without urban centres and with minimal aerial footprints (Lu 1935, 1945).

After World War Two, when the Nationalists were defeated by the Communists in 1949, many urban planners and architects who worked for the former regime retreated to Taiwan. However, even if their names were quickly forgotten in mainland China for political reasons, their anti-air raid strategies took on a new significance in the Third Front Construction project. Just as post-war European cities continued to draw inspirations from the garden city for air raid protection measures, the Third Front Construction also adopted the practice of dispersion by concealing facilities in mountain valleys, natural caves, and artificial tunnels. Since Chinese factories during this time also underscored the need of unifying production and living spaces by combining them into a single entity known as the unit (*dianwei*), most of the Third Front strategic industries were basically small or midsize factory towns. One common layout for these *dianwei*-based towns was to build all the facilities such as industrial plants, worker dormitories, schools,

118 *Tong Lam*

Figure VE 1.3 Factory 300 was one of the several major power plants that supported the heavy industries in the Liupanshui area. Its canteen, market, hospital, school, park, workers' dormitories, as well as power generating facilities were built along a river, making it a typical ribbon-shaped Third Front factory town.
Source: Author photograph.

hospitals, and markets along river valleys deep inside the mountainous areas, giving rise to sprawling factory towns that took the form of a ribbon based on the existing landscape (Figure VE.1.3).

As well, some of the most crucial industrial processes and assemble lines were further hidden in elaborate tunnel systems (Figure VE.1.4). In this

Figure VE 1.4 A sealed hillside entrance to the decommissioned Power Plant 503 in the Panzhihua area. The plant started to operate in the mid-1970s, and was considered as the state of the art at the time with all its facilities concealed in a vast tunnel system.
Source: Author photograph.

Figure VE 1.5 An abandoned factory near Liupanshui that manufactured parts for military aircrafts. Haunting scenes like this, while not uncommon, are generally invisible due to the hidden and remote nature of these secret sites.
Source: Author photograph.

respect, purpose-built Cold War cities such as Liupanshui and Panzhihua were essentially clusters of ribbon-shaped factory towns, even though the development of ICBMs and thermonuclear weapons had increasingly made such spatial arrangements redundant.

As the likelihood of a total war with the superpowers subsided after the normalisation of the Sino-US relation in the 1970s, the Chinese government began to redirect its investment away from the Third Front and to the coastal region. Moreover, by the early 1980s when the project was officially terminated, the concept of dispersion and concealment, which was an extension of World War Two air raid preparations, had become wholly inadequate in light of the latest technological advances such as electronic surveillance, satellites, and precision bombing (Figure VE.1.5).

Still, not all the Third Front factories were deemed irrelevant. Many steel plants and mining operations, for example, have been transformed into profit-seeking state-owned enterprises in the post-socialist era (Figure VE.1.6).

The expansion of these industries has also led to the rapid development of new urban hubs around those industrial plants. Thus, even if many forgotten factories are decaying in the mountains, former purpose-built Cold War cities such as Liupanshui and Panzhihua are experiencing rapid growth and densification, albeit not always in the same ways. In the case of Liupanshui, deindustrialisation of the post-Cold War era has led the city to rebrand its Third Front past as "industrial heritage" in order to attract tourism and investment (Figure VE 1.7).

120 *Tong Lam*

Figure VE 1.6 Formerly known as Factory 671, this petroleum company is located in the Liupanshui area. In the background is a high-rise apartment building under construction for its workers.
Source: Author photograph.

Figure VE 1.7 The post-Cold War Liupanshui is a thriving city trying rebranding itself as China's "Cool Capital" in order to attract tourists and outside investment. In addition to a Third Front theme park and museum, newly completed mega-projects include a new stadium and a new convention and exhibition centre. Recently, the region's high elevation and year-round coolness has also made it an attractive destination for IT companies to set up big data and cloud computing infrastructures.
Source: Author photograph.

In Panzhihua, on the contrary, the Third Front Construction led to the discovery of the area's precious mineral deposits, which has given rise to a vibrant mining industry. In both cases, the reconfiguration of industries and urban spaces in post-Cold War era has resulted in further urbanisation in a rugged geographical terrain that was once widely regarded as uninhabitable.

Figure VE 1.8 A boy viewing an exhibit that celebrates China's success in the detonation of an atomic bomb inside Panzhihua's Third Front Construction Museum. The "museumification" of socialist industrial heritage is also happening in other former Cold War cities such as Liupanshui and Chongqing.
Source: Author photograph.

In contemporary China these Cold War instant cities such as Liupanshui and Punzhihua are dreaming of a new future marked by conspicuous consumption and capital accumulation. Left in their wake, meanwhile, is a dystopian landscape of abandonment and devastation that eerily resembles the scene of what could have happened in the event of a nuclear apocalypse. Buried deep in the mountains with these derelict buildings, too, is a forgotten socialist utopian dream that prevailed, albeit only briefly, at the very moment when the country flirted anxiously with the dystopia of total annihilation (Figure VE.1.8).

References

Baumler, Alan. 2016. "Keep Calm and Carry On: Airmindedness and Mass Mobilization during the War of Resistance." *Journal of Chinese Military History* 5: 1–36.
Lu Yujun. 1935. "Lixiang de fangkong dushi" (The Ideal Anti-Air Raid City). *Kexue de Zhongguo* (Scientific China) 5 (8): 317–319.
Lu Yujun. 1945. "Xinshidai gongyehua zhi yingyou renshi" (Common Knowledge of Industrialization in the New Era). *Gonggong gongcheng zhuankan* (Journal of Public Works) 1: 48–50.
Wang Xingping, Shi Feng and Zhao Liyuan. 2004. *Zhongguo jinxiandai chanye kongjian guihua shejis shi* (Design and Planning History of Modern Chinese Industrial Spaces). Nanjing: Dongnan daxue chubanshe, 46–49.
Xia Fei, 2017. "Mao Zedong zai shang shiji 60 niandai de yige zhongda zhenlue xuanding juece" (An Important Strategic Decision Made by Mao Zedong in the 1960s). *News of the Communist Party of China*. Online at cpc.people.com.cn/GB/85037/8516211.html [accessed September 1, 2017].

Part II
Building the Cold War city

5 The Warsaw metro and the Warsaw pact

From deep cover to cut-and-cover

Alex Lawrey

INDEPENDENT SCHOLAR

Introduction

During the Cold War communist urban planners could approach the possibility of nuclear conflict in two ways; firstly through constructing civil defence infrastructure, including dual-purpose networks like deep-level underground railways, and secondly urban centres could exhibit their vulnerability, making them morally indefensible to attack, detente by reverse deterrence. Cold War urbanism, notably the military-strategic dynamics of transport planning, is a relatively new subject. The history of military use of the railway is the subject of Wolmar's 'Engines of War' (2012); a purposeful but singular example. Wills's (1999, 10–11) study of the development of the London's Fleet/Jubilee line begins with a plan produced by the Ministry of War Transport in 1946, yet does not discuss civil defence and public transport. The complex civilian/military nature of government investment in the aviation industry is tackled in Edgerton's 'England and the Aeroplane', (2013), but it ignores spatial aspects like the dual civilian/military use of airfields. Jakubec's (2014, 145) analysis of the Czechoslovak communist-era railways considers two levels, political and economic, but excludes a third "military-strategic" level, as "less research has been done on this level" which can be considered "as open to investigation". This paper looks at Warsaw's Cold War metro, and the military significance of the national railway, across a broader historical period. Literature on the metro is limited, particularly in English, and mainly focusses on engineering and geology. The Warsaw metro was not completed until after the fall of communism so it is an unlikely choice for a case study into Cold War urbanism. Yet it is precisely because of the super-power conflict that decisions were taken about the purpose of the planned network that stunted its development and left it languishing for years in 'project drift'. For as Burnett and Hamilton (1979, 272–273) have argued "international factors" influenced communist governments, who 'filtered' their actions "towards the fulfilment of the all-Union... economic, political and military objectives of the USSR into their respective urban systems". This compromised "their own national development, trading and spatial strategies", partly due to "contributions to

126 *Alex Lawrey*

Figure 5.1 Metro station sign, 2011.
Source: Author's photograph.

enormous military expenditure". This filtering was achieved through transnational organisations and treaties, notable the Warsaw Pact (military), COMECON (economics) and in the rail industry the OSJD (Organisation for Cooperation between Railways) (Veron, 2016, 249). Roth (2016, 11) noted how "non-economic factors" drove "railway systems forward" in Eastern Europe during the Cold War, with "administrative and military-strategic reasons" often paramount. Metro systems are long-term, capital and resource intensive, projects for city authorities. Moscow's metro required an enormous amount of capital and labour. Abakumov, (1939, 12, 15–16) noted that 12,000 workers from across the nation travelled to Moscow, eventually 65,000 took part, with materials manufactured all over the country. Infrastructure therefore requires, primarily, significant political and economic impetus to achieve realisation.

Humane civilisation and one armed camp

The Polish sociologist and member of the People's Republic State Council Jan Szczepanski wrote of how Socialism was unlike the "materialistic civilization …apparent in the West", for "the technical civilization developing" under the communist urban planning ethos, would "be the humane civilization" (Burnett and Hamilton, 1979, 263). A pamphlet about the Kharkiv metro spoke of how "the level of comfort… influences not only the passengers' mood, but their process of production and social activity", (Hatherley, 2015, 263) and a study of Soviet urbanism in Leningrad noted that reducing commuting times would lead to "overall improvements in living standards and the all-round and harmonious development of the personality" (Trufanov, 1977, 80). Czechoslovak economist Tabery wrote that the "task of railway transportation" was the "uninterrupted growth of socialist manufacturing [a] continuous increase in productivity", increases in passenger "comfort", and it had to "help maximise the satisfaction of the continuously growing material and cultural needs of the people" (Jakubec, 2016, 149–150). Transport, productivity and culture were interlinked, but so too were transportation and military dynamics. A 1980 tourist booklet said that the continued building of Moscow's metro during the war was the "mirror of the Soviet people's gallant fight against fascism and that spirit of unity which made the country into one armed camp" (Hatherley, 2015, 263). This communist nation-state as 'armed camp' idea continued to influence both urbanism and transport policies long after 1945 (Figure 5.2).

Figure 5.2 'Homo Sovieticus' statue at base of Palace of Culture and Science, 2011.
Source: Author's photograph.

Trains, 'homo sovieticus' and pistols under petticoats

In 1917, Lenin had arrived from exile in a sealed train to start the revolution in Russia, and during the Civil War 'agitpoezda' trains were used for propaganda, there were armoured trains, and trains transported Trotsky and the Red Army (Wolmar, 2013, 202–204). The symbol of the locomotive and the railway system were central to the idea of engineering, industrialisation, modernity and progress that the Socialist epoch would fashion (Stefan, 2013, 215). Communist culture would create the New Man and shape the environment. Engineers could rewrite the world according to science and scientific Marxist doctrine, nothing was impossible. Trotsky (2005, 204–207) wrote of how "Communist man" would become "immeasurably stronger, wiser and subtler", with "the average human type" becoming "an Aristotle, a Goethe, or a Marx". "Through the machine" New Man would "command nature in its entirety", pointing "out places for mountains and for passes", changing "the course of the rivers", even laying down the "rules for the oceans". Technology, not faith, would move mountains. This had previously been "done for industrial purposes (mines) or for railways (tunnels)" but "in the future" this would be accomplished "on an immeasurably larger scale… to a general industrial and artistic plan". The 1919 Bolshevik poem 'We' by Mikhael Gerasimov asserted that "We shall know all, We shall take all". 'We' are "Wagner, Leonardo, Titian", walking with Jesus, cutting "the clanging granite", and laying the stones "of the Parthenon" (von Geldern and Stites 1995, 5). Soviet experimental film-maker Dziga Vertov declared "I am the cinema-eye. I am the builder …I create a man more perfect than Adam" (Youngblood, 1991, Chapter 1). The Communist New Man of the Bolshevik imagination was a common trope through Trotsky to Stalin, later parodied as 'Homo Sovieticus'. Polish architect Edmund Goldzamt noted "an architect in a nation building socialism is not only a constructor of streets and buildings, but also of human souls" with opportunities "to influence the masses everywhere and at any time", expressing "the social ideals in the name of which the masses work and live" (Goldman, 2005, 150). Behind the talk of culture, peace and socialism there was a darker, aggressive side. Count Ottokar Czernin, (1919, 222) part of the Austro-Hungarian delegation to the 1917 Eastern Front armistice talks, described the Bolsheviks, like Anastasia Bizenko who had assassinated the Tsarist general Sacharov by hiding her revolver under her petticoats, as "strange creatures" who "talk of freedom and the reconciliation of the peoples of the world, of peace and unity", yet they were "the most cruel tyrants history has ever known …exterminating the bourgeoisie, and their arguments are machine guns and the gallows" (Figure 5.3).

The station marked 'independence': post-war reconstruction, railways and connectivity

Prior to the Great War Poland was divided between three neighbouring empires (Russian, Austro-Hungarian, German Reich) with their own

Figure 5.3 Statue of Pilsudski, central Warsaw, 2014.
Source: Author's photograph.

segregated railways; networks that essentially disturbed Polish aspirations towards nationhood (Wolmar, 2012, 17–18; Przegietka, 2016, 131). In 1846, the Austro-Hungarians dispatched Prussian soldiers to suppress the Polish nationalist rebellion in Krakow. Two years later, Tsar Nicholas I sent troops to the Warsaw to Vienna railway to aid the Austro-Hungarian's in extinguishing the Hungarian uprising (Wolmar, 2012, 18). On 8 November 1918, the former 'Polish Legion' general Jozef Pilsudski travelled to Warsaw, like Lenin in 1917, sent by the Axis military in a sealed train. On arrival he told former 'comrades' that he had ridden the "red tram" but got off at the "stop called independence" (Michta, 1997, 26). On Armistice Day, 11 November 1918 Poland became independent. The Great War had decimated the railways, leaving them incoherent and in need of reconstruction (Przegietka, 2016, 133–134). Post-war assistance came from ARA (American Relief Administration) Polish Mission, under future President Herbert Hoover, who organised the delivery of rolling stock as war reparations from Germany (Fisher, 1928, 196, 199). American help indicated a deliberate strategic re-alignment westwards, forging connections with the victorious Allies. Poland needed friends for there were conflicts with her neighbours in Ukraine, Lithuania, Czechoslovakia and the Soviet Union, who attacked eastern Poland in 1920 reaching the banks of the Vistula. In a dramatic turnaround Pilsudski defeated the Soviet forces, and, after the May 1926 coup, became Poland's leader, establishing the Sanacja ('Sanitation'/Healing) regime, a dictatorship with democratic elements that pursued a modernising agenda, including urban interventions (Kohlrausch,

2014, 213). Initial development of the metro project began in 1925, routes approved in 1927 but the Depression delayed development. The connectivity of place provided a starting point for a re-imagining of urban Warsaw. In 1916, Tadeuz Tolwinski's plan was for the 'Hygienic City', expanding the metropolis beyond the fortified limits set by the Tsarist authorities, (Staniszkis, 2012, 84; Kohlrausch, 2014, 210) followed by Stanislaw Rozanski's 'master scheme' of 1927. In 1930, Oskar Sosnowski, founder of the Polish planning institute, architect and professor, produced an urban study of Warsaw's characteristics and street networks. Kohlrausch (2016, 215) noted how Rozanski and Sosnowski's plans "stressed the geographic assets of Warsaw", namely "its central position at the intersection of international traffic routes". 'Warszawa Funkcjonalna' ('Warsaw Functional') emerged from the 1934 CIAM conference held in Warsaw, produced by the Polish architects Szymon Syrkus and Jan Chmielewski. Their plan was Modernist and Internationalist, emphasising connectivity between east and west, on the route from Paris to Moscow, recalling the 1896 to 1914 Nord Express, travelling from France to Russia in 48 hours (Musekamp, 2016, 121). Syrkus and Chmielewski declared that, "in our conception the scale of the region is interconnected to the scale of central Poland, Europe and even the world" (Kohlsrausch, 2014, 224). As they said, "our plan lies within the realm of utopia" (Crowley, 2010, 107).

Tunnels and war

The utopianism of mooted transport systems co-existed with the dystopian realities of the era; the Spanish Civil War as the prelude to World War II. In 1938, Stefan Starzynski, Warsaw's Sanacja-appointed mayor, gave further impetus to the metro with the establishment of the Bureau for the Study and Projects of the Underground Railroad (ZTM website). Starzynski realised the military value, and dual uses, of the metro project, working with the air defence inspectorate and considering the potential for damage from aerial bombardment and for civil defence (Deszczynski, 2015, accessed 3/9/2017).

Michta (1997, 28) argues that the "Sanacja pattern involved direct interpenetration of the civilian and military branches of government", including urban planning. In Britain, the 1937 'Air Raid Precautions Act' mandated local government with responsibility for providing shelters. During a 1938 debate in the House of Lords, Labour peer Lord Strabolgi asked why "the London Underground Railway" could not be used for civil defence, as had been Barcelona's metro during the Francoist bombardment. French authorities were considering similar ideas for the Paris Metro (Gregg, 2001, 9–10). Moscow's metro was primarily deep level, the first line opening in 1935. Stalin's conversation with senior engineer Pavel Rottert, illustrates his abiding preference for deep-tunnel subways. Stalin liked the young engineer

Makovsky's proposal for deep tunnels, but Rottert thought them too expensive, twice Stalin corrected him, "the government will decide that. We will go ahead and accept comrade Makovsky's plan" (Hatherley, 2015, 308).

Warschau

Both the Nazis and the Soviets invaded Poland in 1939 leading to the second battle for Warsaw in 20 years, which resulted in the deaths of Starzynski, Sosnowski and thousands of others. The German plan for the city, by Pabst, Gross and Nurnburger, intended on the total destruction of Polish Warsaw and its replacement with an ethnically German city, 'Warschau: Die Neue Deutsche Stadt', what Staniszkis (2012, 86) calls the "barbarian urban utopia". In 1943, the city's ancient Jewish community and the built fabric of the Warsaw Ghetto were destroyed. A year later self-described "fearful barbarian" Himmler enacted orders "to burn down and blow up every block of houses" in Warsaw (Bevan, 2016, 130). Himmler noted how Poland had for "700 years blocked our road to the East and stood always in our way", Warsaw "must disappear...and serve only as a transport station for the Wehrmacht... Every building is to be razed to its foundations" (Wituska, 2006, xxii). Before retreating the Wehrmacht suppressed the Warsaw Uprising (with the Red Army camped on the eastern side of the Vistula), decimated the Polish Home Army, killed thousands and destroyed about 80% of Warsaw's buildings. In January 1945, the Red Army 'liberated' Warsaw, finding a wasteland, depopulated, denuded of its built heritage and urban culture and already under new occupation (Figure 5.4).

Underground archives, reconstruction and the Communist take-over

In 1939, Professor Sosnowski and others had secretly recorded much of the city's heritage and hid artefacts in the Technical University's archives and the Piotrkow monastery (Bevan, 2016, 234; Tung, 2001, 81). Urban planning continued in secret throughout the war, and in exile amongst Polish civic design students at Liverpool University, but post-war control of urban spaces was subservient to the new occupiers. The communists were careful to document and publicise both the scale of the destruction and reconstruction, urban salvaging and rebirth became marks of honour. Varsovians had spontaneously begun clearing rubble and rebuilding the city, this enthusiasm was adopted and propagandised by the communists into organised legions and youth groups, and sanctified through the 'Holiday of Reconstruction' on 22 July 1949, celebrating four years of communist party rule, and the launch of the Six Year Plan for the Rebuilding of Warsaw (Crowley, 2012, 36–37). Postwar urban designers took advantage of the ruins by widening streets which allowed for multiple transport modes including buses and trams (Crowley, 2012, 28; Josephson, 2010, 88–89). The communists incrementally took

Figure 5.4 Statue at base of Palace of Culture and Science, 2011. Stalin's name was erased from the list after his death.
Source: Author's photograph.

power through all spheres, including in urbanism and transport. Edmund Goldzamt noted that "the spatial development of Warsaw, and especially its centre, is a critical political, rather than architectural, issue", with decisions "taken solely by our comrades responsible for waging cultural-ideological war, that is to say, by the political leaders". In 1949 communist leader Bolesław Bierut wrote that

> our Party must express itself not only regarding what and how much will be built in Warsaw but also what, where, and for whom....our Party must undertake a determined struggle for a new form for our cities and settlements, and above all, our capital
>
> (Goldman, 2005, 136)

The parameters of Polish urbanism were shaped from the moment of Red Army 'liberation' onwards. In March 1945, Khrushchev wrote to Beirut offering expert help with reconstructing Warsaw including planners, engineers and transport specialists (Kemp-Welch, 2008, 21). In November 1945, Gomulka reported that Stalin told him the Red Army would not stay on in Poland in large numbers except for "small troops guarding the transit railroad", who "would not kill you", and in terms of threats from America and Britain, "there will be no war, it is rubbish", their armies were "disarmed by agitation for peace", for it was "not atomic bombs but armies [that] decide about war". In regard to transport, for Stalin the "most important issue" was the need to have communists in positions to influence policy, such as the economics specialist Hilary Minc. Stalin "was against moving Minc into transportation, but later agreed to it, once he found out that we had no people in transportation" and whilst he "promised to look into our proposals concerning transportation" he foresaw "no possibilities for us to get locomotives and train cars with their help" (Gomulka papers, 117153). The security of the transit route through Poland was the Soviet's abiding strategic objective, Warsaw was only a stop on that route.

The Cold War metro

Initial rapid transport designs, developed from 1945, were for a 64 km network with cuttings as part of the SKM regional rail network. From 1948, an underground metro plan was developed including 26 km of shallow tunnels. This was then adapted, under Soviet influence, to go from 10m depth shallow tunnels to a 30 m to 50 m deep tunnel system. Critical to the design was its potential for use as a shelter from nuclear or conventional bombs, and as a route for military transport. The period which saw the deep-tunnel metro's development, 1950–1956, overlaps with Konstantin Rokossovsky's rule as Marshall of the Polish army. He was Warsaw-born and of Polish noble stock (although his father was a railwayman) but had served as Marshall in the Red Army (when it failed to help the Warsaw Uprising in 1944). He oversaw the near-complete take-over of the Polish army by Soviet officers, and exerted significant control beyond the military. The military imperative behind the amended plans for the metro shows a more aggressive stance taken by Moscow to the independent governance of 'fraternal' nations. The government resolution of December 1950 for the building of the metro lead to works starting in 1951 in Praga, on the strategic east side of the Vistula on a tunnel under the river, which was wide enough to allow regular trains to pass through it (Radio Free Europe Research, 1982, 18). The scheme was for a total of 36 km of rail lines, arranged in Y shape, with works to be completed by 1956. In November 1953, a few months after Stalin's death, the Council of Ministers voted to end major construction works because of rising costs and hydro-geological difficulties. Exploratory drilling and tunnelling continued until 1957, and engineers experimented with a variety of methods to cope with water inundation (Rossman, 1962, 319). Lazar Kaganovich, the leader of the Moscow metro building program in the 1930s, noted that "geology proved to be a pre-revolutionary part of the old regime,

Figure 5.5 Vistula river, 2011.
Source: Author's photograph.

incompatible with the Bolsheviks, working against us" (Hatherley, 2015, 259) (Figure 5.5).

Bierut's 1949 urban plan for Warsaw had included details on the SKM metro system; "the investment crucial for communications in Warsaw will be commenced; the fast urban railway, the metro, will be built, with its first North branch, which will connect the Northern districts with the city centre" (Bierut, 1951, 257). The metro was separated from the general issue of urban transport and outlined in a different plan of the city. There were three line drawings of metro stations, one on the surface and two below the ground. The section on the urban transport system was far more extensive, noting that transport was "currently causing so much trouble to the people of Warsaw" and it would "be greatly improved" with new trams, trolleybuses, buses, 50 km of new tram lines, and "many streets" undergoing "reconstruction", "a new road bridge" connecting the North of the city to Praga, a 10 km long boulevard, and "the New Marszalkowska street" serving as the spine of the reborn metropolis (Bierut, 1951, 249). There is a sense of uncertainty about the plans for the metro, its segregation from wider transport questions in the Plan demonstrates that it was not considered to be part of regular urban development and occupied a special category. In the official history of the Warsaw metro (Rossman, 1962, 345), there is a telling comment regarding "how views of metro's function for defensive purposes in comparison to Moscow and London during WWII, were evolving". The urban priorities of Warsaw's Six Year Plan were to build housing, notably in the city centre, industry, to expand internal transport connections, and to rebuild or conserve Warsaw's decimated heritage. Helena Syrkus, wife of

Figure 5.6 Stare Miasto (the Old City), 2011.
Source: Author's photograph.

one of the authors of Warszawa Funkcjonalna, told the 1949 CIAM conference that the "new Warsaw will conserve its links with the past" with "all that is good in the line of roads, open palaces, and with all the remaining evidences of its ancient culture. In defending and preserving our national culture we defend and preserve international culture" (Crowley, 2010, 115). This mooted defence of culture through links to the past was related to the wider agenda of the pursuit of peace as a strategic direction for the People's Democracies, even in their urbanism (Figure 5.6).

The old city and the Palace of Culture and Science

Two monumental projects came to symbolise the 'Socialist Realist' urbanism in Warsaw: Stare Miasto (the Old City) and the Palac Kultury i Nauki imienia Jozefa Stalina (Josef Stalin Palace of Culture and Science). Although the rebuilding of the Old City has been decried as 'disneyfication' of heritage (Bevan, 2016, 235; Hatherley, 2015, 317), it proudly exhibited Poland's rebirth as a nation. In Warsaw, memories of war and destruction were not erased by rebuilding the Old City but rather inscribed into the landscape by presenting the city as reborn and alive. Stanislaw Lorentz wrote of how it was "our duty to resuscitate" the city, reconstruction being "the last victorious act in the fight with the enemy", with "the decision to reconstruct... made by the highest authorities, who were confident that they were acting according to the wishes of the people" (1984, 71–73). In terms of defence and vulnerability, the construction of soft targets like the Old City, and the purposeful location of residential blocks in the city centre, was in marked

Figure 5.7 Palace of Culture and Science, 2014.
Source: Author's photograph.

contrast to the West's strategy of population dispersal to the suburbs. Warsaw chose concentrated urbanism with high-density residential and civic blocks, and wedges of green space radiating from the centre. The central focus was the Stalin 'gift' of the Palace of Culture and Science, which dominated the skyline, mimicking Bruno Taut's 'City Crowned', and made the city an easy target for potential NATO bombardment (Figure 5.7).

The possible use of the metro for military shelter was a complex issue because of its dual civilian use. Soubry (2013, 508) notes that the public nature of "circulation areas in the stations" and "their openness and nearness to the surface for ease of access is contrary to the needs of resistance to weapons", especially nuclear weapons. Metros were important to the functioning of cities but their civilian usage had to be subservient to a dual function as shelters, so projects had to have a deep tunnel element. When this was an engineering challenge and economically unviable, metro projects were mothballed. Whilst the metro's development was delayed, by 1964 the suburban railways had built 4.5 km of cut-and-cover tunnels, with four tracks on the east-west route, including three passenger stations, (Howson, 1964, 83) and in 1975 Centralna ('Central') station opened, providing an underground hub for rail links to Moscow and Berlin. In the Warsaw region, nuclear bunkers and shelters were located under the Palace of Culture, in the suburbs of Pyry and Janow, and a short section of a secret underground railway between Lasek Bielanski in the west and a park to the east of the Vistula (Ozorak, 2012, 132–133). This is indicative of the relatively limited preparedness of the city for an atomic attack. By 1956, when work on the metro had effectively ended, most of the national railways had been rebuilt after war damage, including bridges over the Vistula (Przegietka, 2016, 138). In Warsaw Pact

The Warsaw pact and metro 137

Figure 5.8 Historic trains (including armed train in background) at the Muzeum Kolejnictwa w Warszawie.
Source: Author's photograph.

terms, the Polish national rail network was (until the 1980s) essential for resupplying garrisons in the GDR and wartime mobilisation plans (Faringdon, 1986, 109–110), and the Warsaw metro, as a civil defence facility, was of secondary importance (Figure 5.8).

Hiatus: project drift

The hiatus of the Warsaw metro project was not atypical in the global histories of metro systems. Extensions to the Central London Railway to Ealing stopped during the Great War, the line opening in 1920 (Hall, 1996, 51). In the USSR, there were breaks in construction for metros in Baku (hiatus 1954–1960, line eventually opened 1967), and in Leningrad and Kiev engineers had developed plans in the late 1930s, work stopped during the war, resuming afterwards, Kiev's opened in 1960, and Leningrad's in 1955, using mainly deep tunnels. As Bennett (2004, 164) noted Moscow's metro is largely "below underground, and has some extremely deep stations. Some were intended to double up as nuclear shelters", notably the Arbatsko-Pokrovskaya line which has a parallel-line deeper than the original, intended for nuclear shelters and which opened in 1953. For Pyongyang's metro, which opened in 1973 and is possibly the world's deepest system except for the St Petersburg (Leningrad) metro, the use of the network "by the public seems to have been only a secondary consideration" to its military nuclear-shelter function (Ozorak, 2012, 125). An east-west line (like Warsaw's going under a river, the Danube) to add to Budapest's existing metro was started in 1950 and abandoned in 1953 to direct labour towards housing and factories, it

Figure 5.9 Detail of Lublin PKP train station, 2010.
Source: Author's photograph.

recommenced in 1963 (Bennett, 2004, 42). Deep-tunnel systems were prohibitively expensive, "the proportion of tunnelling to surface work and elevated construction" becoming "a critical factor" and "the high cost of tunnelling" limiting "construction to the essential minimum through congested city centre areas" (Bennett 2004, 31). Howson (1964, 7) noted how there was then currently, a "boost in underground railways", but that due to the "heavy initial cost" only a "few of these projects can be commercial propositions because of the expense involved". Khrushchev in his memoirs admitted that he had "suspended the construction of subways in Kiev, Baku and Tblisi" to "redirect these funds into strengthening...defences and attack forces" (Hatherley, 2015, 552) (Figure 5.9).

Deep tunnels and the Tsar Bomba: from utility to futility

These 'strengthened' attack forces were intended to counter weapons like those of the American Castle Bravo 15MT thermonuclear test in 1954. Khrushchev met this threat with the 50MT Tsar Bomba, in 1961. By the 1960s nuclear reality required planning for destruction on a scale where if the wind blew in a particular direction a 100 million Western European citizens would die from the NATO nuclear bombing, on a limited 'no cities' basis, of Eastern Europe (Ellsberg, 2017, 137). The Soviets also began to understand the ecological consequences of nuclear war. The utility of deep tunnels shelters became the futility of attempting to survive a nuclear winter outside of the shelters. The proposed deep-tunnel metro would have needed features such as carbon air-filters designed-in from the start to be effective as a fallout shelter. The USSR's global first of a successful ICBM launch in August 1957 hid the reality of the Soviet nuclear capability, Khrushchev's son Sergei, a rocket scientist himself, wrote that Russia made threats with "missiles we didn't have" (Gaddis, 2005, 69). Khrushchev senior boasted about making missiles like sausages in a factory, but this was based on short-range missiles not ICBMs (Ellsberg, 2017, 163). More critical than the perceived 'missile gap' was the potential quantum of destruction, and its fallout legacy.

The thaw and deterrence in reverse

The Khrushchev Thaw in the Soviet Union after death of Stalin in 1953, and Bierut's death during a visit to Moscow in 1956 (just after Khrushchev's 'secret speech' denouncing Stalinism) meant that the once-disgraced Gomulka could be brought back to lead Poland, and the hated Rokossovsky sent back to Moscow. The Warsaw metro project was cancelled under Gomułka, (Hatherley, 2015, 300) whose urban priorities were the construction of housing, and a downgrading of monumental projects like the metro and the Royal Castle. Policy and propaganda were directed towards the pursuit of peace. As Malanekov said "the Soviet government ...is resolutely opposed to the policy of Cold War, the policy of preparation for a new world war" (Roberts, 2008, 31). Soviet Foreign Minister Molotov's swansong was the failed attempt to get the Soviet Union to join NATO in 1954. Despite objections from the Soviets, West Germany finally joined NATO on 9 May 1955. The response, six days later, was that Albania, Poland, Romania, East Germany, Czechoslovakia, Bulgaria, Hungary and the USSR signed the 'Treaty of Friendship, Cooperation and Mutual Assistance', (the 'Warsaw Pact') under the motto of the 'Union of Peace and Socialism'. Warsaw Pact weapons included ICBMs and SLBMs, and the Tupelov Tu-16, the first Soviet-designed long-range bomber. War however, was always the option to be avoided. On 17 November 1956, Soviet Premiere Bulganin proposed a complete nuclear disarmament plan, and after this was refused by America, the Polish Foreign Minister Adam Rapicki proposed a nuclear-free central Europe, which was again rejected (Kemp-Welch, 2008, 125). American Cold War strategist Thomas Schelling (1966, 2) wrote how nuclear weapons "*destroy* value" (his italics), meaning that an enemy can be coerced into a form of behaviour because of the threat of total destruction. In the context of thermonuclear weapons deterrence could also work in reverse: Warsaw

Figure 5.10 Rebuilt city walls, Stare Miasto, 2010.
Source: Author's photograph.

was stronger by presenting itself to the world as a vulnerable, undefended, rebuilt city, a cosmopolitan centre with a recreated historic heart, modern housing and industry, and little in the manner of military infrastructure, other than rebuilt 'historic' city walls. On May Day 1949, children chanted "in reply to atom bombs we are building new houses" (Aman, 1992, 185). The Warsaw Pact's influence was not to militarise the city but to make it more vulnerable and less defended (Figure 5.10).

'The camp of democracy, socialism and peace'

With Sputnik, and Yuri Gagarin, the USSR had conquered outer space the next target was outer earth. Polish town planners and historic building conservators worked in places like Iraq and North Africa, exporting Socialist conceptions of urban Modernity and heritage management. The 1949 National Congress of Polish Architects stated that "two camps" were "realising their contradictory world pictures and ideologies …the camp of democracy, socialism and peace", the People's Democracies, "the Soviet Union" being "the main bastion", and opposing them "the camp of imperialism, economic crisis and warmongering. The contest between these ideologies is also being waged in architecture" (Goldman, 2005, 149; Aman, 1992, 59). The two opposing camps notion was echoed in Hilary Minc's speech to the Polish United Workers Party (communists) in July 1950 when he railed against the

Figure 5.11 Tram, central Warsaw, 2011.
Source: Author's photograph.

"criminal plans" of "bestialized American Imperialism", plans which considered "where and how to spread devastation and war conflagration" and "where and when to drop the atom bomb" (Bierut and Minc, 1950, 65–66). In 1948, Wroclaw hosted the 'International Congress of Intellectuals in Defence of Peace', and Wroclaw's 1956 urban plan for reconstruction replaced the defunct Lessing Bridge with the Peace Bridge (Davies and Moorhouse, 2003, 448–450, 464). Warsaw hosted the Fifth Youth and Students' Festival of 'World Peace and Friendship' in the summer of 1955, described by Applebaum (2013, 474) as a "vast propaganda exercise" (Figure 5.11).

'Landlords in their own country' and 'A million town planners'

Cold War urbanism was often about containing enemies from within as those from without. The Poznan and Hungarian uprisings, and the threat of 'Titoism', communist nations pursuing their own 'path to socialism', resulted in attempts to mollify host populations of satellite states through culture and a degree of nationalism. This was demonstrated through reconstruction of the Stare Miasto and via the adoption of Polish-influenced architectural detailing, a form of nationalist kitsch, added-on to the Soviet architectural vocabulary of Socialist Realism and functional but debased

Modernism (Applebaum, 2013, 371). In 1939, the Armenian architect and planner Karo Alabian, chair of the USSR's Union of Architects, wrote that "Soviet Architecture believes that no progress is possible that is not based on what is progressive and valuable in the past", architecture "like all our culture, is national in form and socialist in content ...an architecture that derives organically from the traditions, customs, life conceptions and climatic conditions of the given nationality" (Hatherley, 2015, 20–21). Goldzamt used the Moscow metro as the exemplar of this idea, where its engineers had "saturated the metro architecture with ideology using rich and easily comprehensible forms", the mass population becoming "landlords in their own country" knowing "that the nation is strong, wealthy, and uses this strength and wealth to improve the citizen's well-being and culture" (Goldman, 2005, 150). The promise of being 'landlords in their own country' because of a metro system in the capital, and then shelving the project did not fit the narrative of a benign proletarian dictatorship pushing forward with technology in the People's Democracy. The metro had once been vaunted as a beacon of the approaching socialist utopia, as the SARP (Polish architects association) wrote in 1953 it was an "assignment from the community" involved in "transforming long underground tunnels into cheerful and monumental palace halls" (Aman, 1992, 81). The 'community' had no say in the decisions, but the myth of the community creating future urbanism and transportation was deeply embedded within Bolshevik culture. Trotsky (2005, 202) wrote of how "in the future ...monumental tasks" such

> as the planning of city gardens, of model houses, of railroads, and of ports, will interest vitally not only engineering architects ...but the large popular masses as well ...the parties of the future for special technology and construction, which will agitate passionately, hold meetings and vote.

The illustrated 'Warszawa Odbudowana/Warsaw Rebuilt', proclaimed that "among us are a million town planners", as "every citizen" held their "own views" on what had been, or should be, built, and how. "Everyone discusses those matters heatedly in the trams, in the streets or at meetings specially organised for the purpose" (Ciborowski and Jankowski, 1963, 53). The book used contrasting photographs of the destroyed city in 1944 and after reconstruction to show how Varsovians had rebuilt their city. By 1963, official views regarding the metro had clearly 'evolved' for it was intended that "by 1975" Warsaw would be "equipped with a rapid urban railway running in a shallow tunnel under the centre, and through the peripheral districts – in cuttings" (ibid, 60). In the ensuing decade, no operational development on the metro took place as Poland coped with economic hardships and the 1970 winter strikes over food prices (Figure 5.12).

Figure 5.12 Construction of metro line II using cut-and-cover technique, 2014.
Source: Author's photograph.

Solidarnosc and gift-giving

Only after the Gdansk shipyard strikes and the Solidarity movement lead by Lech Walesa did Warsaw receive its metro as a fraternal 'gift'. Hatherley (2015, 296–299) calls this "a Metro in exchange for sovereignty", and writes that work on Budapest's underground occurred after the suppression of Hungary's revolution of 1956, and the Prague metro received amplified Soviet support after the 1968 Prague Spring, Czechoslovak author Rybar writing that thanks to the "important technical assistance rendered by the Soviet Union" the metro was to be "marked as a structure of Czechoslovak-Soviet cooperation" (Ctibor Rybar, 'Prague' 1979, cited ibid, 297). After imposing martial law on Poland, General Jaruzelski informed the Sejm (parliament) that "due to generous, comprehensive Soviet help…construction of the Warsaw metro is to start next year" (Radio Free Europe Research, 1982, 17). Tunnelling work, using shallow tunnels, began in April 1983, and in April 1995, after communism had ended, the first north-south line opened (ZTM website). Polish independence after 1990 included membership of NATO (1999) and the EU (2004), the triumph of the "Capitalist Road" writ large; car numbers increased dramatically, and passengers on the State Railway dropped from 951 million in 1989 to 272 million in 2004 and the network shrunk from 27,000 km in the 1980s to 19,000 km in 2004 (Przegietka, 2016, 141–142). Decoupling the metro from its Peoples' Republic past was problematic, only ten of the promised 'gift' of 90 Mietrowagonmasz engines were delivered by 1989, the rest cancelled. Gift-giving fell to the new(/old) Western allies. In 2010, the EU part-funded work on the cut-and-cover east-west line which opened in 2015. In Plac Wilsona, named in 1923 after US president Woodrow Wilson, renamed as Plac Kommuny Parykiej (Paris Commune

144 *Alex Lawrey*

Figure 5.13 Plac Wilsona metro station, ceiling, 2011.
Source: Author's photograph.

square) under the communists, and reverting back to Plac Wilsona in 1991, the metro station design, with its swirling ceiling of light, is both utopian and a retro recreation of the 1930s metro Modernism of Starzyński and Warzsawa Funkcjolnalna (Figure 5.13).

Conclusion

In the 1950s, Warsaw had been at the heart of a military alliance established to oppose and contain NATO, yet was peripheral in terms of receiving its due development for appropriate civilian infrastructure. Stalin pushed plans for metros deeper, expecting them to function as conventional/nuclear bomb shelters. Gomulka and Khrushchev hoped the Cold War would remain 'cold' and the need for nuclear shelters would not arise. Warsaw's urban planning and post-war rebuilding demonstrate not a city attuned to the oncoming nuclear threat but a metropolis geared towards socialist Modernity, confident of its future, and keen to recreate or preserve its past. The rebuilt Old City was not nuclear infrastructure, its only 'value' was in deterrence, yet it was the quintessential example of Polish communist, and military-strategic, Cold War urbanism. Stare Miasto was national in content and form, but primarily Warsaw Pact, not Warsaw, in propaganda purpose. As Aman (1992, 183) says, building on the ruins of the last war showed "faith in the future" and asserted "the cause of peace". The choice to not build a deep tunnel underground system, which could serve as a nuclear shelter, indicates that Warsaw's authorities, in Poland and in Moscow, were not about to ready the city for its next wave of annihilation and destruction (Figure 5.14).

Figure 5.14 Recreated history from 1953, Stare Miasto (detail), 2010.
Source: Author's photograph.

References

Abakumov, Y. 1939. *The Moscow Subway*. Moscow: Foreign Languages Publishing House.

Aman, Anders. 1992. *Architecture and Ideology in Eastern Europe during the Stalin Era: An Aspect of Cold War History*. Cambridge, MA. Architectural History Foundation/MIT Press.

Applebaum, Anne. 2013. *Iron Curtain: The Crushing of Eastern Europe 1944–56*. St Ives: Penguin.

Bennett, David. 2004. *Metro: The Story of the Underground Railway*. London: Octopus Publishing Group.

Bevan, Robert. 2016. *The Destruction of Memory: Architecture at War (Second, Expanded Edition)*. Glasgow: Reaktion Books.

Bierut, Bolesław. 1951. *Szescioletni Plan Odbudowy Warszawy*. Warsaw: Ksiazka I Wiedza. (translated by O. Gradziel)

Bierut, Bolesław, and Minc, Hilary. 1950. *The Six-Year Plan of Economic Development and Building the Foundations of Socialism*. Warsaw: Ksiazka I Wiedza.

Burnett, Alan D. and Ian Hamilton, Frederick Edwin. 1979. Social Processes and Residential Structure. in *The Socialist City: Spatial Structure and Urban Policy*. edited by French, Richard Anthony and Ian Hamilton, Frederick Edwin. 263–304. Old Woking, Surrey: John Wiley & Sons Ltd.

Ciborowski, Adolf, and Stanisław Jankowski. 1963. *Warszawa Odbudowana Warsaw Rebuilt*. Warsaw: Polonia Publishing House.

Crowley, David. 2003. *Warsaw*. Trowbridge: Reaktion Press.

Crowley, David. 2010. Paris or Moscow? Warsaw Architects and the Image of the Modern City in the 1950s. in *Imagining the West in Eastern Europe and the Soviet Union*. edited by Peteri, Gyorgy. 105–130. Pittsburgh: University of Pittsburgh Press.

Czernin, Ottokar. 1919. *In the World War*. London: Cassell and Company Ltd.
Davies, Norman, and Roger Moorhouse. 2003. *Microcosm: Portrait of a Central European City*. St Ives: Pimlico.
Deszczynski, Marek. 8/3/2015. *Nieznany plan warszawskiego metra z 1938 r.* Dzieje. pl website, online at: http://dzieje.pl/content/nieznany-plan-warszawskiego-metra-z-1938-r-przykladem-fachowosci-i-rozmachu
Edgerton, David. 2013. *England and the Aeroplane: Militarism, Modernity and Machines*. London: Penguin.
Ellsberg, Daniel. 2017. *The Doomsday Machine: Confessions of a Nuclear War Planner*. Croydon: Bloomsbury Publishing.
Faringdon, Hugh. 1986. *Confrontation: The Strategic Geography of NATO and the Warsaw Pact*. Padstow: Routledge & Kegan Paul.
Fisher, H. H. 1928. *America and the New Poland*. New York: The MacMillan Company.
Gaddis, John Lewis. 2005. *The Cold War: A New History*. New York: The Penguin Press.
Goldman, Jaspar. 2005. War Reconstruction as Propaganda. in *The Resilient City: How Modern Cities Recover from Disaster*. edited by Campenella, Thomas J. and Lawrence J. Vale. 135–158. New York: Oxford University Press.
Gomulka's Memorandum of a Conversation with Stalin. November 14, 1945, Wilson Centre, History and Public Policy Program Digital Archive, Gomulka papers, in possession of Gomulka Family, translated by Anna Elliot-Zielinska. document number: 117153, online at: http://digitalarchive.wilsoncenter.org/document/117153.
Gregg, John. 2001. *The Shelter of the Tubes: Tube Sheltering in Wartime London*. Singapore: Capital Transport Publishing.
Hall, Peter. 1996. *Cities of Tomorrow: An Intellectual History of Urban Planning and Design in the Twentieth Century*. Updated Edition. Bodmin: Blackwell.
Hatherley, Owen. 2015. *Landscapes of Communism*. St Ives: Allen Lane.
Howson, F. Henry. 1964. *World's Underground Railways*. Weybridge: Ian Allen.
Jacolin, Henry and Ralf Roth (eds.). 2016. *Eastern European Railways in Transition: Nineteenth to Twenty-First Centuries*. London: Routledge.
Jakubec, Ivan. 2016. Transport under Socialism: The Case of the Czechoslovak State Railways 1948–1989. in edited by Jacolin and Roth. 145–156.
Josephson, Paul R. 2010. *Would Trotsky Wear a Bluetooth? Technological Utopianism under Socialism 1917–1989*. Baltimore, NJ: John Hopkins University Press.
Kemp-Welch, A. 2008. *Poland under Communism: A Cold War History*. Cambridge: Cambridge University Press.
Kohlrausch, Martin. 2014. Warszawa Funkcjonalna: Radical Urbanism and the International Discourse on planning in the Interwar Period. in *Races to Modernity: Metropolitan Aspirations in Eastern Europe, 1890–1940*. edited by Behrends, Jan C. and Martin Kohlsrausch. 205–232. Hungary: Central European University Press.
Lorentz, Stanislaw. 1984. Reconstruction of the Old Town Centers of Poland, in *Readings in Historic Preservation. Why? What? How?*. edited by Kellog, Edmund Halsey and Williams, Norman. 70–73. Rutgers University, Center for Urban Policy Research.
Michta, Andrew A. 1997. *The Soldier – Citizen: The Politics of the Polish Army after Communism*. Basingstoke: MacMillan Press.
Musekamp, Jan. 2016. The Royal Prussian Railway (Ostbahn) and Its Importance for East-West Transportation. in *Eastern European Railways in Transition: Nineteenth to Twenty-first Centuries*. edited by Jacolin, Henry. 117–130. London: Routledge.

Ozorak, Paul. 2012. *Underground Structures of the Cold War: The World Below.* Auldgirth: Pen and Sword.

Przegiętka, Marcin. 2016. 1918, 1945 and 1989: Three Turning Points in the History of Polish Railways in the Twentieth Century. in *Eastern European Railways in Transition: Nineteenth to Twenty-first Centuries.* edited by Jacolin, Henry. 131–144. London: Routledge.

Radio Free Europe Research. 1982. *Poland/4: Situation Report.* 3rd March. RFE/RL.

Roberts, Geoffrey. December 2008. *A Chance for Peace? The Soviet Campaign to End the Cold War 1953–1955.* Cold War International History Project: Working paper #57. Woodrow Wilson International Center for Scholars. Online at: https://www.wilsoncenter.org/sites/default/files/WP57_WebFinal.pdf.

Rossmann, Jana (ed.). 1962. *Studia I Projekty Metra W Warszawie 1928–1958.* Warsaw: Arkady. (translated by Olga Gradziel)

Roth, Ralf. 2016. *Introduction.* in *Eastern European Railways in Transition: Nineteenth to Twenty-first Centuries.* edited by Jacolin, Henry. 1–21. London: Routledge.

Schelling, Thomas. 1966. *Arms and Influence.* Fredericksburg, VI: Yale University Press.

Soubry, M. A. 2013. Civil Defence Requirements. in *Civil Engineering for Underground Rail Transport.* edited by Edwards, J. T. 305–330. Bodmin: Elsevier.

Staniszkis, Magdalena. 2012. Continuity of Change vs. Change of Continuity: A Diagnosis and Evaluation of Warsaw's Urban Transformation. in *Chasing Warsaw: Socio-Material Dynamics of Urban Change since 1990.* edited by Grubbauer, Monika, and Joanna Kusiak. 81–108. Germany: Campus Verlag.

Stefan, Adelina Oena. 2016. Passengers' Railway Identity in Socialist Romania during the 1950s and 1960s. in *Eastern European Railways in Transition: Nineteenth to Twenty-first Centuries.* edited by Jacolin, Henry. 213–232. London: Routledge.

Trotsky, Leon and William Keach (introduction), Rose Strunsky (trans.). 2005 (1925). *Literature and Revolution.* Canada: Haymarket.

Trufanov, Ivan (trans. & intro. Riordan, James). 1977. *Problems of Soviet Urban Life: Leningrad 1973.* Newtonville, MA: Oriental Research Partners.

Tung, Anthony M. 2001. *Preserving the World's Great Cities: The Destruction and Renewal of the Historic Metropolis.* New York: Clarkson Potter.

Veron, Paul. 2016. Railway Integration in Europe: UIC – A Key Player of East-West Railway Integration. in *Eastern European Railways in Transition: Nineteenth to Twenty-first Centuries.* edited by Jacolin, Henry. 243–255. London: Routledge.

von Geldern, James and Richard Stites (eds.). 1995. *Mass Culture in Soviet Russia: Tales, Poems, Songs, Movies, Plays, and Folklore 1917–1953.* Bloomington: Indiana University Press.

Wills, John. 1999. *Expanding the Jubilee Line: The Planning Story. Revised Edition.* CW Print Group: London Transport.

Wituska, Krystyna. 2006. *Inside a Gestapo Prison: The Letters of Krystyna Wituska, 1942–1944.* Detroit, MI: Wayne University Press.

Wolmar, Christian. 2012. *Engines of War: How Wars Were Won & Lost on the Railways.* Italy: Atlantic.

Wolmar, Christian. 2013. *To the Edge of the World: The Story of the Trans-Siberian Railway.* Padstow: Atlantic.

Yesterday and Today: Construction History. ZTM website, accessed 20/8/2017, online at http://metro2.ztm.waw.pl/?c=28&l=2

Youngblood, Denise J. 1991. *Soviet Cinema in the Silent Era 1918–1935.* Austin, TX: University of Texas Press.

6 Competing militarisation and urban development during the Cold War

How a Soviet air base came to dominate Tartu, Estonia

Daniel Baldwin Hess and Taavi Pae

UNIVERSITY AT BUFFALO, STATE UNIVERSITY OF NEW YORK;
UNIVERSITY OF TARTU

This chapter traces the demilitarisation of Raadi airbase in Tartu, Estonia, where an early twentieth century aviation facility was transformed, during the Soviet occupation of Estonia, into a Cold War military power centre. Using archival documents and images, the contested urban space that contained this transformation is explored in this chapter through several lenses: the shifting ownership and uses of a key urban site, administrative restraints that limited urban growth and physical barriers that controlled security. The sheer size of the airfield, and the need to 'close' the section of the city in which it was situated, has strongly shaped urban growth throughout the twentieth century. Cultural space and military demands have competed throughout history to dominate this important urban site, formerly a prestigious manor house, occupied by the 'secret' airfield, where a new national museum was recently dedicated on a former runway.

Introduction

With the conclusion of the Cold War and the shrinking of national defence systems in the Western World, most militaries possess excess physical capacity and base infrastructure. Many of these military bases were centrally located, and, if re-development challenges can be overcome, their demilitarisation significantly enriches local land resources that can be used for other opportunities. In recent decades, some military airbases have entered the civilian aviation network or have been adapted for a variety of other uses, including new housing, education, or administrative offices. Others have been the sites of new capital investments or even new towns, but often, former defence space is simply unused.

Demilitarisation of bases has occurred in North America, Western Europe (especially France, Germany, and Great Britain), and throughout heavily militarised Eastern European countries that were part of the

Warsaw Pact or were member states of the former Union of Soviet Socialist Republics (USSR). Reuse of military space affects local, regional, and sometimes national economic, political, and social contexts, as well as the environmental aspects of particular cities and regions. With land at a premium throughout most of Europe, there is more pressure to redevelop former defence sites (Clark, 2016). While multidisciplinary scholarship has examined deaccession of military bases worldwide, notably in the geography (Clark, 1998; Havlick, 2014) and urban planning (Moss, 2003; Bagaeen, 2006, 2016) disciplines, the literature generally focuses on base closure and demilitarisation but, with a few exceptions, fails to address "afterlife" or redevelopment following deaccession (Ashley and Touchton, 2016). For example, Tempelhofer Feld in Berlin, which once housed a Gestapo prison, a concentration camp, and an airport, functions in the last decade as a public park with special uses creating a 'symbol of freedom' (Copley, 2017).

In this chapter, we expand scholarly literature about land use change related to redevelopment of defence sites (Ashley and Touchton, 2016; Bagaeen, 2006) by tracing the demilitarisation and adaptive reuse of a massive Soviet-era airbase. Its redevelopment is unique among usual patterns of urban development: while other military bases have emphasised the cultural heritage of the Cold War (Copley, 2017), no redevelopment has successfully propelled a site of national importance through a full circle life-course like this case study site has. Raadi military air base in Tartu, Estonia, was a traditional manor that became the site for a fledgling aerodrome, subsequently a massive Soviet power centre, then a military wasteland, and finally a national museum. We use a spatio-temporal approach to analyse the continuous friction between cultural land uses and military land uses vying to occupy the same space.

To begin the analysis, this chapter offers a brief history of Tartu, Estonia and one of its most important land and cultural assets, Raadi Manor. Land from this expansive manor, originally owned by Baltic-German nobility, was expropriated for the expansion of a military centre and airbase, which occupying German and Soviet forces used to consolidate their power. We use historical imagery, maps, and archival documents—including a series of reports from the U.S. Central Intelligence Agency, documenting expansion of the defence site during Cold War years, which have recently been declassified and not yet used for scholarly research—to explore how this Soviet military airbase, situated within a closed section of a closed city, stifled regular urban growth patterns. We conclude by describing the prolonged redevelopment of this troubled site and the eventual construction of a signature museum building. Throughout, we compare the processes and outcomes of institutional actions: centralised decision-making in the USSR that designated a historic manor for use as a military centre, and, in recent years, a national monument that has helped to re-invigorate Estonian identity.

Urban history of Tartu, Estonia

The second largest city in Estonia and the cultural and economic centre of southern Estonia, Tartu was first mentioned in 1030 and established as a town in 1230. During the late medieval period, Tartu [population 93,700 (2016), Statistics Estonia] was an important export node and a member of the Hanseatic League. Tartu and its surrounding region serve as the cradle of the Estonian nation, given that significant events—the first national song festival, dedication of important Estonian societies, the first major newspaper—related to the Estonian awakening period (in the latter nineteenth century) happened here.

After the Estonian Independence War (1918–1920), the development of the country as a sovereign state continued for approximately two decades. Estonia's signatory of the Soviet–Estonian Mutual Assistance Treaty in September 1939 marked the beginning of occupations of Estonian Republic territory when about 25,000 foreign troops moved to Estonia. The Molotov-Ribbentrop pact (August, 1939) assigned the Baltic States to the USSR sphere of influence. In 1941, the Second World War became a reality in Estonia when German troops reached Tartu. The front remained in Tartu for several weeks and many parts of the city, especially historic neighbourhoods with wooden tenement housing, were burned (Hess, 2011). Tartu was again on the frontlines of war in 1944, and two-thirds of its buildings were damaged and loss of life was significant. When the Soviet occupation loomed, many Estonians took refuge in Germany and Sweden, and during the first Soviet occupation in 1941, many were deported to Siberia.

Raadi, a manor house at the edge of Tartu

Raadi Manor complex was situated a short distance from the centre of Tartu. Originally a town-owned manor, deriving its Estonian name ("Raadi") from the German *Ratshof*, it was owned by 1751 by the noble Liphart family of Baltic-German descent. The oldest sections of the manor house, depicted in Figure 6.1, date from 1783. Situated on a plateau, the extensive grounds included a manor house and associated service buildings, parks and gardens, agricultural land and orchards and a rotunda pavilion. A landscaped lake beautified the site.

Typical of manors belonging to families with significant wealth, Raadi Manor was expansive.[1] The Liphart family was artistically inclined and possessed extensive art collections. Raadi Manor consequently served as a cultural centre for Baltic-German noble families (Kukk, 2015). The prospering manor also offered ample space to cultivate the aviation interests of the residents. In 1912, Sergei Utoschkin, a Russian aeronautics pioneer, performed demonstration flights, marking the beginning of aviation history in Tartu. In 1914, a small military air brigade unit landed on a specially prepared field near Raadi Manor house, bestowing on Raadi Manor its earliest military importance as an aerodrome.

In 1918, Estonia declared independence and waged war against Soviet Russia. After a short period of Soviet occupation in 1919, a small Estonian military flight unit with two hangars was established on the manor's estate.

Militarisation and urban development 151

Figure 6.1 Raadi Manor house, site of the Estonian National Museum, c. 1930s. Photograph by Eduard Selleke.
Source: Postcard private collection.

A military unit of the Russian czarist army was situated on the urban edge (between Raadi Manor and Tartu town centre) between 1796 and 1918. During the Estonian Independence period (from 1918 to 1940), an Estonian army unit was positioned on the same site. This section of Tartu was always militarised (to various extents), leading Estonian geography scholar Edgar Kant (1927) to label this urban district *kasarmla* ("barracks area") long before the Soviet occupation of Estonia.

An Estonian National Museum dedicated to collecting and interpreting Estonian heritage (both material and oral) was established in Tartu in 1909. Even though Estonia was then part of the Russian Empire, the museum's dedication was one of the most important nationalist movements for the young country (Õunapuu, 2009). Initially, the museum was located in *ad hoc* rented rooms. After Estonia declared Independence in 1918, expansive land reforms occurred in the new republic. Most of the land (including manors belonging to the Baltic German nobility) was requisitioned by the state and then divided. Many manors were converted into schools or cultural centres. Raadi Manor was first assigned to the University of Tartu, but later transferred to the Estonian National Museum (1922).

The museum achieved success at this location despite its intrinsically troubled setting within a Baltic-German manor house in which wealthy Germans employed Estonian peasants as labour. Only half a kilometre away from the manor-turned-museum was the former site of the czarist military unit.

The manor house was heavily damaged during the Second World War and at the conclusion of the war the Soviets appropriated the manor and surrounding land as the territory for an expanded military installation. By 1940, 100 hectares had been requisitioned from Raadi Manor to establish a

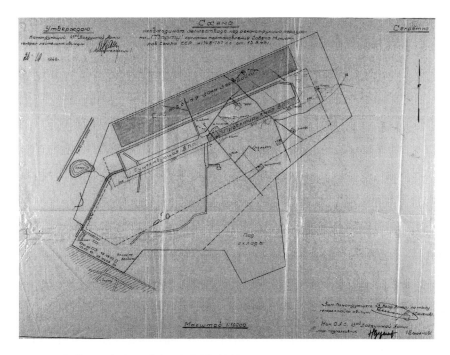

Figure 6.2 General plan of the airfield, 1946.
Source: Courtesy of the Estonian National Archive, ref. ERA.R-1.16.33.

Russian airbase (the Soviets banned private flying). Russians forces began to build the first concrete runway. German troops occupied Estonia in 1941 and extended the concrete runway an additional kilometre in length, to position Raadi airbase as a strategic centre for the *Luftwaffe* (Raukas, 2010). During the Second World War, the Soviets regained control of Estonia, and Raadi Manor house was heavily damaged during battles in 1944. Wisely, the museum collection had been scattered to various places before warfare began. At the beginning of the era of Soviet occupation, Raadi area was heavily damaged and Raadi Manor house and many former military buildings nearby were in ruins. The runway was operable, however. Russian troops began to enhance airfield infrastructure, prepare the airbase for expanded use, and physically restrict the site. These activities are depicted in Figure 6.2, a secret plan from 1946, prepared after the Council of Ministers of the USSR officially assigned the lands to the Tartu Military airfield.

Expansion of Raadi airfield, 1950s

The Soviet occupation triggered extensive militarisation throughout USSR republics. Estonia, together with Latvia and Lithuania, always regarded as geopolitically vulnerable, became critical to the sphere of influence of the

USSR. The Baltic countries were unwilling members of the USSR, and key cities—including Tartu and Tallinn—served as targets for potential nuclear attack during the Cold War, owing to Estonia's strategic location on the western periphery of communism, by American, British, and NATO forces (Burr, 2015; Glew, 2013).

In the early years of the Soviet occupation of Estonia, communist power began to transform Raadi airfield—undertaking a military-regulated redevelopment on land owned by the state and the ministry of defence—into the premiere defence installation in Estonia (Jauhiainen, 2003b). The sprawling land use configuration of the airbase and its environs is depicted in Figure 6.3 on a photograph taken during an aerial reconnaissance mission by the United States Air Force in October 1965.[2] Remarkably, a massive airbase with a runway 3 kilometres in length was located only 2 kilometres from the city centre. Significant expansion was required to remake a regional airport into a military power centre for the Soviet Union on the western

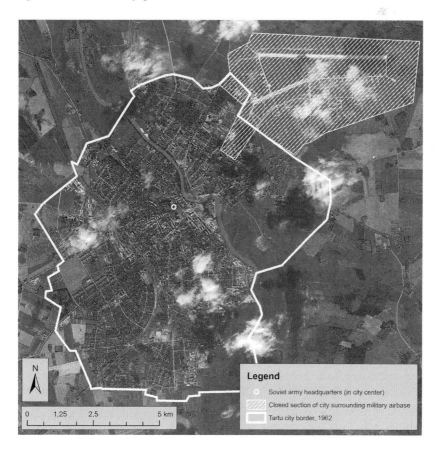

Figure 6.3 Aerial photograph of Soviet-era Raadi airbase, 1965.
Source: Courtesy of the United States Geological Survey, ref. D26 012 4022.

periphery of communism, where air defences could thwart hostile incursion and offensive attacks on Western enemies could be launched from the nearest possible range.

In 1946, farmers in the expansion area were given notice to vacate (CIA, 1951). East of the airfield, a prisoner-of-war camp was dissolved abruptly in early 1947; ground which sloped gently towards the Emajõgi River was levelled in a large earthworks exercise to create more space for runways. Estonian prisoners-of-war performed the labour for the airport expansion (CIA, 1948) and a new east-west runway was completed by 1946 (CIA, 1952). The airport had already been enlarged significantly with the extension of the two original runways and the addition of four more taxiways and aircraft parking spaces (CIA, 1948). A control tower (containing a radio station and weather station), administration building, new hangars, maintenance facility, and barracks were in use; twin-engine bombers were detected on site and four-engine bombers were known to have used the runways (CIA, 1951).

By 1951, more equipment was spotted at the airfield: 25 twin-engine bombers and 250–300 ground attack aircraft. The former manor house was used as a storehouse and the stable was used for parked tanks (CIA, 1954). A jet fighter unit was stationed there by 1952 and consequently a defensive layer was engineered into airplane parking areas through the installation of revetments to surround parked planes lining taxiways. The revetments were 2-meter high curved concrete walls that would inhibit splintering if an aircraft were hit by a bomb.[3] This is not known to have occurred, although in 1954, a four-engine bomber crashed on the Tartu-Võru highway and was dismantled and transported by truck to a secret location (CIA, 1954).[4] Aerial bombs were regularly seen in transport to the airbase by truck (CIA, 1954).

A large Soviet military force was needed to operate the airbase. As early as 1947, there were 800 officers and 400 field workers and ground personnel (CIA, 1952). On the city streets of Tartu, soldiers wearing Air Force uniforms outnumbered any other type of military personnel (CIA, 1954). On the civilian side, 1,200 Estonian workers were employed at the airbase, three-quarters in construction and earthworks and one-quarter in maintenance and warehousing (CIA, 1952).

Secrecy at the airbase was of paramount importance. Intensive use of the airbase included formation flying, and attack drills aimed at ground targets were practiced day and night (CIA, 1952). Preparedness for conflict was critical to military activity at the airbase. For the Soviets, it was important to try to shield these activities from residents. Consequently, road closures occurred during construction periods to avoid civilian viewing of increased military activity, and traffic was forced to make diversions to prevent observation of the airfield by passing motorists (CIA, 1954). Earthen mounds and structures were situated so that they could not be easily seen from offsite, and even the runways were not readily observable (CIA, 1954).

By 1955, the Tartu military airfield appeared on a list of the 30 highest priority targets in the Western USSR (CIA, 1957). Facility development enlarged capabilities at the airbase, and the runway, lengthened to 2,500 meters in 1956, had been extended an additional 500 meters by 1975 to accommodate the largest aircraft (strategic nuclear bombers situated in Tartu included TU-22M "backfire" aircraft). The original airfield was tripled in size and concrete taxi-ways were installed throughout (CIA, 1952). A nearby facility stored nuclear warheads for the "backfire" bombers (Enge, 2000; Jauhiainen, 1997).

The Atomic Weapons Requirements Study of 1959 outlined a plan for American forces to use nuclear bombs to destroy urban-industrial centres in the Eastern Bloc (Burr, 2015). Possessing Soviet bombers capable of striking America, Raadi airbase was part of the Soviet "air power" forces—bases for bombers, missiles, and interception squadrons—that Cold War opponents wished to destroy. Raadi airfield was also the only nuclear target in the USSR Baltic Military District, an extension of the Leningrad air defence zone (Jauhiainen, 1997). Atomic weapons could be used when aggressive action against the USSR was warranted by the United States, and Raadi airbase was listed as 13th among the top 20 installations for an offensive attack [the others were in Belarus (6), Ukraine (7) and Russia (6)] (Burr, 2015).

Housing military personnel at Raadi airfield

As part of the network of Soviet military units capable of strategically managing long distance bombers, the airbase at Tartu eventually employed 9,000 military personnel (Raukas, 1999). Key staff were quartered in a villa on the airport site (CIA, 1951), and other personnel were housed in state-owned apartments in nearby Jaamamõisa (Sommer, 2012). The military participated in the construction of this state-owned housing estate intended for airfield workers; typical of housing estates of its era, it was plagued by low-quality materials and poor-quality construction (Hess and Hiob, 2014). This neighbourhood originally contained about 100 houses that were part of Finnish war reparations to the Soviet Union (Komissarov, 2007). Apartment buildings established there in the 1940s and 1950s eventually deteriorated significantly (Jauhiainen, 2003b).

Tartu functions as a closed city

Elaborate systems were installed in the Soviet Union for protecting state secrets, and the degree of access and visibility afforded cities in this system produced consequences for urbanisation. All Soviet cities can be placed at a point on a continuum of state control, with 'open' cities, where significant restrictions did not apply, on one end, and 'secret' cities on the other end. Sillamäe, for example, in northeast Estonia, was a secret city and did not even appear on maps, since critical production occurred there to support Soviet defence and military operations (Gentile, 2004). Residents of secret Soviet cities had limited or no contact with friends and relatives outside the city and were forbidden to speak about the factories in which they worked.

They were usually compensated for these unpleasant living arrangements by receiving more comfortable conditions than central planning provided in secret cities (better housing opportunities, more diverse consumer goods, and a more developed network of social and cultural services (Gentile, 2004)). Relative dimensions of 'closedness' and 'secrecy' defined a city's place on the continuum, with openness or closure governed by 'varying degrees of rigidity and enforcement' (Gentile, 2004, 264). Activities occurring in a city explained Soviets' desire to designate cities secret or closed, including the strategic importance of a place in terms of the presence of military bases, infrastructures or production capabilities, such as nuclear warheads (Gentile, 2004).

Somewhere in the middle of the continuum was a 'partially' closed city like Tartu, where foreign citizens were forbidden to stay overnight and the movement of inhabitants was restricted (a strict passport system controlled travel within the USSR). Even Soviet visitors could not stay in Tartu overnight without a permit (or *'propusk'*) issued by the local authority and only if warranted. Migration to and from Tartu was closely monitored (Jauhiainen, 1997). With its degree of closure, Tartu did not disappear from maps during Soviet times; however, a 1977 topographic map (see Figure 6.4) deceptively depicts the entirety of the Raadi airbase as agricultural land, with no suggestion of the military uses dominating the space.

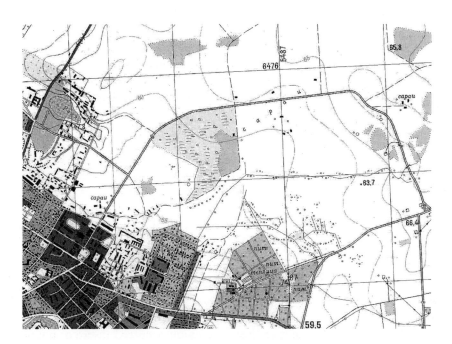

Figure 6.4 Extract from Tartu 1: 50,000 topographical sheet, 1977.
Source: Historical Map Section, Estonian Land Board Geoportal.

Despite the objectives of secrecy, everyone in Tartu knew about the presence of a large military airfield situated near the town centre. In the 1950s, it was possible for curious residents to see the airfield from elevated locations in the city centre, however people were prohibited during the Soviet years from climbing the Tartu cathedral ruins because it offered a view of the airfield (Enge, 2000). Moreover, the presence of military persons was part of everyday life in Tartu, as the airfield headquarters was located in the city centre. Residents heard the unmistakable sound of airplanes breaking the sound barrier, however the existence of the airfield was unacknowledged and operations were shrouded in secrecy (Moore, 2017).

People lived with apprehension. With its main runway situated only 2 kilometres from the city centre, mass civilian casualties would have resulted from a strike on the air base. Raadi airbase maintained its trajectory as a place for important cultural events even in the late Soviet years when, in 1988, there were demonstrations in the manor area, emphasising the contested nature of this space, against environmental degradation cause by Russian activity and military and against the presence of Soviet troops in general (Jauhiainen, 1997).

A secret Soviet airfield shapes urban growth

Underurbanisation and a lack of both physical infrastructure and social infrastructure are hallmarks of Soviet cities (Gentile, 2003; Tammaru, 2001; Leetmaa and Hess 2019; Metspalu and Hess 2018). During the early decades of the Soviet occupation, cities in Estonia grew to support state socialism, but the amplitude and speed of growth varied depending on the nature of activities conducted in particular cities (Tammaru, 2000). By the mid- to late Soviet years, urbanisation had shifted to support socialism in Tartu with the addition of one large housing estate (Annelinn, where construction began in 1973 and about 30,000 people were eventually housed) and other Soviet-era apartment buildings scattered throughout the city. State control of all aspects of life was a hallmark of the Soviet Union, and 76% of Tartu residents lived in state-owned housing in 1989 (Kulu and Tammaru, 2003).

The military airbase and its adjacent closed portion of Tartu strongly affected nearby highway development and urban development, producing the current shape of Tartu, which failed to expand to the east or northeast (Enge, 2000). As a simple monocentric city, Tartu should have expanded evenly in all directions from its centre at the base of the Emajõgi River valley, especially since favourable topographical conditions permitted regular expansion from the centre (soil for buildings was similar in all directions, and land sloped downward from the northeast and southwest on Devonian sandstone plateaus to reach the banks of the Emajõgi River on both the east and west riverbanks). However, the military-driven closure of a large section of town, to protect the secrecy of Raadi airbase, deterred uniform spatial development. The militarisation of Raadi airbase, along with the proliferation of collective farms in its vicinity, led to significant asymmetric mid-twentieth century urban growth (Marksoo, 2005).

During the 1950s and 1960s, limited modifications to the borders of the collective farms permitted Tartu's footprint to extend onto peripheral greenfields. Such growth towards the northeast, however, was impossible due to the location of Raadi airfield. The last revision of the town border, occurring in 1977, enlarged Tartu in every direction except towards the northeast. The military district, which officially belonged in part to the City of Tartu, was considered to be outside the town proper, which is reflected in masterplans. For example, chief architect Raul-Levroit Kivi, who published extensively about the development of Tartu during the twentieth century, did not acknowledge that Raadi airbase had prevented symmetric urban expansion in Tartu (Kivi, 2005).

Afterlife for the airfield

By the 1980s, Raadi airbase was viewed as a controversial place, physically degraded and clouded by 'mental pollution' (Bachmann, 2011, 106). Soviet troops provided a sombre reminder of an unwanted occupation, and the ruins of the former Estonian National Museum, although located in a strictly controlled military area, suggested the possibilities of a new beginning for Estonians in a period of awakening. By the late 1980s, many Estonians dreamed of building an Estonian National Museum at Raadi Manor, on the same site as it existed before the Second World War.

There was low urbanisation pressure in the years immediately following 1991 when the Estonian economy was weak. Although land in Tartu was readily available for development in the 1990s, little occurred because the required foreign capital was unavailable, which rendered market disposal impossible (Jauhiainen, 2003b). Raadi airbase environs did not redevelop, and residential development was impossible on the site because the district was classified as a brownfield due to fuel dump contamination.

Early redevelopment plans in the 1990s called for aviation uses (a local or cargo airport), but the already dilapidated infrastructure made this impossible (Jauhiainen, 1997, 2003a). Despite interest in redeveloping the military airbase, there was no common vision or effective planning strategy (Jauhiainen, 1999). With no other pressing redevelopment plans, new and used car dealers occupied part of the main runway. The Estonian National Museum converted some of the older manor buildings to storage uses.

Environmental clean-up was a priority at Raadi airfield in the 1990s due to a need to eradicate contamination and dispose of tanks, ammunition, and fuel dumps (Auer and Raukas, 2002; Jauhiainen, 2003b; Mander, 1995). Tartu already had a small civilian airport (Ülenurme Airport, 7 kilometers to the south), and a lack of demand and considerable resources needed for modernisation, after the disintegration of the USSR, stopped Raadi airfield from becoming a commercial airport (Enge, 2000).[5]

Raadi Manor was in ruins and surviving structures were deteriorated[6], and the departure in 1993 of the final Russian troops[7] from the former Soviet airbase (and the surrender of its land by the Russian Federation Army) triggered public discussions about the idea of restoring the Estonian National Museum to the manor grounds [along with other proposals for new housing, a university, recreation space, business, and industrial space (Jauhiainen, 1997)]. Exhibition space in the national museum during Soviet times was poor. People recognised two possibilities in the early 1990s for the location of a new Estonian National Museum: renovate Raadi Manor, or construct a new building for the museum in the city centre on Toomemägi (a hill that constituted the historic fortified town). In 1993, an architecture competition was staged for a new museum. Although the competition produced a winning design, inertia and a lack of funding meant that nothing was built (Õunapuu, 2009).

In 2005, a new international competition was launched to design a new building for the national museum that would, at long last, consolidate the collections which for decades were 'dispersed and reassembled' (Moore, 2017, 2) based on the availability of exhibition space.

In a masterstroke of site planning, the winning team (the Paris-based firm Dorell, Ghotmeh & Tane) shifted the museum (from one of several potential sites on the historical manor land in the project brief) to the end of the principal runway on unoccupied space amid the ruins of the airfield. Opened in late 2016, the new 34,000 square metre museum, with a price tag of €70 million, holds 140,000 objects along with exhibition space, storage, education, performance, and community meeting space. It is a large box with a sloping roofline—"taking off" from the runway—and glass walls that provide views of the airfield ruins and manor remains (Moore, 2017) (Figure 6.5). Dubbed a "memory field"—signaling memorialization of this turbulent site and rescue of this post-military landscape from erasure—the design team sought to regenerate the airbase and environs. It is a meaningful adaptive reuse of a troubled place, and as a powerful and explicit extension of the airfield, it confronts the difficult Soviet past in an urban void in Tartu's post-socialist urban space.

Interpreting contested urban space

The prize-winning museum, now the largest in the Baltic countries, celebrates a geo-political shift in the very urban space, shrouded in secrecy for decades, that served as a major Cold War-era military centre. Today, however, locals and international visitors can freely visit the site, which celebrates Estonia's renewed cultural identity and national pride amid the symbolic ruins of the communist occupation (Runnel et al., 2014). The museum confronts Soviet repression, interprets it in the context of Estonian nationalism, and uses artefacts to trace Estonian identity (Runnel et al., 2014). Built on a site important to the establishment of Estonian nationhood—and

Figure 6.5 Estonian National Museum, 2016.
Source: Courtesy of BTH Stuudio.

where, during an unwanted 70 year-occupation in which planes flying over this small university city to a "secret airfield" were a constant reminder of Soviet aggression during the Cold War—the museum interprets the meaning of this contested space and celebrates the nation's past, present, and future. That a runway from the Soviet airbase features prominently in the building's design demonstrates how symbols from the Soviet occupation can be embraced and not buried. A failure to acknowledge the symbolic, ecological, or cultural value of former defence sites is noted as a weakness in many other redevelopment projects (Clark, 2016), however the transformation of Raadi airbase excels in this measure because it avoids the sanitisation of the past.

The presence of a Soviet air base occupied by foreign troops influenced everyday life in Tartu (Jauhiainen, 1997), but it is unclear how residents of Tartu benefitted from Raadi airfield expansion during Soviet times (Sjöberg, 1999), other than to offer military protection for potential conflict by Cold War opponents. With the redevelopment of the space as a new national museum, however, further urban development in this neglected district of Tartu enriches the local economy and people's lives. In this formerly defence-dominated section of a post-socialist city, there is much to be gained from the redevelopment a military site, and benefits can accrue for the city, for the region, and even all of Estonia, with a new national museum and related future development.

Tracing the military and cultural uses of Raadi Manor and environs through the twentieth and twenty-first centuries, cultural uses in this urban district have ascended during certain peacetime and conflict-ridden decades while military uses have descended at opposite times, and vice versa. The

Militarisation and urban development 161

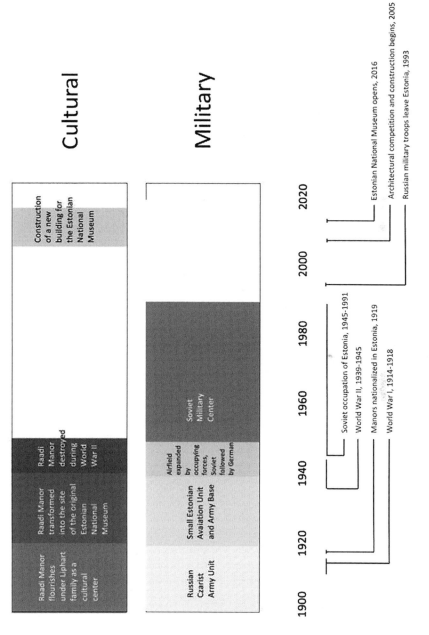

Figure 6.6 Charting the interrelations between cultural and military land uses in the Raadi District.
Source: Authors.

sequence of ownership and use of the contested space is graphically depicted in Figure 6.6. While aeronautics interest was cultivated by a noble family on its expansive manor grounds, military uses and cultural uses were intrinsically incompatible. It is unlikely aeronautics will return to this site, however the memory of early flight in Estonia and war-time expansion of a military base will live on. As potential redevelopment schemes are proposed for the former airbase district, it is critical to balance the memory of the military legacy with contemporary redevelopment aims. This was the conclusion of Dutch architect Winy Maas, who argued that (abandoned) runways are unique elements of the built environment and must not be encroached upon by proposed motorways. Nor should the runway be redeveloped with buildings. Alternative functions must be sought that leave the runways intact (Maruste, 2007). More recently, following the opening of the museum, Raadi has been used as space for large special events: in summer 2019, the rock bank Metallica performed on the runway in front of a crowd of 60,000 people, and in Autumn 2020, the Estonian National Museum was the site of the World Rally Championship

The airbase, located in a closed section of a closed city, suppressed expected urban growth patterns; following demilitarisation in the 1990s, large land resources were relinquished to local governments in a section of the town that was already undeveloped. The land surrounding the redeveloped museum is owned variously by the City of Tartu, the Municipality of Tartu, and private parties. The new Comprehensive Plan for the City of Tartu (City of Tartu, 2017), the Comprehensive Plan for Tartu Parish (Tartu Parish, 2008), and detailed local plans for properties call for new public-oriented investments (hotels, sports facilities) surrounding the Estonian National Museum. This is intended to stimulate urban growth and attract development towards the direction of this former military wasteland, where increased marketability is possible following the opening of the Estonian National Museum. However, recent development discussions suggest tension between competing public and residential land uses in this area. The site is sprawling, and much work remains ahead to reconnect the former manor and airbase with civilian surroundings (Clark, 2016).

Conclusion

A quarter of a century after the disintegration of the USSR, Estonians have achieved an ambition of restoring their national museum to its former location on the Raadi estate, a defence site on contested urban space occupied for two generations by foreign troops. Demilitarisation of Raadi airbase expanded land resources when a centrally located (Hansen, 2004) developable site was vacated near Tartu city centre. The central location of Raadi Manor and airbase is, in the end, advantageous; as profitable commercial activity that strengthens the local and regional economy enhances this former Soviet defence site, it does not displace functions because the central space was 'reserved' when it was occupied by the Soviet airfield. In other words, while the Raadi airbase constrained economic development and urban growth during the Soviet years, it now has the power to facilitate economic expansion and urban redevelopment as it commemorates historical traces of an obsolete airfield on a site of national importance.

The new museum building is an extension—literally and figuratively—of a key runway in this immense military centre, itself a symbol of the terror of communism and an unwelcome occupation that endured for two generations. The historic space of a traditional Estonian manor was scarred by military incursion and persisted as a military wasteland 'polluted both physically and symbolically' (Runnel et al., 2014, 328) until a redevelopment project—motivated by nationalism to enhance the cultural traditions of this re-independent nation—came to successful fruition. The newly dedicated museum—winner of international architectural awards in its own right—now cultivates collective memory and collective identity and dramatically symbolises a skyward ascension.

Acknowledgements

This project has received funding from the European Union's Horizon 2020 research and innovation programme under the Marie Skłodowska-Curie grant agreement No. 655601 and from Institutional Research Grant No. IUT2–17 of the Estonian Research Agency. Ingmar Pastak, Age Poom and Annika Väiko provided invaluable assistance.

Notes

1 A collection of manors preceded Tartu in the Emajõgi River valley, and Tartu's Hanseatic merchant town centre and its surrounding neighbourhoods are built on former manor fields. But Raadi Manor, somewhat preserved at its core but with its extensive grounds assigned to other uses (especially space that grew to become a massive air base) provides us with an anomaly, one that provides an interesting lesson in the formation of Tartu's townscape over various stages of urban development.
2 The base image for Figure 6.3 was part of the CORONA project—developed by the United State Air Force, the Central Intelligence Agency and private industry—to protect United States national security interests during the Cold War by photographing secret places in Asia and Eastern Europe (CIA, 2015). The physical transformation of Raadi airfield is traced using a collection of surveillance reports issued by the U.S. Central Intelligence Agency; this was top secret material during the Soviet era but it has since been declassified. Typically, CIA agents asked visitors from the Soviet Union visiting the United States (or later, people who were in contact with tourists) to describe military activity and installations. The reports do not clarify how the surveillance information was collected, except for one report which states that an informant provided intelligence and that the types of aircraft at Raadi airbase were identified from photographs (CIA, 1954), and, by the 1980s, satellite imagery (CIA, 1982). U.S. Central Intelligence Agency Documents (CIA, 1948, 1951, 1952, 1954, 1957) include CIA-RDP82-00457R001900270008-8, CIA-RDP82-00457R008200490012-9, CIA-RDP82-00457R006500250002-8, CIA-RDP82-00457R014200330007-5,CIA-RDP80-00810A004200890009-9,CIA-RDP80-

00810A005300230005-6, CIA-RDP80S-01540R005000090008-5, and CIA-RDP61S-00750A000500030123-6.
3 By the 1980s, there were 24 large and 30 small revetments at the airfield (U.S. Central Intelligence Agency, 1982).
4 In a later crash in Tartu, in 1991, a Tupolev Tu-16k *Badger* was destroyed and four people were killed.
5 By 1993, the main runway at Raadi airfield was classified as an emergency airfield for aviation emergencies (including commercial) on the Jeppesen chart, although the landing strip is now blocked by an automobile market.
6 Buildings were in disrepair, and Russian troops caused further intentional destruction as they departed in 1993, taking nearly all movable objects with them (Jauhiainen, 1997).
7 The departure of Russian troops was the culmination of a two-year demilitarisation effort following a declaration of Independence by Estonia in 1991 as the Soviet Union disintegrated, and the national government designated responsibility for the Raadi airbase site from the Estonian Defence Ministry to the three jurisdictions it encompasses (Enge, 2000).

References

Ashley, A.J.; Touchton, M. 2016. "Reconceiving military base redevelopment: Land use on mothballed U.S. bases", *Urban Affairs Review*, 25(3): 391–420.
Auer, M.R.; Raukas, A. 2002. "Determinants of environmental clean-up in Estonia." *Environment and Planning C: Government and Policy*, 20(5): 679–698.
Bachmann, K. 2011. "Koha vaimne reostus." Acta architecturae naturalis=Maastikuarhitektuurseid uurimusi. 1. Tallinn, 93–106.
Bagaeen, S.G. 2006. "Redeveloping former military sites: Competitiveness, urban sustainability and public participation," *Cities*, 23(5): 339–352.
Bagaeen, S.G. 2016. "Framing military brownfields as a catalyst for urban regeneration." In Bagaeen, S.; Clark, C. (eds.). *Sustainable Regeneration of Former Military Sites*. 1–18. Routledge.
Burr, W. (ed.). 2015. *U.S. Cold War Nuclear Target Lists Declassified for First Time*. National Security Archive Electronic Briefing Book No. 538. Washington, D.C.: National Security Archive at George Washington University.
City of Tartu. 2017. *Comprehensive Plan 2030+*. Tartu, Estonia: Tartu City Council.
Clark, C. 1998. *Vintage Ports or Deserted Dockyards: Differing Futures for Naval Heritage across Europe*. Faculty of the Built Environment, University of the West of England.
Clark, C. 2016. "Diversity in the transformation of defence sites to new civilian life." In Bagaeen, S.; Clark, C. (eds.). *Sustainable Regeneration of Former Military Sites*. 217–219. London: Routledge.
Copley, C. 2017. "Curating Tempelhof: Negotiating the multiple histories of Berlin's 'symbol of freedom,'" *Urban History*, 44(4): 698–717.
Enge, J. 2000. "Raadi Airport is Tartu's White Elephant," *The Baltic Times*, 6 January, 2000.
Gentile, M. 2003. "Delayed underurbanization and the closed-city effect: The case of Ust'-Kamenogorsk," *Eurasian Geography and Economics*, 44(2): 144–156.
Gentile, M. 2004. "Former closed cities and urbanisation in the FSU: An exploration in Kazakhstan," *Europe-Asia Studies*, 56(2): 263–278.
Glew, C. 2013. "Target: Estonia – Britain's nuclear plan for Tallinn, Tartu and Viljandi," *Estonian World*, 7 October.

Hansen, K.N. 2004. *The Greening of Pentagon Brownfields: Using Environmental Discourse to Redevelop Former Military Bases.* Lanham, MD: Lexington Books.

Havlick, D.G. 2014. "Opportunistic conservation at former military sites in the United States," *Progress in Physical Geography*, 38(3): 71–285.

Hess, D.B. 2011. "Early 20th-century tenement buildings in Estonia: Building blocks for neighborhood longevity," *Town Planning and Architecture*, 35(2): 110–116.

Hess, D. B.; Hiob, M. 2014. "Preservation by neglect in Soviet-Era town planning in Tartu, Estonia," *Journal of Planning History*, 13(1): 24–49.

Hess, D. B., & Tammaru, T. (2019). Modernist housing estates in the Baltic Countries: Formation, current challenges and future prospects. In Housing Estates in the Baltic Countries (pp. 3–27). Springer, Dordrecht, Netherlands.

Hess, D. B., Tammaru, T., & van Ham, M. (2018). Lessons learned from a pan-European study of large housing estates: Origin, trajectories of change and future prospects. In Housing estates in Europe (pp. 3–31). Springer, Dordrecht, Netherlands.

Jauhiainen, J. 1997. "Militarisation, demilitarisation and re-use of military areas: The case of Estonia," *Geography*, 82(2): 118–126.

Jauhiainen, J. 2003a. "Urban planning and land-use in post-socialist Estonia: The case of Tartu." In Lars, R. (ed.). *Sustainable urban patterns around the Baltic Sea.* Uppsala: Baltic University Press. 36–43.

Jauhiainen, J. 2003b. "Development of a former military district: Jaamamõisa at the former Raadi air base in Tartu." In Lars, R. (ed.). *Sustainable urban patterns around the Baltic Sea.* Uppsala: Baltic University Press. 51–56.

Jauhiainen, J.S. 1999. The Conversion of Military Areas in the Baltic States. *North European and Baltic Integration Yearbook 1999*: 327–334. Springer Berlin Heidelberg.

Kant, E. 1927. "Tartu. Linn kui ümbrus ja organism." Tartu (koguteos). Tartu, 192–433.

Kivi, R.L. 2005. *Tartu planeerimisest ja arhitektuurist: artikleid ja mälestusi.* Tallinn: Eesti Arhitektuurimuuseum.

Komissarov, K. Sämplima ja skrätsima. Koha praktikad ehk Tartu Hiinalinna kujunemisest. *Maja*, 52: 24–27.

Kukk, I. (ed.) 2015. "Unistuste Raadi : Liphartite kunstikogu Eestis : kataloog = Raadi of our dreams: The Liphart family and their art collection in Estonia: Catalogue." Tartu: Tartu Ülikool kirjastus.

Kulu, H.; Tammaru, T. 2003. "Housing and ethnicity in Soviet Tartu." *Yearbook of Population Research in Finland*, 39: 119–140.

Leetmaa, K., & Hess, D. B. (2019). Incomplete Service Networks in Enduring Socialist Housing Estates: Retrospective Evidence from Local Centres in Estonia. In Housing Estates in the Baltic Countries (pp. 273–299). Springer, Dordrecht, Netherlands.

Mander, Ü. 1995. "Environmental Pollution in the Former Soviet Military Airfield in Tartu." Presented at "Reuse of Former Military Bases," Tartu, Estonia, December 1995.

Maruste, M. 2007. Linna loomine – Raadi piirkonna workshop. *Maja*, 52: 42–45.

Metspalu, Pille and Daniel Baldwin Hess. 2018. Revisiting the Role of Architects in Planning Large-Scale Housing in the USSR: The Birth of Three Large Housing Estates in Tallinn, Estonia. Planning Perspectives. vol. 33. no. 3. pp. 335–361. [doi: 10.1080/02665433.2017.1348974]

Moore, R. 2017. "Estonian National Museum Review: Touching and Revealing," *The Guardian*, January 1.

Moss, T. 2003. "Utilities, land-use change, and urban development: Brownfield sites as 'cold-spots' of infrastructure networks in Berlin." *Environment and Planning A*, 35(3): 511–529.

Õunapuu, P. 2009. *Eesti Rahva Muuseumi 100 aastat.* Tartu: Eesti Rahva Muuseum.

Raukas, A. 1999. *Endise Nõukogude Liidu sõjaväe jääkreostus ja selle likvideerimine.* Tallinn: Keskkonnaministeerium.

Raukas, A. 2010. "The birth and unglorious end of the Raadi military airfield." In Siimets, Ü.; Madisson, S.; Liiv, J. (eds.). *The Raadi Book: Pictures of Things Gone by.* Tartu: Eesti Rahva Muuseum: Eesti Rahva Muuseumi Sõprade Selts. 212–213.

Runnel, P.; Tatsi, T.; Pruulmann-Vengerfeldt, P. 2014. "Who Authors the Nation? The Debate Surrounding the Building of the New Estonian National Museum." In Knell, S.; Aronsson, P.; Amundsen, A.B. (eds.). *National Museums: New Studies from Around the World.* London: Routledge. 325–228.

Sjöberg, Ö. 1999. "Shortage, priority and urban growth: Towards a theory of urbanisation under central planning," *Urban Studies*, 36(13): 2217–2236.

Sommer, S. 2012. *Soviet housing construction in Tartu: The era of mass construction (1960–1991).* Master's Thesis, Department of Geography, University of Tartu, Estonia.

Strömberg, P. 2010. "Swedish military bases of the Cold War: The making of a new cultural heritage," *Culture Unbound*, 2: 635–663.

Tammaru, T. 2000. "Differential urbanisation and primate city growth in soviet and post-soviet Estonia," *Tijdschrift voor Economische en Sociale Geografie*, 91(1): 20–30.

Tammaru, T. 2001. "Urbanization in Estonia in the 1990s: Soviet legacy and the logic of transition," *Post-Soviet Geography and Economics*, 42(7): 504–518.

Tartu Parish. 2008. *Comprehensive Plan.* Tartu, Estonia: Tartu Parish Council.

U.S. Central Intelligence Agency (CIA). 1948, 1951, 1952, 1954, 1957. Information Report. Washington, D.C.: U.S. Central Intelligence Agency.

U.S. Central Intelligence Agency (CIA). 1982. Status of Soviet Military Transport Aviation, National Photographic Interpretation Center. (CIA-RDP82-T00709R000100260001-8). Washington, D.C.: U.S. Central Intelligence Agency.

U.S. Central Intelligence Agency (CIA). 2015. CORONA: Declassified. *CIA News and Information.* Washington, D.C.: U.S. Central Intelligence Agency.

7 In-between the East and the West
Architecture and urban planning in 'Non-Aligned' Skopje

Jasna Mariotti

QUEEN'S UNIVERSITY BELFAST

Introduction

The development of Skopje after WWII was an ambitious project that ran parallel to Yugoslavia's processes of modernisation and mass urbanisation. Its architecture and urban planning followed these dynamics and imaginaries. These radical transformations of the city's built tissue were possible in part because of the particular position of Yugoslavia in the post-war era, actively engaged in cooperation alternative to the Cold War divides and adopting a policy of non-alignment and peaceful coexistence of various ideological systems. This paper focusses on two themes and links architecture and urban planning with the shifting political paradigms represented at a national level, focussing on the city of Skopje in the period following WWII. The first theme studies the reconstruction of Skopje after the earthquake in 1963, focussing on the discipline of urban planning and the second one explores the role of architecture in the making of the city of solidarity. These themes contribute to the exploration of how the reconstruction of Skopje contributed in part to the shaping of Yugoslavia's project of socialist modernisation.

* * *

The development of Skopje after the Second World War was an ambitious project that ran parallel to Yugoslavia's processes of modernisation and mass urbanisation. Its architecture and urban planning followed these dynamics and imaginaries. The radical transformations of the city's built tissue during this period were possible in part because of the particular position of Yugoslavia in the post-war era. After the break between Yugoslavia and the Soviet Union in 1948, the Truman administration adopted a policy of 'keeping Tito afloat' (Lees, 1997), hoping that the Yugoslav influence would spread to other communist states. During this period, Yugoslavia became a key ally of the United States in their attempt to create fissures in the communist world. Through a 'wedge' strategy, involving military and economic assistance, the United States aimed at undermining Soviet control and creating divisions between the Soviet Union and other communist countries (Lees, 1997, pp. xiii–xviii). Yet, Yugoslavia remained neutral and actively engaged in cooperation alternative to the Cold War divides. The country adopted a policy of non-alignment with either of the great powers in the Cold War, based on a peaceful coexistence of various ideological systems.

Architecture and urban planning played an important role in the framework of shifting political orientations. The city of Skopje, an important administrative centre during the Ottoman rule, which ended in 1912 with the Balkan Wars, was transforming fast in the period that followed - its changing form and appearance paralleled changes in those who ruled. Skopje's planned transformation began in 1914 when the first regulatory plan for the city was officially ratified. The First World War stopped its implementation and prompted the development of a new plan. The plan that followed, prepared in 1929 under the leadership of architect Josif Mihailovic, was the acknowledged framework for development until WWII (Popovski, 1968). During the inter-war, Skopje had its own established architectural scene and reputable urban planning practice. In the period after WWII, international influences helped in the transformation of the city and manifested as modern architecture.

The post-war brought significant changes in the patterns of urban development in Skopje. The Macedonian capital city grew alongside the processes of political and societal restructuring. The formation of the Federal People's Republic of Yugoslavia in 1945, within which Macedonia was one of the six republics, contributed to the instilment of socialist ideals at a national level and attendant rapid urbanisation. The urbanisation of Yugoslavia after WWII was linked with a decline in agrarian production, or a transfer from agriculture to other economic activities – in particular to industry – and, in this process of rapid socialist restructuring and industrialisation, its cities grew.

Skopje developed rapidly after WWII. The city was transformed into a modern metropolis with a rapidly growing industry and its population rose quickly from 68,880 in 1931 to 120,130 in 1953 and 166,870 in 1961 (The Population of Yugoslavia, 1974, p. 54). In fact, its growth rate was among the largest recorded among the Yugoslav cities (Musil, 1980). Immediately after WWII, the reconstruction of existing factories and the construction of new ones commenced. The adoption of the first Five-Year Plan for Socialist Industrialisation and Electrification, provided conditions for the rapid development of industry in the country, and hence, the rapid development of Skopje as a capital city. The pace of industrial development prompted an increase in the number of inhabitants in the city, due to an influx of workers, which in turn almost eradicated unemployment. During the first Five-Year Plan, 20 new factories were opened in the city (Gradski odbor na Nariodniot front – Skopje, 1949). Through this transformation, from a small town to a centre for metallurgical, chemical and pharmaceutical industries, the city of Skopje became the third largest city in Yugoslavia, after Belgrade and Zagreb (Stefanovska and Koželj, 2012).

After 1945, planning paradigms for the city also changed to accommodate Skopje's continuing growth and the projected visions for its future expansion. In the newly established context of the socialist country, the Macedonian authorities invited Luděk Kubeš, a Czechoslovakian modernist architect and urban planner, and his team, *Atelier Arhitektu,* to draft a new masterplan for the socialist city of Skopje that would serve as a base for the post-war transformation of the city. His plan for the city from 1948 showed strong influences from, and references to, Le Corbusier's "Radiant City" of 1935 (fr. *La Ville Radieuse*). *Atelier Arhitektu* outlined a new ground for city planning, which started from a *tabula rasa* and disregarded the patterns of earlier settlements. The plan followed the principles that '[i]n the new socialist society, the regulatory plan serves as a basis for a lively and healthy development of future generations to which the socialist society gives them all political, cultural and economic development opportunities' (Gradski odbor na Nariodniot front – Skopje, 1949, p. 31). In an attempt to accommodate the city's spatial and demographic growth, new interconnected neighbourhoods with freestanding buildings were proposed and, following the functionalist principles of the modern movement, dwelling, working, recreation and transport were separated. Industries were carefully arranged in clearly defined functional zones in the city's periphery, repetitive patterns of mass housing in-filled the inner city territory, while large areas of the urban fabric were set for recreation. This formed the basis for the development of a socialist Skopje. The city's boundaries expanded and its area grew, influenced by its adjacency to the River Vardar and its relationship to surrounding urban agglomerations. The plan was amended in 1955 to propose that in the following 30 years, the city was planned to house 300,000 people.

The earthquake of July 1963 brought about a marked change in urban planning in Skopje. In August 1963, less than one month later, the existing masterplan for Skopje was declared void. The authorities signalled the need for the formulation of alternative models of growth and for the future development of the city. Adolf Ciborowski, who was previously Chief Architect in Warsaw and coordinated city's post-war reconstruction, was appointed as Project Manager of Skopje's new master plan (UNDP, 1970). In the period that followed, under the auspices of the United Nations, the Institute of Town Planning and Architecture of Skopje (ITPA) led the process of drafting the new master plan for the city with the help of foreign experts, Doxiadis Associates from Athens and Polservice from Warsaw.

In this chapter, two themes that link architecture and urban planning with the shifting political paradigms represented at a national level are examined, focussing on the city of Skopje in the period after 1945. The first theme studies the reconstruction of Skopje after the earthquake in 1963, focussing on the discipline of urban planning and the second one explores the role of architecture in the making of the city of solidarity. These themes consider the 'non-aligned' territory of the city of Skopje and its position in-between the East and the West, analysing how the reconstruction of Skopje contributed in part to the shaping of Yugoslavia's project of socialist modernisation.

Architecture and urban planning in the Non-Aligned world

Yugoslavia's global geo-political position after the Second World War was distinctive. In 1948, the Tito–Stalin split (or Yugoslav–Soviet split) resulted in Yugoslavia's expulsion from the Communist Information Bureau and banished the country from the international association of socialist states. This event offered the country the opportunity to construct its own path of socialism, through the formation of a socio-political system that introduced socialist self-management and decentralisation of the state power as credible policies for domestic reform (Siani-Davies, 2003, pp. 1–31). In the following years, Yugoslavia stood firmly between the capitalist West and the communist East, and little more than a decade later, in 1961, the country became one of the six founding members of the Non-Aligned Movement (NAM). This organisation advocated a middle course between the Western and the Eastern blocs, effectively placing Yugoslavia in a unique position between them and at the forefront of tracing an alternative third way alongside the newly independent and decolonised countries of the Third World (Ashcroft et al., 2007, p. 212).

The idea for the NAM was at first conceived at the conference of the Afro-Asian countries, held in 1955 in Bandung, Indonesia. There were 29 participant countries, which represented over half of the world's population. The 'ten principles' adopted at the Bandung Conference promoted world peace and co-operation, respected the sovereignty, territorial integrity and equality of all nations, proposed abstention from intervention in the internal affairs of another country, promoted co-operation and respected justice and international obligations (Final Communique of the Asian-African conference of Bandung, 1955). Josip Broz Tito, President of Yugoslavia, was among the observers at the conference in Bandung and the only European leader present. Tito was interested in the position of Yugoslavia in-between the blocks and in building alliances with the newly decolonised countries, an approach that would further reinforce country's international standing.

Six years after the conference in Bandung, in September 1961, at the initiative of Tito, the First Conference of Heads of States or Government of Non-Aligned Countries was held in Belgrade. Tito became the first Secretary General of the movement, whose founding fathers and emblematic leaders together with the Yugoslav president were Nasser of Egypt, Nkrumah of Ghana, Nehru of India and Sukarno of Indonesia. At the opening of the conference Tito proclaimed "[w]e are the conscience of the humanity" (Nova Makedonija, 1961a), and pushed further for the awakening of the suppressed, addressing the issue of colonialism and the unresolved issues in the relations amongst the great powers. The conference in Belgrade lasted five days and the programme covered a wide range of topics, including strengthening of world peace and security, liquidation of colonialism, non-interference in the internal affairs of other states and a ban on nuclear tests (Nova Makedonija, 1961b). The conference received wide international interest and coverage of the conference was published on the front pages of newspapers around the world. Most reports praised the conference as a first attempt towards the peaceful and positive resolution of international problems, and for providing an opportunity for the voices of Asian and African peoples to be heard in Europe (Nova Makedonija, 1961c).

Twenty-five countries, from Afghanistan, Ceylon and Ghana to Ethiopia, were represented in Belgrade and there were three countries represented as observers: Bolivia, Brazil and Ecuador. In a period of global Cold War tensions, the NAM was seen as an alternative to the Cold War divide, calling for a world of hope, peace and cooperation but also for nuclear disarmament and peace; "[a]ll nations have the right of unity, determination and independence by virtue of which right they can determine their political status and freely pursue their economic, social and cultural development without intimidation or hindrance" (The Belgrade declaration of the Non-Aligned countries of 1961, article 13).

In his opening speech at the beginning of the conference, Tito outlined the importance of the conference and the responsibility of the leaders present in Belgrade:

> People are tortured by the Cold War which takes a severe form, and they are afraid of a possible catastrophe which would be caused by the new World War. Accordingly, I believe that we will do a great service to the world if we clearly and adamantly point the road which leads towards peace in the world, towards freedom and equality and peaceful cooperation between nations.
>
> (Nova Makedonija, 1961d)

Yugoslavia's role in the NAM was an instrument to vocalise its progressive views to the world and to solidify its political standing. It also served as an opportunity to intensify cultural exchanges and economic collaborations and to distribute knowledge and technology to other Non-Aligned countries (Sekulić, 2012). As part of his leadership in the NAM, Tito travelled on goodwill missions to more than 60 countries, from India and Burma, as well as to the United Kingdom following an invitation from Winston Churchill. The trips, as Tito explained "greatly benefited the country, by enabling leaders abroad to gain a greater understanding of Yugoslavia's internal development and outlook on international issues" (Rubinstein, 2015, p. 53).

Architecture and urban planning became one of Yugoslavia's most significant exports thanks to the connections established through the NAM, a process through which architects and planners from Yugoslavia took part in the modernisation of the newly independent and decolonised countries

(Stanek, 2012). They were also important tools in constructing the nation's distinction from the communist bloc and the gradual orientation towards the West (Kulić, 2009). The newly decolonised countries were urbanising rapidly and the Yugoslav architectural firms found new markets in the Non-Aligned world and contributed to their growth.

The transfer of architecture between Yugoslavia and Africa and Asia during the Cold War was through Yugoslav construction companies that were responsible for the planning, design and the building of numerous key projects in the Non-Aligned world. Energoprojekt and Mostogradnja from Belgrade, Mavrovo and Pelagonija from Skopje, Energoinvest and Centroprojekt from Sarajevo, Industrogradnja and Ingra from Zagreb were just some of the Yugoslav companies that had projects in other countries from the NAM. They built trade fairs, hotels, conference centres, ministerial complexes, hospitals and administrative offices across Asia and Africa, including Libya, Algeria, Nigeria, Zimbabwe as well as Iraq (Sekulić, 2012). Through these architectural projects and planning policies they were also involved in the transformation of the newly decolonised societies, which enabled Yugoslavia to promote its model of socialism to other countries from the NAM (Kulić, 2017). Yugoslavia's internationalism was mirrored in the assistance provided by other countries in the recovery and reconstruction of Skopje after the earthquake of 1963.

Urban planning in post-1963 Skopje

On the 26th July, a devastating earthquake struck the city of Skopje, causing tremendous damage to the built fabric of the city. The earthquake ravaged about 85% of its buildings, and took the lives of more than 1,000 people. Shortly after the earthquake, national and international parties offered aid to the city, which made its recovery an unprecedented action of international solidarity and co-operation (Ladinski, 1997). As a result, the transformation of the city intensified after the earthquake.

The city centre of Skopje, which the earthquake of 1963 had almost completely destroyed, required immediate attention. In 1964 the International Board of Consultants, chaired by Ernest Weissmann of the United Nations (UN) and comprised of national and international experts in seismology and town planning, recommended the organisation of an international competition for its reconstruction, possibly with help from the UN. Acknowledging that the city centre should be 'radically replanned to the highest possible standard', the Board urged the most talented urban planners and architects to participate, marking another new beginning for the planning of Skopje and earmarking the city as a testing ground for new modern planning concepts and ideas, but also as a site for collaboration among experts from different countries (UNDP, 1970, p. 297). Eight teams were invited to submit proposals, four foreign teams: Luigi Piccinato (with Studio Scimemi) from Rome; Maurice Rotival from New York; Kenzo Tange from Tokyo; J. H.

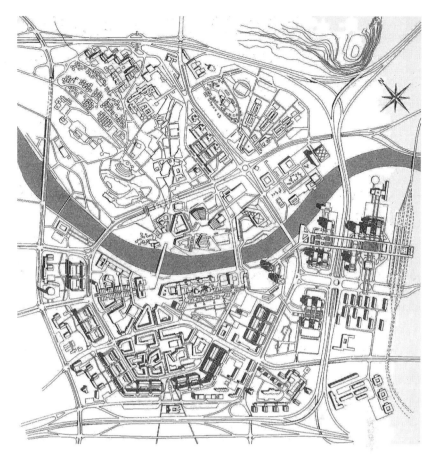

Figure 7.1 The 'Ninth Project' for the reconstruction of the central city area of Skopje, 1966. The project illustrates the formation of the City Wall and the City Gate as prominent design elements of the masterplan.
Source: UNDP, 1970, p. 331.

van den Broek and Bakema from Rotterdam; and four Yugoslav teams: Radovan Miščević and Fedor Wenzler from the Croatian Institute of Town Planning in Zagreb; Eduard Ravnikar and associates from Ljubljana; Aleksandar Djordjević and his colleagues from the Belgrade Institute of Town Planning from Belgrade; Slavko Brezovski and his associates from *Makedonija-proekt* from Skopje.

In July 1965, the Jury awarded the first prize to two teams. Three-fifths ($12,000) was given to the team of Kenzo Tange from Japan, and the other two-fifths ($8,000) to the team of Miščević and Wenzler from Zagreb. The jury reached this unconventional distribution of awards on the basis that '[...] there was no one entry which should be the single basis for implementing the reconstruction of the centre of the city of Skopje. Each of the entries contained a variety of promising ideas and proposals' UNDP, 1970, p. 370). In its report, the jury noted that Kenzo Tange's entry '[...] has dealt with many aspects of the plan in a serious, original and inspired way', pointing particularly to '[t]he architectural interpretation of larger structures and planning and design of the urban ensembles' as being 'of high quality' (UNDP, 1970, p. 373). The entry of Miščević and Wenzler in turn was praised for '[...] the opportunity it provides for realization in stages and thus for flexibility' (UNDP, 1970, p. 373).

Shortly after the winners were announced, the international team with members from Japan, Zagreb and Skopje began collaborating on a new master plan for the city centre of Skopje which aimed to 'contain all the positive aspects of all the competition entries' and would – according to Adolf Ciborowski – '[...] be a base for a beautiful city for happy citizens' (Boškovski, 1965). The 'Ninth Project', which was the name given to the master plan that the team of Kenzo Tange, Miščević and Wenzler, Polservice traffic engineer S. Furman and Skopje's ITPA drafted, was approved in 1966 (UNDP, 1970, p. 122) (Figures 7.1 and 7.2).

The new masterplan proposed a relocation of the railway station and a formation of a transportation centre, a project for which Kenzo Tange developed the initial design, and a series of new roads that penetrate through its built fabric. Six storey buildings would encircle the city centre to form

Figure 7.2 Kenzo Tange and his team in front of the model for the reconstruction of the eastern area of the city centre of Skopje. The model displays the new Transportation centre. The architect Kenzo Tange is in the centre of the picture.

Source: UNDP, 1970, p. 327.

the 'City Wall', a dynamic and inhabited spatial structure, whose construction in the built tissue of Skopje still represents an exemplary piece of the visionary ideas for the city post-1963. The 'City Gate' is another element of his proposal, a nucleus and a centre for administrative and commercial activities in the city.

The adoption of the masterplan for the city centre of Skopje opened the road to the intensive reconstruction of Skopje that started from its city centre. In its final layout, deriving from Tange's competition entry, the masterplan represented a synthesis of an intense international involvement and was a product of an international collaboration of a group of experts who transferred their ideas about the future development of the 'world city' with help from the UN.

Making of the city of solidarity: building in Skopje after 1963

After the earthquake in 1963, help for Skopje came from 82 countries, including the United States and Russia as well as from Afghanistan, Ethiopia, Indonesia, China, Uganda and Sri Lanka. They offered financial and material aid and, thanks to this immense display of international aid, the city recovered quickly (Skopje – grad na solidarnosta, 1975). These expressions of international solidarity helped in reshaping the destroyed city. Blagoj Popov, president of the city's executive council and mayor, said:

> For the deadly wounded city, its citizens fighting to rescue precious human lives and against the terrible consequences of this disaster, aid immediately was available, we have witnessed extraordinary humanity, solidarity and sacrifice. This help was from all over Yugoslavia, from all over the world and from different sources continuously and efficiently, knowing no barriers or obstacles for this noble cause.
>
> (Vasa pomoc Skopju, 1964)

Skopje's housing structure suffered heavily in the earthquake. Of the city's 36,518 housing units, 3,411 were completely destroyed, 11,891 were heavily damaged beyond repair, 14,194 were badly damaged, and only 7,082 were partially damaged (Jordanovski, 1993). The major reconstruction priority in the wake of the earthquake was new housing. Between 1963 and 1965, 22,250 new housing units were built in the city and around 16,000 were refurbished. As a result, during this period, 18 new neighbourhoods emerged in Skopje, the traffic network expanded and the water supply system was improved (Skopje – grad na solidarnosta, 1975). Both the locations and the layouts of new homes were under the guidance of a centralised planning system and in socialist Skopje, a state controlled agency, the *Self-governing Interest Association* (Mk. *Samoupravna interesna zaednica*, SIZ), was in charge of their realisation. Compared to the pre-earthquake conditions, the overall standard of housing also improved. The average habitable area available per person rose from 8.5 to 14.7 sq meters. It was projected that by 1981 all the housing units in the city will have a kitchen and a toilet as compared to the 24% that had such facilities in 1961 (Popovski, 1968). These radical transformations facilitated Skopje's growth rates, which increased spectacularly post-1963. As forecast, between 1961 and 1971, the city's population almost doubled: from 166,870 in 1961 to 314,552 a decade later (The Population of Yugoslavia, 1974).

Shortly after the earthquake, prefabricated housing units started to emerge in Skopje, and Levittown-like neighbourhoods started to shape the city (Figure 7.3). In the immediate aftermath of the earthquake, the prefabricated housing seemed to represent an ideal way to accelerate the construction process through quick and cheap assembly, while responding to the housing shortages and providing city dwellers with an opportunity to have their own home. USSR, Romania, Czechoslovakia and Poland, as well as Italy, Norway, Finland and France donated prefabricated houses to the different neighbourhoods of the city. The USSR built a factory for prefabricated concrete panels used for the construction of apartment buildings. The new factory could build up to 1,200 apartments per year and in the process of recovery after the earthquake around 14,000 units in the city were prefabricated.

The construction of prefabricated housing units in socialist Skopje benefited from the technological development of the country and, in the context of severe housing need, prefabrication appeared to be the right way forward. Prefabrication processes accelerated the reconstruction

Figure 7.3 Finishing work on the prefabricated houses of the "Taftalidze" neighbourhood. Still existing today, these houses were donated by the government of Czechoslovak Socialist Republic.

Source: Vasa pomoc Skopju, 1964.

process to such an extent that in October 1963 Edvard Kardelj, President of the Federal Parliament, pondered if 'this type of temporary and rapidly constructed unit should become a standard element of housing policy' (Fisher, 1964, p. 48). Such construction methods were positively endorsed by Yugoslav politicians as well as by Khrushchev (First Secretary of the Communist Party of the Soviet Union), who visited the city of Skopje on 22 August 1963.

Besides housing, other buildings were constructed in Skopje. In the 'non-aligned' territory, designers from both East and West were invited to develop their urban visions for the devastated city. As a result, Skopje's architectural and urban culture was exposed to international influences, through architects and urban planners from all over the world working in the city, a condition certainly facilitated by Yugoslavia's foreign policies in the 1960s. Poland, Romania and Czechoslovakia built hospitals, Bulgaria, Czechoslovakia and Switzerland built schools, and Denmark and East Germany built kindergartens. The UK aided with the construction of a theatre and USSR and Algeria together donated the youth cultural centre 25th May. The Universal Hall was built with a donation from 35 countries, as well as donations made by foreign tourists who visited Yugoslavia after the earthquake.

The Museum of Contemporary Art in Skopje was part of the Polish contribution for the city after the earthquake, together with Ciborowski's involvement in the planning processes. In the help that was offered to the city, Poland provided conceptual ideas as well as technical documentation for this public building. The Museum of Contemporary Art would house the art collection and donations – around 900 works, from the foreign states to the ravaged city. The project for the Museum of Contemporary Art in Skopje was based on a national competition organised in 1966 by the Society of Polish Architects in which 89 proposals were submitted. In May 1966, the winners were announced, the group 'Tigers': Waclaw Klyszewski, Jerzy Mokrzynski, Eugeniusz and Wierzbicki. The building is an impressive spatial construct, positioned on the top of Kale Hill, overlooking the city, and immediately after it was built, it became a visually dominant edifice in the city (Figure 7.4). The director of the museum in Skopje, Boris Petkovski, in an article from 1966 wrote that the building represented "not only a functional architectural solution, but also a symbol of the architecture of this century and the solidarity that enabled its birth" (Vasa pomoc Skopju, 1964).

These projects exposed Skopje to international influences, leaving a permanent stamp over its built form. Donated by various countries from both sides of the Iron Curtain, these monumental structures in the city can be seen as expressions of international solidarity and an effort of cooperation on equal terms, alleviating Cold War divides and testing complex spatial configurations.

Figure 7.4 Museum of Contemporary Art in Skopje by "Tigers", a Polish group of architects. The museum is positioned on top of the Kale Hill overlooking the city.
Source: Jasna Mariotti.

Skopje in-between the East and the West

The reconstruction of Skopje after the earthquake of 1963 was an act of international solidarity and the multifaceted nature of its architecture came out of the state of mediation between global influences and complex local conditions. In the aftermath of destruction, the city of Skopje was receptive to new ideas. Ernest Weisman, the President of the International Board of Consultants for the reconstruction of the city after the earthquake of 1963 commented:

> I am deeply impressed with what has been done in Skopje after the earthquake. I visited Skopje immediately after the disaster and I had the opportunity to get to know the horror of the tragedy. It is now that Skopje received another character. The results presented at the meeting of the International consultative committee are outstanding. I think that this is the result of the people from Skopje, but also the international solidarity.
> (Vasa pomoc Skopju, 1964)

Architecture and urban planning in Skopje after the earthquake operated in the complex framework of politics, ideologies and cultural exchanges and the political meanings attached to them are visible in the reconstruction of the city after 1963. Yugoslavia's position in the NAM certainly influenced these complex relationships.

After 1963, Skopje became a world city, aided by knowledge and technology from both sides of the Iron Curtain and facilitated through Yugoslavia's unique 'non-aligned' position between the rigid East and the capitalist West. Shortly after, the city became a showpiece of the achievements of the architects both from the East and the West, but also through the growing influence of the local architects, a young generation of Macedonian architects that were trained abroad, some of them even working in offices of renowned international architects. Upon their return to Yugoslavia they introduced their foreign experiences to Skopje. Georgi Konstantinovski, for instance, designed the Student Dormitory 'Goce Delčev' (1971–1975) and Skopje's City Archives (1966–1968) after studying at Yale under Paul Rudolph; Janko Konstantinov, who worked for Alvar Aalto in Jyväskylä in 1957, in turn built Skopje's Telecommunications Centre (1974) and Post Office (1979), sculptural buildings from *beton brut* that still dominate the cityscape (Konstantinovski, 2001). Petar Muličkoski designed the building of the Central Committee of the League of Communists of Macedonia in 1970 after his studies in the USA, and Živko Popovski, who worked in the office of Van den Broek and Bakema in Rotterdam, contributing to their competition entry for the reconstruction of Skopje after the earthquake in 1963, following his return to the city designed the City Trade Centre (Figure 7.5).

> After the devastating earthquake, the West and the East competed in Skopje with money and ideas, it was an interesting international game. Western civilization wanted to compete with the East on a different level, on the level of solidarity, with money and teams. The projects for the reconstruction were not solely on a level of architecture and design. The experts thought both sociologically and philosophically, and made a concept for a city that was supposed to be representing what Western civilization was considering to be a democratic and libertarian culture. Skopje needed to be prepared for the XXI century.
>
> (Popovski, 2004)

The earthquake, although disastrous, provided the city of Skopje with the ground for testing novel concepts in accordance to the international tendencies of that time, incorporating knowledge and culture into the projects for the city. Politics and architecture intersected in Skopje after 1963, and its transformation following the earthquake was in itself a modern project, a post-war display of modernity projected over the 'non-aligned' territory

Figure 7.5 Projects for Skopje a) Student dormitory 'Goce Delčev' by Georgi Kontantinovski, b) Telecommunications centre by Janko Konstantinov, c) Central Committee of the League of Communists of Macedonia by Petar Muličkoski, d) City Trade Centre by Živko Popovski and Associates.

Source: Sojuz na drustvata na arhitektite na Makedonija. (1978). *15 godini novo Skopje* Photo: Rumen Kamilov.

180 Jasna Mariotti

of the city. The architectural discourse was a ground for experimentation and at the forefront of international knowledge exchanges in Skopje; its exceptional quality was illuminated through the specific global networks of Yugoslavia during the Cold War and its unique position of in-between.

Bibliography

Ashcroft, B., Griffits, G. and Tifflin, H. (2007). *Post-colonial Studies. The Key Concepts.* London: Routledge.

Boškovski, V. (1965). Skopje so 63,000 Novi Stanovi: Devettiot Proekt ke gi Sodrzi Site Dobri Resenija na Dosega Predlozenite Proekti [Skopje with 63,000 New Apartments: The Ninth Project will Contain all the Good Solutions of the so far Proposed Projects]. *Nova Makedonija,* 30 July 1965, p. A6.

Final Communique of the Asian-African conference of Bandung. (1955). Available at: http://franke.uchicago.edu/Final_Communique_Bandung_1955.pdf [Accessed 20 August 2017].

Fisher, J. (1964). The Reconstruction of Skopje. *Journal of the American Institute of Planners,* 30(1), p. 48.

Gradski odbor na Narodniot front – Skopje. (1949). *5 godini slobodno Skopje* [5 Years Free Skopje]. Skopje: Goce Delcev.

Jordanovski, K. (1993). *Skopje: Catastrophe, Reconstruction, Experience.* Skopje: Matica Makedonska.

Konstantinovski, G. (2001). *Graditelite vo Makedonija XVIII–XX Vek* [Architects in Macedonia XVIII–XX Century]. Skopje: Tabernakul.

Kulić, V. (2009). "East? West? Or Both?" Foreign Perceptions of Architecture in Socialist Yugoslavia. *The Journal of Architecture,* 14(1), pp. 129–147.

Kulić, V. (2017). Building the Socialist Balkans: Architecture in the Global Networks of the Cold War. *Southeastern Europe,* 41(2), pp. 95–111.

Ladinski, V. B. (1997). Post 1963 Skopje Earthquake Reconstruction: Long Term Effects. In: A. Adenrele, ed., *Reconstruction after Disaster: Issues and Practices,* Aldershot: Ashgate Publishing, pp. 73–107.

Lees, L. M. (1997). *Keeping Tito Afloat: The United States, Yugoslavia, and the Cold War.* University Park, PA: The Pennsylvania State University Press, pp. xiii–xviii.

Musil, J. (1980). *Urbanization in Socialist Countries.* White Plains, NY: M. E. Sharpe.

Nova Makedonija. (1961a). Sovest na covestvoto [Conscience of the humanity]. 2 September 1961, p.A1.

Nova Makedonija. (1961b). Deklaracija na Konferencijata na sefovite na drzavite i vladite od vonblokovskite zemji [Declaration of the Conference of Heads of State and Government of the Non-aligned countries]. 7 September 1961, p. A2.

Nova Makedonija. (1961c). Odzivi vo svetot [Responses in the world]. 1 September 1961, p. A3.

Nova Makedonija. (1961d). Tito: Vreme e da se likvidira podelbata po blokovi [Tito: It is time to eliminate the division into blocs]. 2 September 1961, p. A2.

Popovski, J. (1968). Skopje: vcera, denes, utre [Skopje: yesterday, today, tomorrow]. Skopje: Nova Makedonija.

Popovski, Ž. (2004). Mozevme, no ne Napravivme Svetski Grad [We Could Have, but we Didn't Do a World City], *Dnevnik,* 27 July 2004. Available at: http://okno.mk/node/35128/ [Accessed 17 January 2015].

Rubinstein, A. Z. (2015). *Yugoslavia and the Nonaligned World.* Princeton, NJ: Princeton University Press.

Sekulić, D. (2012). Constructing a Non-aligned Modernity: The Case of Energoprojekt, In: M. Mrduljaš & V. Kulić, eds., *Unfinished Modernizations: Between Utopia and Pragmatism*, Zagreb: Croatian Architects' Association, pp. 122–133.

Siani-Davies, P. (2003). Introduction: International Intervention (and Non-intervention) in the Balkans, In: P. Siani-Davies, ed., *International Intervention in the Balkans Since 1995*, London: Routledge, pp. 1–31.

Skopje – grad na solidarnosta [Skopje – city of solidarity]. (1975). Skopje: The City Assembly of Skopje.

Sojuz na drustvata na arhitektite na Makedonija. (1978). *15 godini novo Skopje* [15 years new Skopje]. Skopje: NIP "Nova Makedonija".

Stanek, L. (2012). Introduction: the 'Second World's' architecture and planning in the 'Third World'. *The Journal of Architecture*, *17*(3), pp. 299–307.

Stefanovska, J. and Koželj, J. (2012). Urban Planning and Transitional Development Issues: The Case of Skopje, Macedonia. *Urbani izziv*, *23*(1), pp. 91–100.

The Belgrade declaration of the Non-Aligned countries of 1961. Available at: http://namiran.org/wp-content/uploads/2013/04/Declarations-of-All-Previous-NAM-Summits.pdf [Accessed 5 September 2017].

The Population of Yugoslavia. (1974). Belgrade: Demographic research center, Institute of Social Sciences.

United Nations Development Programme. (1970). *Skopje Resurgent: The Story of a United Nations Special Fund Town Planning Project*. New York: United Nations.

Vasa pomoc Skopju [Your help to Skopje]. (1964). Skopje: Komitet za obnovu i izgradnju Skopja.

8 Atomic urbanism under Greenland's ice cap
Camp Century and cold war architectural imagination

Kristian H. Nielsen

AARHUS UNIVERSITY

City under the ice

One of the 1961 episodes of *The Big Picture*, a television show produced by the United States Army Signal Corps to promote the armed forces and its activities around the world, featured an extraordinary urban settlement, Camp Century in Greenland, presented as city under the ice (*City under the Ice*, 1961). Camp Century, in reality, was a small but strategically significant military research facility built beneath the surface of the ice cap in Northwestern Greenland on Danish territory. Together with four other bases on the ice cap, Camp Century formed part of the US Army Corps of Engineers' Greenland Research and Development Program initiated in 1954 to perform studies of the elements of military engineering in regions of snow, ice, and permafrost (Doel *et al.*, 2016). Camp Century was the first base located entirely below the ice. It provided year-round accommodation for up to 200 soldiers and was powered by an experimental portable nuclear generator developed by the US Army (Suid, 1990). The narrator of the *The Big Picture* described the project in the following terms:

> On the top of the world, below the surface of a giant ice cap, a city is buried. [...] This is an ideal Arctic laboratory, for more than 90% of Greenland is permanently frozen under a polar ice cap, which covers all but a few coastal areas of the island. [...] In this remote setting, less than 800 miles from the North Pole, Camp Century is a symbol of man's unceasing struggle to conquer his environment, to increase his abilities to live and fight, if necessary, under polar conditions (*City under the Ice*, 1961).

A city, an ideal Arctic laboratory, a symbol. These were the terms used to introduce Camp Century in the US Army's campaign to "sell" the camp and the Army's other research and development activities in Greenland to American audiences (Kinney, 2013). Camp Century was an American outpost constructed by Army engineers as part of the effort to afford protection to America's homeland, but also a symbol of Western modernity based on science, technology, and continual frontier exploration and conquest.

Atomic urbanism under Greenland's ice cap 183

Figure 8.1 Cutaway view of Camp Century and its nuclear power plant below icecap by artist and illustrator for National Geographic magazine, Robert C. Magis (used with permission).

As *The Big Picture* episode on Camp Century unfolded, the viewers would learn that the basic concept consisted of 23 trenches dug into the ice cap and then covered with steel arches and snow. There was a main trench, referred to as Main Street, with a series of lateral trenches housing research, laboratory and test facilities, modern living quarters and recreation areas, and a complex of support facilities, including the medium-sized nuclear power plant. Camp Century was, in the words of the *Big Picture* show's narrator, a "complete modern community deep under the ice" (*City under the Ice*, 1961).

Why would the US Army and others refer to Camp Century as a "city" under the ice? Clearly, Century was not a city as one would normally understand the term, but a military base built to accommodate research and development projects related to military polar engineering. As such, it formed part of the geopolitical polar concept developed by US military strategists after World War Two, based on the recognition that the shortest air route between the Soviet Union and the United States crossed the Arctic. It seems reasonable to suspect that the US Army presented Century as a city (and research facility) in order downplay its military aspects and thus make it easier for the Danish Government to accept the presence of the relatively large military structure outside the designated US defense areas in Greenland such as Thule Air Base. In addition, and this is where the symbolic dimensions become important, by using the city metaphor for Century, the US Army offered cultural connotations of homeliness and safety, but also urban domestication of hostile environments such as the ice cap by means of (American) ingenuity and heroic (Army) engineering.

This chapter explores additional meanings attached to the "city" at the time. The main thesis to be developed is that the very notion of the city was fraught with Cold War anxieties. If we consider those anxieties, the resulting image of Camp Century will become very different to the one shown in the *Big Picture* show. Communication officers in the US Army may have seen—and for obvious reasons—Camp Century as a snug and scientific miniature "city" dug into the ice cap. To be sure, the footage from Century showed men working hard and relaxing in comfortable milieus, all heated and supplied with electricity by the most modern and most technically sophisticated energy supply source, the nuclear reactor. By portraying Century as a city, however, the US Army—probably unwittingly—tapped into a highly contested and problematic concept. The story of Camp Century, as a result, was not only a triumphant story about strategy, conquest and security, but also a more paradoxical, even paranoid one involving nuclear deterrence and pervasive nuclear fears (Weart, 2012). To understand how deterrence and fear entered into the urban architectural imagination, Farish's (2003, 2004) notion "anxious urbanism" and Monteyne's (2011) "subsurface urbanism" will be explored.

The city in an age of nuclear annihilation

During an early Cold War period otherwise characterised by unprecedented economic growth, city centres in the United States paradoxically began to decline. Beauregard (1993, 2006) noted that in this period cities underwent "parasitic urbanization" in contrast to the "distributive urbanization" characterising the prewar industrial city. Parasitic urbanisation entailed that suburbs emerged as the most powerful and prosperous areas partly due to government programmes subsidising suburban development by re-siting education, retail and recreation facilities, locating new sources of employment in suburban office parks and industrial estates, and funding new roadways and highways. In addition, post-war economic liberalism promising consumer affluence and increased prosperity led to promises of personal and social mobility tied primarily to the suburbs and not the city centre. The suburbs, moreover, were where the white middle-class could escape the threat posed by the cities and their black inhabitants. Anxieties affiliated with the city grounded in perceived racial differences as well as social class. Suburbs first appeared as para-sites located outside the city and then gradually transformed prevalent ideas about the city itself.

Farish (2003, 2004) contributes to our understanding of the complexities involved in postwar urban development by addressing the consequences of nuclear weapons of mass destruction for urban spaces. He introduced the notion "anxious urbanism" to refer to the tension induced by strategies of containment and nuclear deterrence. US military strategists saw atomic weapons as a means to control large parts of the globe, containing Communist influence as far as possible and deterring the Soviet from attacks on Western capitalist countries. At the same time, after the first successful atomic bomb test by the Soviets in August 1949 and with advent of the even more powerful thermonuclear bombs in 1951, atomic weapons became a palpable threat to the West. The doctrine of massive retaliation, announced by Secretary of State John Foster Dulles in 1954, made military analysts and, almost at the same time, nearly all segments of American society realise that the Cold War could easily release worldwide nuclear Armageddon. Bernard Brodie, one of the most influential strategists of the early Cold War, calculated that 55 thermonuclear bombs alone, easily deployed in just one overnight attack, would destroy the 50 largest Russian cities killing around 35 million people. Moreover, the same calculation applied to Soviet bombs and American cities, which was why Brodie already in 1954 suggested the term "national suicide" to characterise massive retaliation strategy (Steiner, 1991).

The anxieties produced by nuclear strategies such as containment, nuclear deterrence, and massive retaliation inspired many developments in urban planning. New towns across the United States were built to accommodate scientists, engineers, and others working on projects sponsored by the military-industrial-scientific complex (Hales, 1997). A coordinated network of military bases, often planned as small-scale American communities and designed to project US power across the globe were established (Gillem, 2007). And, most perspicuously, the decentralisation of major cities, suburban sprawl, began to take its hold (Beauregard, 2006). The common denominator between such otherwise unrelated developments was the cruel fact of a pending nuclear holocaust and what that meant for US urban development at home and city planning across the world. One immediate response to such atomic fears was urban planning for civil defence, which, as Farish (2010) notes, included everything from suburban dispersal to planning evacuation routes and designing supposedly safe zones in the city and beyond.

The civilian fallout shelter became particularly important to architects and urban planners as they prepared for what Monteyne (2011) calls "hypothetical Hiroshimas" in the United States. The Kennedy Administration in the early 1960s launched The National Fallout Shelter Program, intended to provide "shelter for all". Up until then, efforts in designing and building atom bomb shelters were largely aimed at individual homes. Individual, usually well-off, families provided their own protection from nuclear blasts in sheltered spaces. The national programme targeted entire cities including residential blocks, public and commercial buildings, and public spaces. Cities were being surveyed; they were marked with the characteristic fallout shelter signs and stocked with supplies, sanitation kits, and more. Even though it can be debated how serious this planning for shelters really was and whether it would have effective in the event of nuclear war, urban planners and architects increasingly thought about designing cities for the atomic age and atomic warfare.

The Schoharie Valley Townsite project, originating as a graduate school project at Cornell in 1959, attracted attention as probably the first comprehensive study for protective planning and construction for an entire community. The students and their professors collaborated with IBM and state and national civil defence authorities to select the site and then plan and design for a town of 9,000 people to service an electronic manufacturing facility. The project, which never materialised, included hardened underground areas for dual use. Normally, they were to accommodate daily activities, but in case of a nuclear emergency, they would become "buttoned up" and converted into refuge living spaces. As another precaution against what was called "radical external developments", such as "a nearby explosion of a 20-megatons bomb", the town's power source needed to be as independent as possible. A nuclear reactor was thought to satisfy the criterion of independence and another important consideration, "fail-safe" (Edmondson, 1962).

During the early Cold War, the notion of the city changed radically. As cities became targets for atomic weapons with increasing destructive powers, civil defence and sheltered underground facilities gained in importance in urban planning. Fallout shelters and protected public areas served as safe city spaces, while at same time reflecting fears of total urban destruction.

The very idea of planning a city safe from atomic attacks ultimately indicated that its inhabitants lived at the brink of a nuclear holocaust. Nuclear fears, feeding off perennial themes of transmutation, fiery destruction, transforming rays, and a planet laid waste, were widespread (Weart, 2012). With suburban sprawl and underground shelters, nuclear fears became integral features of the city. Suburbs and underground shelters promised safety and comfort, but also carried an undertone of danger and violence due to their more or less explicit connotations to nuclear holocaust. The atomic city's layout and architecture embodied both optimism and paranoia. Paraphrasing Vanderbilt (2002), we might say that the city had become "a paradoxical netherworld of both security and insecurity".

Militarisation and modernisation in Greenland

There are clear parallels between the paradoxical atomic subsurface urbanism represented by the Schoharie Valley Townsite project and fallout shelters, and Camp Century (Monteyne, 2011). Conceived at about the same time as the Cornell project, Century was much more than a military-scientific base. In the propaganda published by the US Army and in the news media, Century stood for a utopian version of urbanism profoundly shaped by nuclear fears. Camp Century was born out of nuclear deterrence and massive retaliation strategies, which placed Greenland at the geopolitical heart of the Cold War. Yet, Century also demonstrated that life underground (under-ice) could be comfortable and sustainable. The Army even argued that Century would enable its underground inhabitants to survive against the many parameters of atomic attacks, protected by the depth of the icecap. Although the very existence of Camp Century reminded people that nuclear war was imminent, Century, like shelters and other architectural measures against atomic bombs, seemed to contain Cold War threats. In case of a devastating nuclear war, safe and stable communities might even still be possible.

The rise of the Cold War in the late 1940s changed the geopolitical status of the whole of the Arctic and specifically Greenland. Although Greenland during World War Two had served as an island stepping stone for troop movements across the North Atlantic and US weather stations in Greenland had supplied crucial weather information to the Allied forces—allowing them to plan D-day, among other things—Greenland had remained peripheral in the war. Soon after the war, US military geographer Paul Siple wrote to Major General Henry Aurand that

> in reality the Arctic Ocean is a Mediterranean Sea in the middle of the populated land masses of the northern hemisphere. [...] Technically, the research and development to cope with the Arctic will not be insurmountable if we put our minds to it.
>
> (quoted in Nielsen and Nielsen, 2016, p. 197)

Time Magazine on 27 January 1947 published an article with the heading "Deep-freeze defence", showing Alaska and Greenland as joint Arctic fortresses. This idea provided the basis for the polar concept or strategy, which later became enforced by the 1951 agreement between the US and Denmark with respect to the defence of Greenland for the benefit of NATO (Petersen, 2011).

Also in 1951, the United States established Thule Air Base in northeast Greenland. The large-scale engineering operation required about 12,000 men and 300,000 tons of cargo. Thule Air Base strategically served the U.S. Air Force's Strategic Air Command and its first-generation bombers of near-continental range, B-36s and B-47s. Thule was an important operational base for in-flight refueling of B-36s and B-47s until the late 1950s and for strategic reconnaissance missions against the Soviet Arctic. In 1958, as it became clear that the Soviet Union was developing intercontinental ballistic missiles, the US Air Force built the Ballistic Missile Early Warning System (BMEWS), including a major radar facility at Thule Air Base. At the same time, with the deployment of B-52 continental-range bombers, Thule Air Base no longer was required as a refueling base. Greenland, however, maintained an indirect strategic offensive role in the polar strategy with its airspace used on a daily basis by B-52s on Airborne Alert (Petersen, 2011). Thule Air Base remains the northernmost base of the United States, operating as part of a global network of sensors providing missile warning, space surveillance and space control.

The Cold War transformed prevailing ideas about the Arctic. Despite the work of archaeologists and historians demonstrating the continued interconnectedness of Arctic areas with the rest of the world throughout human history, there has been a long tradition for thinking about the Arctic as "a world apart" (McGhee, 2005). Large-scale military engineering project such as Thule Air Base and the Distant Early Warning (DEW) Line, a system of radar stations in Alaska, Canada, Greenland and Iceland, changed all that. The Arctic had become a nuclear frontline, and the threat of total nuclear destruction was pervasive, perhaps even more so than anywhere else. In addition, the amount of pollution generated by military and industrial activities reached new levels, threatening the fragile ecosystem even more. On 17 March 1959, the US nuclear submarine Skate broke the pack ice near the North Pole. The image of military intrusion into the pristine Arctic wilderness ranked with images of mushroom clouds as the most enduring images of the early Cold War (McCannon, 2012).

The early Cold War militarisation of the Arctic was accompanied by efforts to modernise civil society and often the processes were more or less closely linked (Farish and Lackenbauer, 2009). In the case of Greenland, the Danish and the Greenland authorities in 1948–1950 agreed that Greenland should undergo complete and rapid modernisation in order to raise standards of living for ordinary Greenlanders to Danish levels. The decision was influenced by the unsuccessful American attempt to buy Greenland from Denmark in 1946, and as a response to international calls for decolonisation.

In 1953, Greenland officially became part of the Danish realm and equated with other counties in Denmark. The modernisation programme encompassed all areas, from education and health to culture and industry. The capital of Greenland, Nuuk (until 1979, Godthåb), and other settlements on the west coast flourished, not the least because of booming cod and shrimp fisheries. Modernisation ultimately affected most of Greenland (Nielsen, 2017).

When British writer Geoffrey Williamson visited Greenland in the summer of 1952, he was impressed by the speed with which Greenlanders seemed to be embracing modern ways of life (Williamson, 1953). In the course of a decade, it seemed, they had abandoned traditional seal hunting and adapted to factory conditions, motorboats and workshops. A genuine enterprising spirit permeated Greenland with new fish processing plants being built and mining consortia exploring the extraction of minerals. Williamson observed with some amusement that so many Greenlandic girls preferred Western commodities such as nylon stockings, high-heeled shoes, cigarettes and chewing gum. He was surprised at the level of education and language skills in Greenland, where he noticed Greenlandic translations of many Western books, including his own childhood favourites. Williamson concluded that an Arctic renaissance was happening. It would no longer be possible to dismiss Greenland as a remote, frozen territory "out of this world", because Greenland was changing by tapping into international forces of economic, technological and cultural change (Nielsen, 2016). Greenland was becoming modern.

Nuclear Greenland

When the US Administration in late 1958 and early 1959 first approached Denmark about its plans to build Camp Century as a subsurface, nuclear-powered camp, Greenland already had become the focal point for discussions about nuclear issues, mostly military ones. Greenland was Danish territory and thus governed by Danish laws on nuclear weapons and nuclear power, which were in the process of being worked out at the time. Danish nuclear policies, as it turned out, were highly ambiguous. Denmark, one of the founding members of NATO, supported nuclear deterrence as a military strategy. Yet, in the course of 1957, the Danish Government adopted a policy of not permitting nuclear weapons "under the present circumstances" (Agger and Wolsgaard, 2006). Denmark was being pressured by the Premier of the Soviet Union, Nikolai Bulganin, who in an open letter to the Danish Prime Minister H.C. Hansen warned Denmark strongly against allowing nuclear weapons on Danish territory. Also, anti-militarist ideologies represented by the Social Liberal Party in the Government were opposed to nuclear armament. As regards peaceful uses of nuclear power, Denmark got off to a late start as nuclear research and development activities generally lacked support in the Danish innovation system, and in 1985 the Danish utilities and the Danish Parliament withdrew their support of nuclear power in Denmark (Nielsen and Knudsen, 2010).

Ambiguity also permeated questions relating to nuclear issues and Greenland. In 1957, after the Bulganin letter and the first announcement of the Danish no-nuclear policy, the US Administration made an informal inquiry to the Danish Prime Minister H.C. Hansen whether the Danish Government would wish to be informed in the event that the United States should place nuclear weapons in Greenland. It was a very sensitive issue, and Hansen decided to respond in an informal, personal, highly secret written statement that would be limited to one copy each on the Danish and American side. He did not inform or consult either the Danish Parliament or The Foreign Policy Committee—the latter, according to Danish law, must be consulted by the Government on matters of major importance to foreign policy. Hansen simply told the Americans, "I do not think that your remarks give rise to any comments from my side". The US Administration, and probably rightly so, interpreted it as full acceptance of US deployment of nuclear weapons at Thule Air Base (Petersen, 1998).

The 1951 agreement about the defence of Greenland said that Denmark must approve of all US activities outside Thule Air Base and a few other designated defence areas. Very few people, if any, at the time imagined that less than ten year later, a nuclear reactor would be installed in tunnels under the icecap. In 1954, the US Army Corps of Engineers launched its Greenland Research and Development Program to consider the engineering problems involved in operating on the icecap. The programme included many military-scientific activities on the icecap, all of which had been approved by the Danish authorities prior to their implementation. Proper field investigations on the Greenland Ice Sheet began at Site 2, about 350 km east of Thule Air Base, where Army engineers studied the feasibility of building military camps under the ice. Using Peter Snow Millers, snow removal machines originally used to clear roads in the Swiss Alps, they excavated trenches in the ice and experimented with various types of roofing. In 1957, the first subsurface camp, now designated Camp Fistclench, was completed at Site 2 with a capacity to support about 150 persons (1st Arctic Task Force, 1955; U.S. Army Engineer Task Force, 1957).

Based on the positive outcome of the feasibility study at Camp Fistclench, the US Army planned for a larger subsurface camp with a proposed location about 100 miles east of Thule Air Base and therefore named Camp Century. In the spring of 1959, just over a year after Prime Minister H.C. Hansen, going against the official Danish policy, had secretly allowed US nuclear weapons to be deployed at Thule Air Base, the Army applied for permission to build an extended subsurface camp with its own nuclear reactor installed. The nuclear reactor, part of the Army's experimental nuclear power programme, was to be small and portable and at the same time would be the first nuclear reactor in Denmark. The Danish authorities were concerned.

They worried that since the Army clearly was investing heavily in the subsurface concept there might be more to it. They also feared that the very mention of the "concept of atoms" in relation to Greenland might induce new public discussions about Denmark's commitment to NATO's strategy of nuclear deterrence and Denmark's proper path to peaceful uses of nuclear power (Nielsen and Nielsen, 2017).

Although the Danish authorities did try to persuade the US Administration not to go ahead with building Camp Century in Greenland and rather to conduct the experiment in Alaska, there was too much at stake to stop the project. The US Army even went ahead and dug the first trenches in the summer of 1959 before Denmark had officially granted permission. When the Danish authorities realised that eventually news about Camp Century would reach the Danish public, they announced in late-August 1959 that there were Danish plans to use nuclear power in relation to the lead-zinc mine in Mestersvig located on the east coast of Greenland. Therefore, when the news about Camp Century, the nuclear-powered camp under the icecap, broke in the Danish media in November 1959, the Danish Foreign Ministry confirmed the information, adding that nuclear power was to become an important part of the power supply in Greenland. This turned out to be the first instance of the Danish authorities' sustained attempts to "contain" news coverage of Camp Century by always emphasising the peaceful uses of nuclear power in Greenland and downplaying the camp's military significance (Nielsen and Nielsen, 2013).

Building Camp Century

The 100-man crew arrived at the planned location of Camp Century—for a number of reasons it turned out to be not 100 miles, but 138 miles or 220 km east of Thule Air base—in June 1959. They cut trenches about 10 meters deep, covered them with corrugated steel arches, and finally used the Peter Snow Millers again to seal in the trench roofs with loose snow that soon would harden enough to allow for removal of the steel arches (Figures 8.2 and 8.3). At Camp Fistclench, the Army engineers had learned that flat structural roofs over snow trenches were incompatible with the ever-increasing snow load for which they had to be designed. It was also here that they developed the so-called undercut trench concept, according to which the trenches were wider at the floor than at the bottom. Undercut trenches provided maximum floor space with minimal overhead clearance. The main trench and the trenches containing the nuclear power plant were wider than other trenches, and steel arches were in place throughout the operation of Camp Century (Clark, 1965).

Originally, the question of whether to construct camps beneath or above the surface was not an easy one to answer for the US Army engineers. The harsh environment with severe storms and extremely low temperatures presented obvious challenges for conventional construction methods. The

192 *Kristian H. Nielsen*

Figure 8.2 The Peter Snow Miller in operation cutting a trench in the ice. US Army Photos (public domain).

Figure 8.3 Trench partially covered with steel arch roofs. The US Army Signal Corps' Pictorial Service team is filming in the trench. US Army Photo (public domain).

inevitable accumulation of drifting snow would mean that above-surface structures would soon become subsurface ones. This was a condition for which they were not designed, and ultimately they would fail due to the stresses of an ever-increasing overburden. Subsurface camps, on the other hand, while providing some shelter from the cold winds and drifting snow, would also have limited life span due to tunnel deformation caused by the viscoelastic properties of snow. Maintenance of the original geometry and cross section of the tunnel could be achieved with a continuous snow-trimming regime, but eventually the camp would become buried to a depth where the deformation rate was so great that the trimming task would become unacceptably burdensome and costly. Nevertheless, subsurface camps were preferred since they were less expensive to build and easier to camouflage (Clark, 1965).

The designers of Camp Century planned for a series of parallel trenches housing living quarters, kitchen and mess hall, toilets and showers, scientific laboratories, the nuclear reactor power plant, and other essential services. The camp was a more or less enclosed system with separate sumps for sewage, located about 45 meters from trench 19 (living quarters), and for radioactive liquid waste disposal (up to 50 millicuries per year according to the Danish-American agreement over the operation of nuclear reactor). All solid radioactive waste was to be removed from Greenland and disposed of in accordance with national regulations provided by the Atomic Energy Commission. In addition, the camp had its own water well supplying melted glacial ice, and it boasted the world's largest deepfreeze. Most of these techniques, except for the disposal of radioactive liquid waste were previously tested at Camp Fistclench (Clark, 1965).

Camp Century was designed to protect its inhabitants from the hostile natural environment, but also from hostile enemy forces by means of effective camouflage under the icecap. On the icecap, visible facilities such as the DEW Stations, built using telescopic structures, were clearly visible against the white and featureless icescape. Although the ramps leading to the surface and the meteorological hut located "topside" revealed the location of Camp Century, future subsurface camps deliberately and systematically could be camouflaged for the safety of the military personnel who might take shelter there. Moreover, it was demonstrated that the ice sheet could provide some protection from missile attack. At Camp Century, scientists and engineers from the US Army Engineer Waterways Experiment Station performed high-explosive tests to simulate the effects of nuclear weapons on terrain and structures on and inside the ice sheet. As expected, the snow did seem to attenuate the blast from above-surface detonations effectively (Joachim, 1967). The experiments did not consider the effects of radiation.

The most pressing engineering design problem was tunnel deformation. The Army engineers were fully aware that the Greenland icecap is not a fixed structure, but steadily flows from the centre towards the sea. Trenches dug into the icecap, therefore, would gradually deform and overhead clearance would be gradually lost. Eventually, the trenches would collapse. It was also known that complex interactions between several variables such as temperature, stress, density and depth determine deformation rates. These variables meant that the expected life span of Camp Century was ten years. After this period, the camp would be buried to a depth where the viscous flow of the icecap would be so great that it would be too expensive to maintain the trenches by trimming. For various reasons, including the fact that heat accumulated in the trenches and accelerated snow deformation, the rate of trench closure was higher than expected (Clark, 1965).

The nuclear city under ice

As mentioned in the opening of this chapter, the US Army eagerly sought to present Camp Century as an actual city under the ice. The Army's own film about Century showed the men living comfortably inside heated and spacious buildings where generous portions of food were being served to satisfy the enormous appetite that working in the cold climate produced, and leisure activities such as ping-pong, film screenings and music were available. Although the film did emphasise that this was a unique Arctic research installation inhabited by military scientists and soldiers, it also portrayed Century as a completely modern community. Except for the fact that the men lived deep under the ice with no windows and in near-isolation from the rest of the world, they were able to enjoy facilities for every modern convenience, in part due to the nuclear reactor at Camp Century (*City Under the Ice*, 1961).

The Army cast Camp Century as a technological spectacle, similar in many ways to the ways in which the Navy's new nuclear submarines were publically presented at the time. The first operational nuclear-powered submarine, USS Nautilus, completed a submerged transit of the North Pole in the summer of 1958. The Navy carefully avoided mention of the military ramifications of Nautilus' mission, known as Operation Sunshine. Rather, Operation Sunshine was described as a scientific expedition and as an opening up of a new commercial seaway benefitting shipping enterprises everywhere. Reports mentioned the relative comforts experienced by Nautilus' crew such as private bunks, expanded areas for mass and recreation, and closed circuit TV, even as the submarine glided through icy Arctic waters. To avoid the risk that the crew was seen as mere extensions of the amazing submarine, the Navy's press reports made sure to highlight the men's technical expertise and courage (Griffin, 2013).

Camp Century was a nuclear-powered city and thus reflected some of the ambiguities and anxieties involved in early Cold War atomic urbanism. Century's nuclear power plant was an experimental, modular nuclear reactor developed in the course of the nuclear power programme led by U.S. Army Engineer Reactors Group (Suid, 1990). Like other US Army nuclear power plants, it was designed to generate heat (10 MW) and electricity (2 MW) at remote, inaccessible places where delivery of oil or coal would be logistically difficult to accomplish. The plant, known as PM-2A, cost nearly $6 million out of Camp Century's total $8 million budget. Although the plant has many start-up problems, including "unacceptably high" radiation levels, which were lowered by means of additional shielding of the reactor, the PM-2A demonstrated the validity of the Army's nuclear power programme. Its successor, the PM-3A, powered the McMurdo base at Antarctica from 1962 to 1972, suffering numerous malfunctions (Suid, 1990, pp. 67–69).

Initially, most of the men at Camp Century were unhappy about living in ice tunnels near to a nuclear reactor. The fact that soon after the plant had been put into operation the crew discovered radiation levels exceeding design limits probably increased their concerns. The first of the many journalists who visited Camp Century, Walter Wager, attributed the men's uneasy notions about radiation to normal fears of atomic blasts shared by everyone at the time. However, he added, as the men learned, from trained nuclear experts, about all the safety devices built into the reactor and as they became busy with their daily concerns in the "tunnel town", they soon put their anxieties to rest. In Wager's account, the men soon learned to appreciate that they lived next to a "modern miracle, a remarkable achievement in which all Americans can take pride". The combination of subsurface living and nuclear power was not only key to opening up the entire Arctic and Antarctic, it would also make the idea of settlements in space less fanciful (Wager, 1962, pp. 81, 131).

Wager's, and many other journalists', stories about Camp Century made it clear that Century was a nuclear outpost on the Arctic frontline of the Cold War. Even if the Danish authorities, who under agreement with the United States were entitled to comment on all visiting journalists' drafts before publication, tried their best to get the journalists to emphasise Century's scientific dimensions rather than the military ones, they were not successful (Nielsen and Nielsen, 2013). Wager, for example, noted that, despite its subsurface location, Camp Century was visible from the air. Since there was nothing between Greenland and the missile and bomber power of the Soviet Union, all US bases in Greenland were a prime target. "The stark but strangely beautiful island may be ravaged by an aggressor's modern weapons of mass destruction," he concluded (Wager, 1962, p. 121).

In response to the nuclear threat from the Soviet Union, the US Army and US Air Force, according to Wager (1962), were considering future Century-like "combat holes". Such Polar holes could be used for radar detachments, squadrons of jet fighters, defensive anti-ballistic missile batteries, airborne commando teams and retaliation missiles. Wager said that it was all pure notion. What he probably did not know and what very few people knew was that the US Army in fact were proposing to build mobile deployment of intermediate-range ballistic missiles in tunnels under the ice cap. The so-called Project Iceworm, conceived by the Planning Studies Division in the Army's think-tank, The Engineering Studies Center, but never realised, was presented as an alternative to basing the Navy's Minuteman missiles in hardened silos within the United States. Project Iceworm would base a modified version of Minuteman known as the Iceman in a large network of tunnels dug in the icecap using the construction techniques tested at Camp Century. Located closer to the Soviet Union, Iceman missiles would be able to retaliate much faster and with greater accuracy than Minuteman. Dug into the ice cap and its critical element, Iceman missiles, moveable, Project Iceworm would be hidden, elusive and relatively difficult to target (Baldwin, 1985; Petersen, 2008).

Anxious atomic urbanism under the ice

Walter Wager, one of the journalists who stayed for a brief period with the iceworms as the men of the subsurface installation were called, saw Camp Century as an exemplar of future cities. As the most remarkable scientific and technological achievement in Polar Regions, Camp Century produced knowledge that, it was argued, would benefit all of mankind. With the rapid increase in global population, it would not be too farfetched to imagine domed or even subsurface communities on empty land in the Arctic. Larger versions of Century's PM-2A nuclear power plant would be able to generate enough heat and electricity, including power for melting off enough of the icecap to supply water for the needs of tomorrow's future polar towns. The only thing missing to sustain proper settlements in such remote and hostile settings, Wager half-jokingly continued, was women. "With plenty of atomic power and lots of feminine company—a combination that seems highly promising—there may be a continuous encroachment of civilization upon the icecaps" (Wager, 1962, pp. 98–100).

Wager's extrapolation, as we now know, never came true, but it seemed to have held some currency at the time. In 1964, the General Motors interactive exhibit Futurama II at the New York World's Fair included a succession of techno-colonies in space, in the deep sea, in remote deserts, and on and under the Antarctic ice sheet (Figure 8.4). Futurama II was one of the main attractions of the fair and told visitors that establishing communities in such extreme environments would be necessary to accommodate an ever-growing human population, and was made possible by scientific research and advanced technologies. In the future, people would live wherever they wanted with neither terrain nor distance to deter "the man of the city" from building his home. Thanks to nuclear power in particular, Antarctica in the future would become a land of "growing communities". As with its predecessor, the Futurama designed by Norman Bel Geddes for the 1939 New York World's Fair, Futurama II stressed the virtually unlimited capabilities of American technology, which would transform wastelands into habitable environments (Vanderbilt, 2002).

Atomic urbanism under Greenland's ice cap 197

Figure 8.4 Scenes from Futurama II depict "growing communities" in under-ice laboratories and settlements in Antartica. Uncredited/AP/Ritzau Scanpix (used with permission).

The very idea of building subsurface, nuclear-powered cities in inaccessible and hostile environments such as Greenland or Antarctica was a Cold War techno-fantasy evoking ambiguous and even anxious urban connotations. The community at Camp Century was cozy and comfortable, and the men living there were brave pioneers of tomorrow's cities. However, the nuclear technology that made such communities possible also generated fears of radiation and possible annihilation. Camp Century existed for the sole purpose of extending NATO's nuclear deterrence strategy to the Arctic battlefront, and the Army made sure to test Century for its potential capability to withstand nuclear attack. Camp Century was a highly ambiguous, even paranoid architectural and engineering accomplishment in the sense that it was designed to contain or reduce anxieties about nuclear war, yet at the same time served to produce or reinforce those very anxieties.

Today, Camp Century is buried deep under the ice. The Army engineers left it in 1965 after having removed the nuclear reactor the year before. They thought the remains of Camp Century would remain under the ice for eternity sinking deeper and deeper into the Greenland ice cap as more snow would fall. What they did not—and could not—take into account was the accelerating global warming that someday would threaten to reverse that development. In 2016, a team of researchers led by William Colgan (2016) published climate model simulations and different soundings of the Camp Century site indicating that in about 80 years' time melting of the icecap would penetrate down to the depth of Century. All other things being equal, this would mean that polluted meltwater could get into the sensitive Arctic Ocean. Camp Century, once "a symbol of man's unceasing struggle to conquer his environment, to increase his abilities to live and fight, if necessary, under polar conditions", now has become a symbol of the unintended consequences of military-technology developments under the Cold War.

Bibliography

1st Arctic Task Force (1955) *After Operations Report. Research and Development Program, Greenland*, Fort Belvoir, VA: Defence Technical Information Center. Available at: http://www.dtic.mil/docs/citations/AD0123338 (Accessed: 18 March 2018).

Agger, J.S. and Wolsgaard, L. (2006) "All Steps Necessary: Danish Nuclear Policy, 1949–1960," *Contemporary European History*, 15(1), 67–84.

Baldwin, W.C. (1985) *A History of the U.S. Army Engineer Studies Center 1943–1982*, Fort Belvoir, VA: US Army Engineer Studies Center.

Beauregard, R.A. (1993) *Voices of Decline: The Postwar Fate of US Cities*, Oxford and Cambridge, MA: Blackwell.

Beauregard, R.A. (2006) *When America Became Suburban*, Minneapolis, MN: University of Minnesota Press.

City under the Ice (1961) film produced by the US Army Signal Corps Pictorial center, 30 mins. Available at: https://archive.org/details/gov.archives.arc.2569752 (Accessed: 18 March 2018).

Clark, E.F. (1965) *Camp Century: Evolution of Concept and History of Design, Construction and Performance*, Hanover, NH: U.S. Army Research and Engineering Laboratory. Available at: http://www.dtic.mil/dtic/tr/fulltext/u2/477706.pdf (Accessed: 18 March 2018).

Colgan, W., Machguth, H., MacFerrin, M., Colgan, J., van As, D. and MacGregor, J. (2016) "The Abandoned Ice Sheet Base at Camp Century, Greenland, in a Warming Climate," *Geophysical Research Letters*, 43(15), 8091–8096.

Doel, R.E., Harper, K. and Heymann, M. (eds.) (2016) *Exploring Greenland: Cold War Science and Technology on Ice*, New York: Palgrave Macmillan.

Edmondson, F.W. (1962) "Design of a Nuclear City," in *Design for the Nuclear Age. Proceedings of a Conference Held as Part of the 1961 Fall Conferences of the Building Research Institute, Division of Engineering and Industrial Research*, Washington, DC: National Academies of Science, 74–89. Available at: https://www.nap.edu/catalog/21569/ (Accessed 18 March 2018).

Farish, M. (2003) "Disaster and Decentralization: American Cities and the Cold War," *Cultural Geographies*, 10(2), 125–148.

Farish, M. (2004) "Another Anxious Urbanism: Defense and Disaster in Cold War America," in Graham, S. (ed.) *Cities, War, and Terrorism: Towards an Urban Geopolitics*, Oxford: Blackwell, pp. 93–109.

Farish, M. (2010) *The Contours of America's Cold War*, Minneapolis, MN: University of Minnesota Press.

Farish, M. and Lackenbauer, P.W. (2009) "High Modernism in the Arctic: Planning Frobisher Bay and Inuvik," *Journal of Historical Geography*, 35(3), 517–544.

Gillem, M.L. (2007) *America Town: Building the Outposts of Empire*, Minneapolis, MN: University of Minnesota Press.

Griffin, C.J.G. (2013) "'Operation Sunshine:' The Rhetoric of a Cold War Technological Spectacle," *Rhetoric & Public Affairs*, 16(3), 521–542.

Hales, P.B. (1997) *Atomic Spaces: Living on the Manhattan Project*, Urbana and Chicago: University of Illinois Press.

Joachim, C.E. (1967) *Shock Transmission through Snow and Ice*, Technical Report No. 1–794, Vicksburg, MI: U.S. Army Engineer Waterways Experiment Station, Corps of Engineers.

Kinney, D.J. (2013) "Selling Greenland: The Big Picture Television Series and the Army's Bid for Relevance During the Early Cold War," *Centaurus*, 55(3), 344–357.
McCannon, J. (2012) *A History of the Arctic: Nature, Exploration and Exploitation*, London: Reaktion.
McGhee, R. (2005) *The Last Imaginary Place: A Human History of the Arctic World*, Chicago, IL: University of Chicago Press.
Monteyne, D. (2011) *Fallout Shelter: Designing for Civil Defense in the Cold War*, Minneapolis: University of Minnesota Press.
Nielsen, H. and Knudsen, H. (2010) "The Troublesome Life of Peaceful Atoms in Denmark," *History and Technology*, 26(2), 91–118.
Nielsen, H. and Nielsen, K.H. (2013) "Inddæmning og tilbagerulning: Om danske myndigheders censur af presseomtale af amerikanske militære aktiviteter i Grønland, 1951–63," *temp - tidsskrift for historie*, (7), 141–162.
Nielsen, H. and Nielsen, K.H. (2016) "Camp Century: Cold War City Under the Ice," in Doel, R.E., Harper, K.C. and Heymann, M. (eds.) *Exploring Greenland: Cold War Science and Technology on Ice*, New York: Palgrave Macmillan, pp. 195–216.
Nielsen, H. and Nielsen, K.H. (2017) *Camp Century: Koldkrigsbyen under Grønlands indlandsis*, Aarhus: Aarhus Universitetsforlag.
Nielsen, K.H. (2016) "Small State Preoccupations: Science and Technology in the Pursuit of Modernization, Security, and Sovereignty in Greenland," in Doel, R.E., Harper, K.C. and Heymann, M. (eds.) *Exploring Greenland: Cold War Science and Technology on Ice*, New York: Palgrave Macmillan, pp. 47–71.
Nielsen, K.H. (2017) "Cod Society: The Technopolitics of Modern Greenland," in Körber, L.A., MacKenzie, S. and Stenport, A.W. (eds.) *Arctic Environmental Modernities: From the Age of Polar Exploration to the Era of the Anthropocene*, New York: Palgrave Macmillan, pp. 71–85.
Petersen, N. (1998) "The H.C. Hansen Paper and Nuclear Weapons in Greenland," *Scandinavian Journal of History*, 23(1–2), 21–44.
Petersen, N. (2008) "The Iceman that Never Came – 'Project Iceworm,' the Search for a NATO Deterrent, and Denmark, 1960–1962," *Scandinavian Journal of History*, 33(1), 75–98.
Petersen, N. (2011) "SAC at Thule: Greenland in the U.S. Polar Strategy," *Journal of Cold War Studies*, 13(2), 90–115.
Steiner, B.H. (1991) *Bernard Brodie and the Foundations of American Nuclear Strategy*, Laurence: University Press of Kansas.
Suid, L.H. (1990) *The Army's Nuclear Power Program: The Evolution of a Support Agency*, New York: Greenwood Press.
U.S. Army Engineer Task Force (1957) *Greenland Research and Development Program. 1957 After Operations Report*, Fort Belvoir: Department of the Army, Corps of Engineers.
Vanderbilt, T. (2002) *Survival City: Adventures among the Ruins of Atomic America*, New York, NY: Princeton Architectural Press.
Wager, W.H. (1962) *Camp Century: City Under the Ice. Story of Our Incredible Polar Base Below the Greenland Ice Cap*, Philadelphia, PA and New York: Chilton Books.
Weart, S.R. (2012) *The Rise of Nuclear Fear*, Cambridge, MA: Harvard University Press.
Williamson, G. (1953) *Changing Greenland*, London: Sidgwick and Jackson Ltd.

Welfare or Warfare?
Civil defence in Danish urban welfare architecture

Rosanna Farbøl

UNIVERSITY OF SOUTHERN DENMARK

From June 1951 and once a week throughout the Cold War, the testing of more than 1,000 sirens reminded Danish city dwellers that the Cold War could turn hot any minute. Once more, the country risked occupation, this time not by Germany, but by the Soviet Union.[1] From the 1950s, Warsaw Pact war plans included the conquest of Denmark by Soviet and Polish forces in order to control the only shipping route from the Baltic Sea to the North Sea, and after 1961, the use of nuclear weapons became a constant element in the attack rehearsals (Jensen 2014, p. 86). As World War II had demonstrated, civilians had become a target on par with military installations, and total war in the age of nuclear weapons threatened impending doom for entire populations.

In Denmark, the threat of nuclear war led to the creation of the Civil Defence Service in 1949, which established a network of shelters, warning systems, emergency stockpiles, ambulance and firefighting services, planned activities such as evacuation drills, and created educational material. Gradually from the late 1950s, the authorities developed a protocol for civil emergency planning that described the maintenance of essential public utilities, including electricity, water, food supplies, transportation, and communication. While Civil Defence was public and dependent on volunteers, civil emergency planning was mostly carried out in secrecy.

The Cold War was, however, not just a time of tension and nuclear anxiety in Denmark. It was also an optimistic period characterized by affluence and political, social, and cultural transformation. The economic boom of the late 1950s through to the 1960s facilitated the expansion of the universal welfare system covering health care, social security benefits, old-age pensions, child care, and free education (Wium Olesen 2017). The Social Democratic governments' Keynesian doctrine of fiscal stimulus accelerated the building of new schools, hospitals, day-care institutions, and social housing.

In Cold War Denmark, these visions of nuclear dystopia and welfare utopia dovetailed in urban architecture. Buildings central to the welfare state were typically designed to include nuclear civil defence features, meaning that Cold War security policy and welfare policy overlapped in concrete form: as Jennifer Light has demonstrated (2003), Cold War civil defence was a nexus where welfare and warfare met. David Monteyne has examined the architecture and aesthetics of civil defence designs and buildings (2011). He argues that welfare and civil defence institutions alike were concerned with sheltering, nurturing, and healing people (one could add educating), and that both were designed to produce and preserve valuable citizens for (re)building the nation (p. 187).

This visual essay presents a selection of separate cases of Danish urban civil defence architecture, all of them constructed in ordinary welfare state institutions for living, working, learning, recreation, and convalescing. Whereas civil defence architecture in the USA, as Monteyne shows, was an important ideological tool for the government to demonstrate that "... plans for protecting citizens in the imagined aftermath of nuclear attack were based on the material realities of building construction and everyday spaces" (p. xii), in Denmark this endeavour was more subtle. In Danish buildings, there were no signs telling residents or passers-by that they were now in a fallout shelter: such signs would only be put up in times of imminent crisis. Civil defence architecture, moreover, would often be integrated – perhaps even hidden – in already existing and accepted welfare institutions, making civil defence less visible, offensive, and alarming.

This essay illustrates that through civil defence, the Danish state planned to not only control but also protect, and *care for*, its citizens in the event of nuclear war. This was in contrast to, for example, the British government, who – while simultaneously creating a secret network of government bunkers to run the country in the event of nuclear war – advised citizens to 'protect and survive' in improvised shelters at home (Grant 2010; Cordle 2012; Stafford 2012; Deville, Guggenheim & Hrdlicková 2014, pp. 190–191). The Danish welfare state, through placing emphasis on protecting its citizens through a system of civil defence and civil emergency planning, designed buildings that served as unobtrusive but constant reminders of nuclear danger; solidifying and setting in stone (or rather concrete) the welfare state's promise that it would protect its citizens from cradle to grave.

The city as target

Extensive preparations were made to take care of the inhabitants in Aarhus, if the city was attacked, as demonstrated by Figure VE.2.1. Civil defence efforts would be led from the municipal control centre (*Kommandocentral*, abbreviated KC on the map). Besides the control centre, ten observation posts (*Observationsposter*) were established at different strategic locations, for instance the city hall tower. Should it be necessary to enact a mass civilian evacuation, more than 40 aid stations (*Hjælpestationer*, abbreviated HJST) would be ready to provide safe place for a short rest, refreshments and medical care. Nine turn-out stations (*Udrykningsstationer*, URS) were prepared, and here the civil defence personnel would assemble in ambulance, firefighting, and salvage corps teams. Depots were places at various schools around the city,

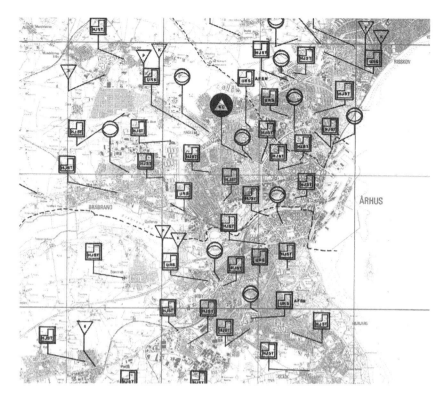

Figure VE.2.1 Aarhus civil defence plan 1983.
Source: Aarhus Stadsarkiv/Erhvervsarkivet.

storing among other things stretchers, firefighting equipment, gas masks, and 135 kilometres of fire hose. Besides equipment for firefighting and rescue, civil defence depots also contained everything people made homeless by the attack would need to survive, such as food, mattresses, blankets, clothes, candles, matches, soap, tableware, infant formula, and even romper suits.

Aarhus, the second largest city in Denmark, had an important role to play as it housed the Western regional control centre from where the military and civil defence would be organized, if Copenhagen was occupied or destroyed by bombing (Hansen, Tram Pedersen & Stenak 2013, p. 45). The municipal control centre in Aarhus was constructed between 1952 and 1954. Here, the mayor, police, local civil defence forces, an ABC-officer (the officer with special knowledge of Atomic, Biological and Chemical weapons) as well as the directors of the energy and water distribution networks would assemble to try to manage the area during crisis or war. Their main task was to keep society functioning as routinely as possible for as long as possible, and to minimize the impact of war on the population. In every city with more than 10,000 inhabitants, such a control centre was established, and most were built in the 1950s and 1960s (Holt Pedersen 2013, p 108; Holt Pedersen & Pedersen 2014, pp. 260–275).

The map does not show the 35 large public shelters that could accommodate an estimated 10% of the population, or the more than 300 smaller public shelters, accommodating 50 persons each, or the unknown number of private shelters (Stenak 2013; Holm 2014). By the end of the Cold War, Aarhus Civil Defence planned to have a surplus of almost 20,000 seats in various types of shelters. Many of these, however, were designed and projected but never actually prepared.

From 1949, all public buildings in Denmark were required to have public air-raid shelters. The construction was financed, in whole or in part, by the state. In the wake of the Korean War, more shelters were built and existing World War II shelters were renovated, but they were not adapted to thermonuclear warfare before 1964–1965; the wooden doors were replaced with doors of steel, and new ventilation systems with sand filters were installed. They were supplied with the most necessary survival equipment such as radios, earth closets, and water containers as well as hammers and chisels in case the exit were destroyed (Stenak 2013; Hansen 2015, p. 42). In total, the various forms of public and private shelters could accommodate 4.2 million people of a population of a little less than 4.6 million people in 1960 (danmarkshistorien.dk 2015).

Housing and sheltering citizens: *Gellerupplanen*

In Aarhus, the social housing project *Gellerup Plan* (Figure VE.2.2) is one example of how urban ideals of "the good life" had to incorporate the prospect of atomic warfare. It was a prestigious project, conceived in the late 1950s, although the building only began in 1967 (Høghøj & Holmqvist 2018). Around 3,000 apartments were built in precast concrete construction of four- and eight-storey blocks. The overall design was in a Le Corbusian-inspired style of modernism and functionalism. In Denmark, this architectural style was closely associated with ideals of internationalism, equality, solidarity, welfare, modernity, and improved public health (Bæk Pedersen 2005).

The non-profit housing association behind the project, *Brabrand Boligforening*, offered spacious and technologically sophisticated apartments, along with good collective sports and leisure facilities, playgrounds and a library, as well as a shopping centre with shops, banks, a post office, pubs, and restaurants. Surrounding the apartment blocks were open spaces and a system of paths for pedestrians and cyclists.

In the basements of the blocks were cycle stores that had been fitted with emergency exits. An architectural blueprint from one of the low-rise blocks demonstrates that the cycle store had a dual purpose: storerooms in peacetime, and shelters in case of war.

Figure VE.2.2 Gelleup 1975.
Source: Den Gamle By/Århus Luftfoto.

The construction of private *reinforced rooms*, in effect provisional shelters, had become mandatory by law in newly built housing for more than two families in cities.[2] These were rooms in the basement that could be transformed into an air-raid shelter within 24 hours. In older buildings, supplementary provisional shelters were appointed. Boosted by the general building boom in the 1960s, the number of seats in provisional shelters had reached 3,743,822 across Denmark at the end of the Cold War.

This particular block in Gellerup (Figure VE.2.3), and the ones identical to it, had six *reinforced rooms*. In total, the shelters could accommodate 245 people, corresponding to the number of inhabitants in the block. The welfare utopia of Gellerup had, literally, concrete preparations for a nuclear apocalypse built into its foundations.

Schoolchildren and refugees

The children shown in Figure VE.2.4 at *Nørrevangsskolen* in Slagelse, some 100 kilometres south-west of Copenhagen, had no idea that their school was built to provide more than their welfare state-sponsored education. Below the classrooms, provisional shelters were established for the children and local residents. Schools all over the country were built with basements that

206 *Rosanna Farbøl*

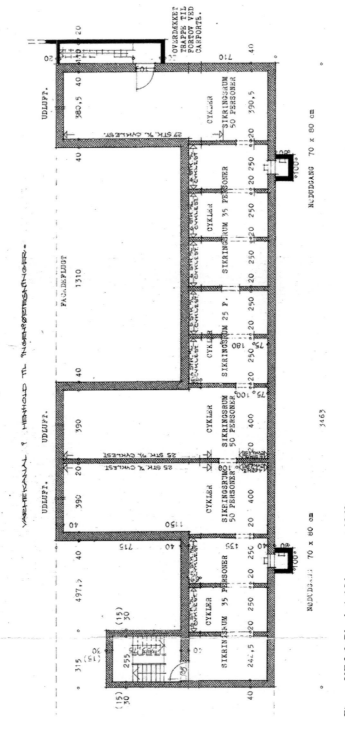

Figure VE 2.3 Block A 15 basement 1969.
Source: Brabrand Boligforening.

Figure VE 2.4 Steen Rasmussen, published in Sjællands Tidende 24-01-1970.
Source: Printed with the permission of Slagelse Stads- og Lokalarkiv.

were double reinforced and bombproof with 50 cm concrete between floor plan and basement (Hansen 2015, pp. 42–43). In Slagelse, a purpose-built ramp from the outside made it easy to get to the underground shelters. The building, however, had another special role in the regional Cold War preparedness programme, as it was designed to function as a refugee camp in a time of crisis.

Danish authorities expected that a large number of people, particularly from the major cities, would flee at the prospect of atomic war. Moreover, forced evacuations of particularly vulnerable areas, including large parts of the capital, were planned (Tram Pedersen 2013). Streams of refugees would need to be accommodated, and *Nørrevangsskolen* was one such facility. While not acknowledged publicly at the time, the architectural blueprints for *Nørrevangsskolen* detail the dual design (Figure VE.2.5).

The school building covered around 6500 square meters (Trier 2017); quite a large structure for less than 1,000 pupils. It also had a strikingly simple layout: one floor (besides a reinforced basement), wide corridors in straight lines, and classrooms to each side. Large but with no hiding places, it was easy to manage and control. The spacious classrooms were all fitted with running water and a sink, and the corridors were extremely wide to accommodate a large number of people. A total of 12,000 Copenhageners

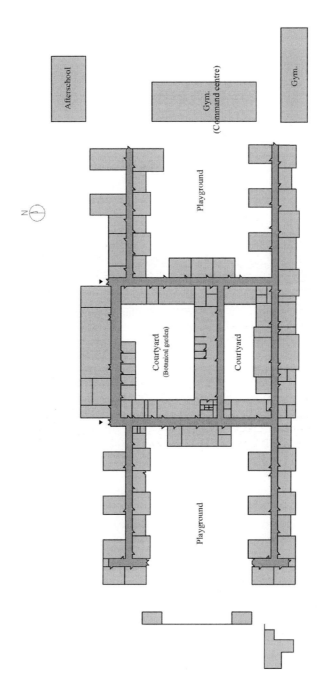

Figure VE 2.5 Illustration based on 'Ny folkeskole. Orienteringsplan', Arkitekter Hermann Steudel & J. Knudsen Pedersen, 1965 (Blueprint).
Source: By courtesy of Peter Møller.

were meant to be evacuated to the school if nuclear war was imminent. The school library was designed to double as an emergency kitchen so local Civil Defence personnel could feed the evacuees (Holt Pedersen & Pedersen (n.y.) pp. 42–45; Tram Pedersen 2013). The local control centre for the city of Slagelse was built underneath the school's gymnasium. Public schools played an important part in Civil Defence planning, as they were often designated or even designed to function as refugee camps, field hospitals, or depots.

Health care: Glostrup emergency hospital

One of the core features of the post-war welfare state in Denmark was the universal health care system. The expansion of the hospital services meant many new facilities were built, and they came to symbolize the national sense of pride in the welfare state (Paulsen 2008; Wium Olesen 2017, p. 463).

The county hospital in Glostrup, a suburb of Copenhagen, was opened in 1958, and it was at the time one of the most modern hospitals in Europe. It was also one of the architecturally most iconic in Denmark; it was even printed as a postcard (Figure VE.2.6). The building also boasted clear characteristics of the *bunker style* (Monteyne 2011) such as

Figure VE 2.6 Carl Stenders Kunstforlag A/S/Pictura and Byhistorisk Hus/Glostrup Arkiv.

210 *Rosanna Farbøl*

bold, rectilinear masses in exposed, rough concrete, moderate fenestration (compared to the dominant Modernist style of glass curtain walls), and fortress-like details.

The Cold War also cast its shadow over the Danish health care sector, and hospitals like Glostrup. After nuclear attack, a large number of wounded civilians were expected, and in consequence, emergency hospital services were organized. According to the Civil Defence Act of 1949, all hospitals were required to have an emergency section that could feasibly operate during war. Sixty hospitals, including the one in Glostrup, established bomb-proof surgery facilities (Paulsen 2008, p. 30). In 1963, the popular magazine *Ude og Hjemme* featured Glostrup's emergency hospital facilities.

The hospital director took the reporter though heavy steel doors to the underground emergency section where doctors and nurses could continue their vital work protected by walls of 50-centimetre armoured concrete (Figure VE.2.7).

Figure VE 2.7 Ude og Hjemme, 25 January 1963.

Figure VE 2.8 Ude og Hjemme, 25 January 1963.

The self-contained medical bunker had beds for 120 patients (Figure VE.2.8), an x-ray department, a laboratory, and stockpile of essential drugs. It had its own independent power generator, and even a well for water supply (Hammelev 1963; see also Beck 2017).

The reporter was clearly impressed, and concluded the article by stating that "None of us hopes that the hospitals underground will ever be needed. But one cannot accuse the authorities of not having taken all eventualities into consideration. All that is possible to do, has been done" (Hammelev 1963, p. 39). This rational response emphasized the preparedness of the authorities, while making no attempt to disguise the existence of an unprecedented nuclear threat. The nuclear anxiety that readers of *Ude og Hjemme* possibly felt may – or may not – have been assuaged by this fusion of welfare and warfare.

In the mid-1960s, civil defence officials and the National Health Service began planning to expand the existing hospital capacity in case hospitals above ground were incapacitated by atomic attack. Thus, existing peace-time capacity was supplemented with 50 field hospitals complete with medical equipment and 200 beds. These were packed in boxes and secretly stored away in the basements of public buildings such as schools all over the country (Tram Pedersen 2013; Hansen 2015). One of the hospitals was set up for a test run at Felsted Central School in 1976, where consultant doctor H. J. Hansen was photographed inspecting the equipment (Figure VE.2.9).

Besides preparing for the care of the injured and sick, the authorities also planned pragmatically for a large number of civilians that would die during or after an attack with atomic weapons. In 1960, all municipalities were

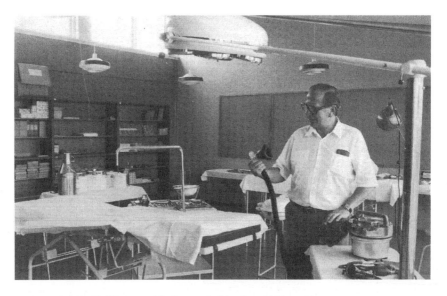

Figure VE 2.9 Civilforsvarsbladet nr. 5, Oktober 1976, p. 11.

ordered by the Ministry of Ecclesiastical Affairs to designate and prepare emergency cemeteries; preferably hidden by trees, close to a church, and far from ground-water tables and neighbours. Primitive wooden crosses and paper bags in two sizes (for adults and children) were to be stored in depots nearby (Sørensen 2012; Tram Pedersen 2013).[3]

Conclusions

From 1949, nuclear civil defence became incorporated into the architecture and ideology of the expanding welfare state in Denmark. Welfare and warfare merged: civil defence was a welfare-security policy strategy. *Society* was the fortress when the *state's* sovereignty was threatened. Cold War architecture in Denmark became the material and social manifestation of nuclear danger.

On the face of it, Danish Cold War civil defence seemed quite impressive. It was extensive and planned in minute detail, yet, there were obvious gaps between these plans and the likely consequences of a nuclear strike. Danish civil defence officials realized as early as 1951 that it was almost impossible to construct public shelters that could withstand a direct hit, and for socio-economic reasons, this was not considered feasible

(Sylvest 2018). If Aarhus was at the epicentre of a nuclear attack, would the provisional shelters in Gellerup and elsewhere in the city, in fact, become tombs? A study made by the civil defence authorities in 1957 estimated that a surprise attack on the capital with a 1MT nuclear bomb would result in 41,000 wounded in Zone A, closest to ground zero, alone. In the Copenhagen metropolitan area, there would be 292,460 injured (Sylvest 2018, p. 26). Clearly, even an ambitious expansion of hospital facilities would not come close to the capacity necessary to cope with this many casualties.

Scholars have argued that the nuclear age disrupted state-citizen relations, because the state could no longer absolutely guarantee the welfare of its citizens (Masco 2008; Monteyne 2011; Hogg 2016; Sylvest 2018). Indeed, it is tempting to view Cold War civil defence efforts as merely a 'necessary façade' (Grant 2010) to ensure enough public support for NATO's deterrence policy. Yet, while Danish civil defence realized that in the event of total nuclear warfare, it was impossible to save all citizens, they refused to give up the ambition of saving as many as possible. Despite hopeless odds, the Danish state actively attempted to extend the social contract to provide security and comfort in the nuclear aftermath. Welfare principles were applied in a serious, systematic, and discreet manner via civil defence principles, and architecture for civil defence – designed and built into ordinary welfare buildings – successfully reproduced and materialized the welfare state narrative, and encouraged the idea that welfare principles could survive and endure at any cost.

Architecture for civil defence was 'politics in matter' (Weizman 2007): Danish citizens were exposed to the risk of possible nuclear warfare, and were encouraged to exhibit the correct behaviours in response to that risk (for instance obeying authorities, remaining vigilant, showing community spirit, and taking responsibility). Crucially, citizens were assured that nuclear war was possible to plan for, prepare for, and most importantly overcome. In Cold War Denmark, the materiality and architecture of the welfare-warfare buildings offered safety and comfort in war and peace, gave substance to the possibility of catastrophe, and promoted the idea of state-led preparedness.

Notes

1 The Danish island Bornholm in the Baltic Sea was in fact occupied by Soviet forces from 9 May 1945 to 5 April 1946.
2 "Lov bygningsmæssige civilforsvarsforanstaltninger", lov nr. 253 af 27. maj 1950 (chapter 2, section 6); Bekendtgørelse nr. 314 af 28. juni 1950 om udførelse af sikringsrum.
3 Also American authorities dealt with the questions of how to handle casualties after a nuclear attack in the pamphlet "Mortuary services in Civil Defense" (1956), Wellerstein (2012).

Bibliography

Århus Civilforsvar, 1983, *Civilforsvarsplan for Århus Kommune*, Århus, Århus Civilforsvar, Aarhus Stadsarkiv/Erhvervsarkivet 865.85 'Civilforsvaret'.

Bæk Pedersen, P 2005, *Arkitektur og plan i den danske velfærdsby 1950–1990. Container og urbant rester*, Aarhus, Arkitektens forlag.

Beck, MA 2017, 'Hospitalsbyggeri i paddehattens skygge', *Rigshospitalets medarbejdermagasin Indenrigs*, p. 16.

'Bekendtgørelse om udførelse af sikringsrum', 1950, bekendtgørelse nr. 314 af 28. juni 1950.

Beredskabsstyrelsen, 2002, Rapport om beskyttelsesrumsberedskabet – herunder udviklingen på sikringsrumsområdet.

Brabrand Boligforening, 1969, 'Gjellerupplanen. Plan af underkælder med sikringsrum' K. Blach Petersens tegnestue.

Carl Stenders Kunstforlag A/S/Pictura and Byhistorisk Hus/Glostrup Arkiv 1959, 'Københavns Amts Sygehus i Glostrup'.

Civilforsvar 6/83, pp. 4–5 "12 kontante spørgsmål om CF besvaret af Britta Schall Holberg".

Civilforsvarsbladet nr. 5, oktober 1976 "Skole omdannet til fuldt funktionsdygtigt sygehus".

Civilforsvarsstyrelsen "Beretning om arbejdet indenfor Civilforsvaret Oktober 1961 - oktober 1962", november 1962, Beredskabsstyrelsen, Civilforsvarsdirektør E. Schultz' embedsarkiv, Årsrapporter fra Civilforsvarsstyrelsen 1947–83, ks 231, Rigsarkivet.

Cordle, D 2012, 'Protect/protest: British nuclear fiction of the 1980s', *British Journal for the History of Science*, vol. 45, no. 4, pp. 653–656.

Danmarks købsteder, 2010, *Århus – Historiske befolkningstal*, (Accessed 30 August 2018) http://ddb.byhistorie.dk/koebstaeder/befolkning.aspx?koebstadID=74

Den Gamle By/Århus Luftfoto, 1975, Gellerupplanen. Udsigt fra sydvest over kvarteret omkring Hejredalsvej og Edwin Rahrs Vej.

Deville, J, Guggenheim, M & Hrdličková, Z 2014, 'Concrete governmentality: Shelters and the transformations of preparedness', *The Sociological Review*, vol. 62, no. S1, pp. 183–210.

Grant, M 2010, *After the bomb. Civil defence and nuclear war in Britain, 1945–68*, Basingstoke, Palgrave Macmillan.

Hammelev, J 25 January 1963, 'Hospitalet under jorden', *Ude og Hjemme*, vol. 4, pp. 38–39.

Hansen, PH 2015, 'Kold krig i kælderen: Hvis katastrofen ramte Ærø', *Årbog for Ærø Museum*, vol. 22, pp. 40–47.

Hansen, PH, Tram Pedersen, T & Stenak, M 2013, *Den Kolde Krigs anlæg, baggrundsnotat*, København, Kulturstyrelsen.

Hogg, J 2012 ,'"The family that feared tomorrow": British nuclear culture and individual experience in the late 1950s', *British Journal for the History of Science*, vol. 45, no. 4, pp. 535–549.

Hogg, J 2016, *British nuclear culture. Official and unofficial narratives in the long 20th century*, London/New York, Bloomsbury.

Høghøj, M & Holmqvist, S 2018, 'Da betonen blev belastende. Den emotionelle kamp om Gellerupplanen i 1960'erne og 1970'erne', *Temp – tidsskrift for historie*, vol. 8, no. 16, pp. 124–145.

Holm, H 2014, 'Hvis krigen kommer', *Aarhus Stiftstidende* 9 January 2014.

Holt Pedersen, P 2013, 'Kommunale kommandocentraler i Skanderborg, Odense og Hadsund', in M Stenak, T Tram Pedersen, PH Hansen & M Jespersen (eds), *Kold krig. 33 fortællinger om Den Kolde Krigs bygninger og anlæg i Danmark, Færøerne og Grønland*, København, Kulturministeriet, pp. 108–112.

Holt Pedersen, P & Pedersen, K 2018, 'Kommunale kommandocentraler (KC) i region V, pp. 42–45, (Accessed 10 April 2018) http://reganvest.dk/Andre_bunkere.html

Holt Pedersen, P & Pedersen, K 2014, *Danmarks dybeste hemmelighed. REGAN VEST – regeringens og kongehusets atombunker*, Værløse, Billesø og Baltzer, pp. 260–275.

Jensen, B 2014, *Ulve, får og vogtere*, København, Gyldendal.

Light, J 2003, From *warfare to welfare: Defense intellectuals and urban problems in Cold War America*, Baltimore, John Hopkins University Press.

'Lov bygningsmæssige civilforsvarsforanstaltninger', 1950, lov nr. 253 af 27. maj 1950.

'Lov om civilforsvaret', 1949, lov nr. 152 af 1. April 1949.

Masco, J 2008, '"Survival is your business": Engineering ruins and affect in nuclear America', *Cultural Anthropology*, vol. 23, no. 2, pp. 361–398.

Monteyne, D 2011, *Fallout Shelter: Designing for civil defense in the Cold War*, Minneapolis, University of Minnesota Press.

Nørrevangsskolen, 1965, 'Ny folkeskole i Slagelse, stue og kælderplan', Arkitekter Hermann Steudel & J. Knudsen Pedersen, (Blueprint).

Nørrevangsskolen, 1965, 'Ny folkeskole. Orienteringsplan', Arkitekter Hermann Steudel & J. Knudsen Pedersen, (Blueprint).

Paulsen, S 2008, Glostrup Hospital gennem 50 år, 1958 – 2008, Glostrup Hospital.

Petersen, JH, Petersen, K & Christiansen, NF 2010–2014, *Dansk Velfærdshistorie*, Odense, Syddansk Universitetsforlag.

Sørensen, LM 2012, 'Atomkirkegården i Rødovre', *Politiken* 24 August 2012.

Stafford, J 2012, 'Stay at home: The politics of nuclear civil defence, 1968–83', *Twentieth Century British History*, vol. 23, no. 3, pp. 383–407.

Stenak, M 2013, 'Arkitektur og byggeri', in M Stenak, T Tram Pedersen, PH Hansen & M Jespersen (eds), *Kold krig. 33 fortællinger om Den Kolde Krigs bygninger og anlæg i Danmark, Færøerne og Grønland*, København, Kulturministeriet, pp. 58–63.

Sylvest, C 2018, 'Atomfrygten og civilforsvaret', *temp – tidsskrift for historie*, vol. 16, pp. 16–39.

Tram Pedersen, T 2013, 'Beskyttelsen af civile', in M Stenak, T Tram Pedersen, PH Hansen & M Jespersen (eds), *Kold krig. 33 fortællinger om Den Kolde Krigs bygninger og anlæg i Danmark, Færøerne og Grønland*, København, Kulturministeriet, pp. 26–32.

Trier, H 2017, 'Nørrevangsskolen 1967–2012', *Slagelse leksikon*, (Accessed 10 April 2018) http://www.slagelseleksikon.dk/mod_inc/?p=itemModule&id=403&kind=11&pageId=169&tit=

Vanderbilt, T 2002, *Survival city*, New York, Princeton Architectural Press.

Virilio, P 2009 [1997], *Bunker archaeology*, New York, Princeton Architectural Press.

Weizman, E 2007, *Hollow land: Israel's architecture of occupation*, London, Verso.

Wellerstein, A. 2012 '"Mortuary services in civil defense" (1956)' Restricted Data, the Nuclear Secrecy Blog, (Accessed 28 June 2018) http://blog.nuclearsecrecy.com/2012/02/29/weekly-document-16-mortuary-services-in-civil-defense-1956/

Wium Olesen, N 2017, 'Velfærd og kold krig', in J Fabricius Møller, M Fink-Jensen & N Wium Olesen (eds), *Reformation, enevælde og demokrati, Historien om Danmark*, København, Gad, pp. 420–558.

Part III
Culture and politics in the Cold War city

9 Urban space, public protest, and nuclear weapons in early Cold War Sydney

Kyle Harvey

UNIVERSITY OF TASMANIA

Introduction

In Sydney in the early 1960s, tentative experiments with public demonstrations of opposition to nuclear weapons were an early indicator of how Sydney's peace movement thought about the spaces of the city it inhabited. A small, but vibrant cultural and political community, the peace movement had traditionally favoured private spaces to discuss and debate the impact of war, political conservatism, and injustice on Australia and its citizens. In the early Cold War years, however, peace activists campaigning against nuclear weapons started to do so publically, using Sydney streets, parks, plazas, and also the spatial iconography of the city itself. The spaces of the city, its suburbs, and the idea of its geographic expanse served multiple functions for these peace activists. On one hand, the city was the backdrop to demonstrations seeking public support and engagement, a familiar component of most social movements. On the other, though, ideas about the city and its spatial dimensions were themes to be mobilised and contested. What was public space and how could its traditional purposes be used by anti-nuclear campaigns? How would Sydneysiders think about the mapping of nuclear destruction over their city? And how might the peace movement use the city's geography in ways that might visualise their opposition to nuclear weapons in new and interesting ways?

These approaches made use of familiar ideas about nuclear weapons and their destructive potential, echoing similar responses of activists, artists, and others in Europe and North America to the magnitude of devastation a nuclear weapon would cause if exploded over a major city (see Boyer 1994; Grant 2016; Weart 1988). Sydney was somewhat removed from cities of similar size in the Northern Hemisphere, not least due to its remote location in the southern hemisphere and Australia's status as a non-nuclear armed nation. Nevertheless, Sydney's peace movement, along with local civil defence bodies, utilised the spatial iconography of city maps, and the physical spaces of the city and its suburbs, to convey their attitudes about the dangers of nuclear weaponry to the public. What makes Sydney an interesting case study is its nature as an antipodean capital city, seemingly far removed from

Cold War hotpots in Britain, Western Europe, and North America. Low population density and a conservative political culture meant that its peace movement struggled to find much public support. However, these conditions led, in the early 1960s, to a creative approach to the spatial conception of how activists could use the city – and various ideas about its geography – to advance their aims. These approaches are akin to what Farish (2003, p. 127) calls "anxious urban imaginaries," in that the mobilisation of fear of nuclear weapons and an imagined nuclear attack were the centrepieces of strategies of urban activism and engagement with public space.

This chapter responds to a varied literature on Cold War protests and the western city. Histories of urban social organisation, geographic and spatial histories of activism, and the urban environment's centrality to political and cultural expression in the Cold War era have contributed to a rich body of scholarship exploring the intricate relationships between activism, urban environments, and the concept of public space. What Schregel (2015) calls "municipal interventions in defence" are instructive in the spatial dimension given to local responses to the threat of nuclear weapons, but also in other local and regional manifestations of anti-nuclear and peace sentiment based on a spatial approach to activism (see also Bennett 1987; Cutter, Holcomb and Shatin 1986; Hobbs 1994; Schregel 2017). These histories align with other research on the energising potential of urban environments in fostering social movement activity, from studies of counterculture and performance (Brown 2013), to suburban anti-nuclear activism (Cameron 2014; O'Mara 2006), to a broad literature on youth and student protest cultures in urban areas (see Castells 1983; Schildt and Siegfried 2005.) In Australia, however, it is somewhat difficult to translate many of these European and North American historical antecedents. Australia's cities developed from British colonial power structures, and public access to urban spaces was historically restricted and policed (Douglas 2004). In response, protest movements, youth culture, and demands for access to public space have similarly demanded what Mitchell (2003) and others have called "the right to the city." Similarly, as geographic social movement research has suggested, social movements and the geographic environments in which they operate are mutually constitutive (Gregory and Urry 1985; Miller 2000). In Sydney, we can apply these ideas to the early Cold War years and the unique interplay of urban culture, youth, modernity and evolving social movement activity that permeated these early experiments at public, urban protests against nuclear weapons.

This history sits squarely within Australia's Cold War entanglements. Australia played a significant part in the allied powers' Cold War nuclear weapons programmes: British atomic bombs were tested on Australian soil from 1952 to 1963, Australian territory was used for US strategic and defence facilities, Australian mined uranium was vital in the production of nuclear weapons, and Australian ports welcomed US nuclear-armed warships and submarines (Ball 1980; Reynolds 2000; Tynan 2016). Expressing its dissent, Sydney's peace movement utilised those opportunities that were available to it within the central and suburban space of the city. Public halls, union meetings, churches, and speakers' corners in public parks were popular locations for discussion groups, lectures, and meetings. Larger congresses were also held every three to four years in Sydney and its southern counterpart, Melbourne. Such events were typical for political activity in mid-century Sydney, but it was not until the early 1960s that peace activists began to experiment with a modern variety of public protest, akin to the kinds of experiments with direct action that would radicalise later activists against the Vietnam War.

In the politically conservative 1950s, public displays of anti-nuclear dissent were often considered dangerous or risky. Anti-communism was rife, Sydney's mainstream media was hostile to progressive ideas, and Australia's conservative federal government meant that concerns about nuclear weapons were relegated to the fringes of political life. The overt expression of dissent – via graffiti, marches on city streets, outdoor rallies, or any other activity that broached the limits of permissive public behaviour – was rare (Harvey 2018). Additionally, the small numbers involved in peace movement meant that large mobilisations in city squares were not possible. Even the numbers of demonstrators willing to risk arrest by impeding traffic in sit-downs, something that became a divisive, radical strategy of the anti-war movement from 1965, were not present in the anti-nuclear movement in the early 1960s. In spite of this, as Irving and Cahill (2010, pp. 4–5) argue, radical movements throughout Sydney's past have sought to use public space in the city centre to express their demands, and did so at sites of symbolic significance, such as the heart of Sydney's civic and commercial district. Never numbering more than a few thousand people, these campaigns were small and their influence muted, yet their perseverance demonstrates an evolving willingness to air demands in a public, rather than a private fashion.

In the early 1960s, Australia's anti-nuclear advocates took tentative steps towards rectifying their reticence to demonstrate publicly. Building on the rapid development of anti-nuclear protests in Britain and the United States, Sydney's peace movement began to experiment with broadening its public support by staging marches and rallies. As Irving (2017, pp. 32–38) has argued, much of this experimentation aimed to build an international community of solidarity based on imagined connections and an implied cultural closeness. In practice, this was a translated exercise, and organisers attempted to mobilise Australians unfamiliar with political marches to accept this newer style of public demonstration. Doing so would use the space of the city and its extensive suburbs as a means to validate the authenticity and public nature of a movement opposed to nuclear weapons. It would also, organisers asserted, help to show Sydneysiders that the threat posed by nuclear weapons was one that enveloped the whole city, and this very city-centric style of protest – with its accompanying iconography – was the best way to elicit additional public support. This attitude indicates an openness to challenging the co-optation of urban space by various interests – both public and private – that sought to modernise and commodify urban space, private space, and infrastructure in Sydney's inner city, particularly in the 1970s (Anderson and Jacobs 1999). This chapter argues that the tentative experiments of the early 1960s in thinking about public protest in the context of the urban environment of Sydney are a crucial aspect of Australian thinking about protest, the spaces of the city, and the significance of nuclear weapons to a nation embroiled in the Cold War.

Atomic fears in the early 1960s

Visions of cities destroyed by nuclear weapons were nothing new to Australians in the early 1960s. Since the bombings of Hiroshima and Nagasaki in 1945, Australians were familiar with descriptions of atomic destruction and fascinated with the uses of atomic technology (Sherratt 2003). As elsewhere in the world, John Hersey's *Hiroshima*, an issue-length article that appeared in the *New Yorker* in August 1946, was republished in Sydney's *Sun* newspaper later that year. The *Sun* published the article in four parts "because it helps plain human beings to understand the reality of atomic destruction in terms of everyday life – and death." Hersey's account, the *Sun* editorial argued, "brings the atom bomb to us as closely as if it had fallen on [Sydney suburbs] Ashfield or Balmain instead of Hiroshima" (*Sun* 1946, p. 1). The following decade, Nevil Shute's novel *On The Beach* became a domestic best seller in 1957, and a hit film upon its release in 1959. Although both novel and film were set in Melbourne, Sydney peace activists used the opportunity to leaflet cinemagoers in late April 1960, using the final words uttered in the film – "There is still time, brother" – to advertise attendance at an upcoming Sydney peace march (*Tribune* 1960a, p. 2).

The reality of Sydney's destruction by a nuclear attack, however, was minimal at best. As Steinbach (2002, p. 92) has written, "As long as any postulated threat to Australia, real or imaginary, remained distant, in Europe or Asia, its impact on day-to-day life was effectively minimal." British nuclear tests that had occurred in the interior of the country, as Gerster (2014, p. 61) noted, "caused barely a ripple on the placid surface of Australian life." In the early 1960s, Sydney remained a small outpost in Britain's former empire, and by many accounts, a conservative, sleepy city of just over 2 million inhabitants (Spearritt 1999, p. 3). Significantly, "Australians were more likely to go on nuclear alert by alarming happenings elsewhere," and growing unease in the United States, Britain, and Western Europe in the late 1950s and early 1960s over atmospheric nuclear testing and the unpredictability of Cold War tensions certainly influenced those in Australia prone to criticism of nuclear weapons and the potential disaster of their use (Gerster 2014, p. 62). At the same time, Australia was in the middle of an unprecedented period of conservative government that lasted from 1949 to 1972, corresponding with a rapid expansion of urban sprawl, car ownership, and affluence

(Spearritt 1999, pp. 151–52). Early in this period, Prime Minister Robert Menzies had remarked that Australia would need to prepare for war within three years, and the sorts of Cold War conservatism and paranoia that gripped Britain and North America played out in similar ways in Australia (Lowe 1999). Civil defence bodies and a small but vibrant peace movement responded accordingly, fixating on the grim possibility of a nuclear attack on cities such as Sydney, and what this meant for the city's residents, and their political, social, and cultural relationship with nuclear weapons.

In the 1950s, civil defence propaganda became a fixture of the Australian government's efforts to assure its citizens that defence preparations would lessen the expected 120,000 casualties, should a nuclear weapon hit an Australian city (Steinbach 2002, p. 94). As the power of these weapons grew, Australian academics, newspapers, and indeed many politicians, agreed that there was no defence for Australia's highly urbanised population against an atomic attack (Steinbach 2002, pp. 104–105). The message of preparedness advocated by the New South Wales (NSW) Civil Defence Organisation, a state government branch, attracted predictable degrees of ridicule and confusion. In a 1961 sponsored article appearing in the popular magazine the *Australian Women's Weekly*, the Organisation emphasised that "there is no uncertainty as to the destruction and devastation which could be caused by a war in which modern weapons may be used" (*Australian Women's Weekly* 1961, p. 14). Detailing preparedness measures, warning alerts, and suggestions for stockpiling, the article prompted severe criticism from a group of progressive scientists, amongst others, who argued "[i]t is an example of living in a "fool's paradise" to perpetuate schemes for personal survival in total nuclear war" (Blackwood *et al.* 1961, p. 30). As in nations such as Canada, the illusion of preparedness raised by civil defence was a key aspect of the peace movement's response to the idea of nuclear weapons and their impact on urban environments (Burtch 2012).

In the early 1960s, when the extent of Australia's defence relationship with the United States was kept secret from the public, the idea of Australian cities as nuclear targets was an often abstract matter, influenced by rumour and assumption as much as it was by a sense of antipodean removal, compounded by vast distances between Australia and the Cold War's hotspots in Western Europe. How central was Australia to a potential nuclear war between the superpowers? How would Australians be affected? Would radioactive fallout from nuclear war in the northern hemisphere still affect Australia? Would Australians have sufficient warning if an attack were to occur? Such questions were at the heart of the peace movement's early thinking, as was the approach taken by government agencies that in part echoed British and North American civil defence programmes, designed to encourage preparedness in the event of a nuclear attack. Civil defence films – the most famous of which was 1952's *Duck and Cover* – were produced and widely screened in the United States and reinforced the idea of a sudden attack (Matthews 2012), but even though Australian defence planners felt somewhat differently about civil defence, these films were still screened in Australia. For example, the Civil Defence Organisation branch based in Penrith, in Sydney's western suburbs, screened both the 1955 film *A New Look at the H Bomb* and the 1957 film *A Day Called X* at a 1960 event for local fire brigades, industrial firms, and members of local civic organisations such as bowling clubs and veteran servicemen's leagues (*Nepean Times* 1960, p. 14). In this sense, Cold War culture in Sydney operated in an interesting manner. On one hand, Australians' lack of exposure to the kinds of civil defence propaganda common in the United States meant Australians remained somewhat isolated from the same kinds of fear of nuclear attack that mobilised peace movements elsewhere in the world. That the public had little idea of just how entwined Australia was with its nuclear-armed allies also made Australian anti-nuclear activists' task much more pronounced.

Official civil defence publications also emphasised the severity of destruction wrought by a potential nuclear attack on an Australian city. The NSW Civil Defence Organisation prepared numerous pamphlets for distribution to citizens detailing the magnitude of a nuclear attack, and the precautions that individuals and families could take to ensure their safety should an attack occur. In Sydney's leafy harbourside suburb of Mosman, a 1961 pamphlet illustrated the destruction by using the familiar visualisation of blast and radiation radii superimposed over an aerial photograph of the city. The heat, blast impact, and radiation produced by nuclear weapons, "irrespective of size" were "facts to be faced" in times of crisis, yet it remained the citizen's own responsibility to be informed, to fill in household registers of occupants, display house numbers clearly, and to volunteer their services to their local civil defence body (NSW Civil Defence Organisation 1961).

Evacuation drills in 1961 continued to emphasise the unusual nature of civil defence preparations, contrasting with mainstream news of severe tensions between the United States and the Soviet Union. In February 1961, "Operation Picnic" took place, where 629 people participated in a mock evacuation from metropolitan Sydney to the foothills of the Blue Mountains on the western fringes of the city, where a local children's health centre had been converted into a decontamination and first aid post (Australian Broadcasting Corporation 1961; *Nepean Times* 1961, p. 5). This was the first exercise of its kind to take place in Sydney, and was ridiculed by the left-wing press as "remote from reality" and "essentially miss[ing] the whole point of the threat of nuclear weapons" (*Tribune* 1961, p. 10). Similarly, residents of western Sydney felt that "a Civil Defence Organisation would be practically useless against atomic attack; prevention of such an attack is the only answer" (Lambert 1961, p. 4). These attitudes resonate with Burtch's (2012) examination of the failures of Canadian civil defence measures in the Cold War, where an incomplete and insufficient civil defence program resulted in frequent criticism and ridicule. As in Canada, Australians challenged what Oakes (1994, p. 79) calls "the Cold War conception of nuclear reality" in order to reject governmental efforts to legitimate the world in which nuclear weapons and their threats existed.

The geographical expanse of Sydney hinted at by these evacuation drills was also a key component of the geographical and cartographical imagining of a nuclear bomb's impact on the city. Sydney's suburbs radiate north, west, and south over some 50 kilometres from its city centre, and population density over this vast area has been historically low compared to British or European cities. In the nuclear age, city maps with theoretical blast radii from a nuclear explosion superimposed were commonly used by Civil Defence bodies and peace organisations alike, and emphasised the wide swathe of destruction caused by the blast impact, heat, and radiation produced by a nuclear attack (Boyer 1994, p. 14; Farish 2003, pp. 131–133). In Sydney, such maps appeared in local newspapers showing varying degrees of devastation for the city and its surrounds (Titterton 1951, p. 2; *Tribune* 1954, p. 1). Peace groups occasionally repurposed these maps – and created their own – for leaflets and pamphlets that emphasised, for example, a "potential Death Zone of radio-activity" from a blast in which "Sydney would vanish" (NSW Peace Council [mid-1950s]). The emphasis on total urban destruction, combined with a rhetoric of urgency amidst international crises and tensions between the nuclear powers, influenced the peace movement in its new experiments in the early 1960s with public protest. The space of the city and its suburbs, accordingly, became a part of the peace movement's public display of dissent, just as social organisation around geography and space has historically been a key part of the political and cultural demands of various social movements (see Castells 1983; Miller 2000). Responding to new trends in protest from Britain and the United States, and to an attitude that treated the spatial scales of the city akin to the spatial nature of nuclear destruction (see Miller 2000, p. 139), Sydney activists began to think in new ways about how to protest, combining these ideas with experiments in public engagement that were, in the early 1960s, tentative steps toward the kind of disruptive, performative direct action that would soon become a staple of modern social movements in Australia and across the western world.

Early experiments with public protest

By the beginning of the 1960s, Australia's peace movement was experiencing the beginnings of substantial change. The traditional alliances of left-wing unions, progressive academics, churches, and cultural groups with the Communist Party of Australia (CPA) had faltered, due in part to the CPA's split in the mid-1950s over the Party's rigid adherence to Stalinism. What emerged were various groupings of a diverse socialist alternative left that was dedicated to non-alignment and democratic ideals, and that was beginning to think about indigenous rights, direct action, gender, and, soon, conflict in Vietnam (Murphy 1993, pp. 122–126). The beginnings of this 'New Left' were visible in growing Labour Clubs at the major universities, the

Youth wing of the Australian Labor Party, in Fabian and Rationalist societies, and in Campaign for Nuclear Disarmament (CND) groups founded in four cities, including Sydney. More broadly, there existed a growing population of young Sydneysiders interested in anti-nuclear protest. Their presence, along with a style and attitude influenced by the development in youth protest in Britain and the United States in recent years, combined to shake up the existing peace movement in Sydney.

Perhaps most significant was the emerging trend from Britain of large anti-nuclear marches. From 1958, increasing numbers of demonstrators had marched for four days at Easter between London and the Atomic Weapons Research Establishment at Aldermaston in Berkshire. The tens of thousands taking part in these Easter marches were, as Irving (2016) and Scalmer (2002, pp. 12–13) argue, a substantial inspiration to Australians desperate to mobilise such numbers of demonstrators at home. In the Sydney socialist magazine *Outlook*, for example, a 1962 editorial spoke of the marked differences between the numbers involved in the British protests and their counterparts in Melbourne and Sydney, which although numbering several thousand in total, appeared "small fry indeed" (*Outlook* 1962a, p. 2). In Sydney, outdoor marches and demonstrations had a lengthy heritage in the labour movement, yet in this new era of youth participation and non-aligned socialist ideas, anti-nuclear protests involved a different crowd. Marking the anniversary of the 1945 atomic bombings, Hiroshima Day demonstrations in Sydney grew rapidly, from 400 braving heavy rain in 1960, to 6,000 participants the following year (*Tribune* 1960b, p. 1; *Peace Action* 1961, p. 1).

In 1960 and 1961, the traditional peace organisations recognised that young Sydneysiders were a key target for this evolving movement. Young people were unlikely to attend meetings and discussion groups; instead, peace organisations aimed to provide a more social, festive environment to attract young participants. The Peakhurst peace group in Sydney, for example, held a barbecue that attracted 93 young people, complementing half-day "car picnics" and other social gatherings. However, the group's representative Jim Lambert lamented, "we lost these young people because we couldn't … keep in touch with them and draw them closer to the movement" (NSW Peace Committee 1961). The public nature of peace demonstrations, though, had an enduring appeal, connecting social life, leisure, and the outdoors in various urban and suburban spaces with a genteel politics of dissent. At a meeting of Sydney area peace group representatives in March 1961, for example, most recognised the value of a visible public appeal. The South Head Peace Committee, for example, recommended "a large picnic, with cars going through the streets of Sydney, well decorated with peace banners etc. This is very good publicity and shows outside people that we are a large mass organisation" (NSW Peace Committee 1961). A representative of the Sutherland Shire group agreed: "there should be more singing of peace songs at our peace rallies" to attract younger members, she argued. "And instead of having too many meetings, we should try to have more social activities, picnics, cultural functions, etc." (NSW Peace Committee 1961). Such an attitude contrasts with May's (1999) examination of the private domain of the home as the key site of Cold War cultural containment and resilience. Instead, Sydney activists looked to the spaces of leisure the city afforded them, combined with strategic use of public political demonstrations, to enrich a movement that had hitherto existed in private spaces – homes, churches, and meeting rooms – that lacked any effective engagement with the spaces of the city.

The idea that cultural and civic engagement could be aligned with peace movement activity on Sydney's city streets found favour with organisers. In October 1960, a large NSW Peace Committee meeting was held the same day as the annual Waratah Spring Festival in central Sydney. Held annually since 1956, this was a large cultural festival, replete with vibrant parades through city streets, and in 1960 attracted some 850,000 spectators (*Sun-Herald* 1960, p. 2). Stan Hatton, attending this event, had seen the large crowds at the Festival earlier that day, and recommended – "from a propaganda point of view" – that the Peace Committee organise a float in the following year's procession to capitalise on the growing numbers of spectators in what was becoming a major fixture in Sydney's civic calendar (NSW Peace Committee 1960). What these ideas suggest is how the peace movement imagined the possibilities of public demonstrations within the confines of acceptable civic engagement. As Miller (2000, p. 166) explains, "social movement processes ... are constituted through space, place, and scale, and that constitution affects how they interact, articulate, and play out." In Sydney, we can observe tentative expressions of thought about how the city's geography, its cultural and social hubs, and the distances between them could invite new forms of anti-nuclear protest.

The idea that anti-nuclear concerns could be expressed publicly in such civic, cultural and social ways was not an inevitability for many suburban peace groups lacking access to the urban, social and political conditions favourable to such activity in the inner city. The Sutherland Shire District Peace Committee in the city's south, for example, reported plans for a march from surrounding suburbs to the popular beach suburb of Cronulla, "ending with a public protest meeting", a relatively new activity for the group (NSW Peace Committee 1961). In the northern beach suburbs, the Manly-Warringah Peace Committee proposed something even simpler. Monthly street corner stalls, selling the periodical *Peace Action* and a variety of other literature, would serve as an opportunity to talk directly to passers-by, encourage them to sign up to mailing lists, and join the cause (NSW Peace Committee 1961). Such an innocuous tactic might seem pedestrian by modern standards, but in conservative Sydney in the early 1960s, and especially in its suburbs, suspicions of communist affiliation still hindered the peace movement's ability to canvass publically (see Summy 1988). Without access to sympathetic trade unions, too, a local peace group would have struggled to find new members, elicit additional support, or even engage in fundraising. The conservatism of suburban life, compounded by what Lowe (1999, p. 103) calls "popular anxieties about change", contrasted dramatically to the radical potential of the inner city, or at least its romantic ideal (see Brown 2013, p. 827). There were, of course, many diverse spaces in a geographically large city, and in some ways, Sydney peace activists adopted a "spatial hierarchy of risk" (Farish 2003, p. 126) as they sought to prioritise the kinds of public demonstration that reaped the most reward.

The nascent Sydney CND group was less hindered by these concerns. Younger in age and comprised of Trotskyists, socialists, and others with a shared aversion to the CPA's official, dogmatic communism, the group advocated nonalignment and unilateral disarmament, like British CND. From the outset, its members endeavoured to use the city's public space for public demonstrations and leafleting drives. Only several months old in 1962, the group had, according to *Outlook*, "periodically startled the lunch-hour crowds in Martin Place with the sight of leafleters clad in black calico, mystic, wonderful, bearing the semaphore sign in white" (*Outlook* 1962b, p. 20). The sign was the ubiquitous peace symbol, first used by British CND in 1958 and, in the early 1960s, almost exclusively allied with the anti-nuclear cause (Rigby 1998). More than simply a radical alternative to the larger NSW Peace Committee and its various suburban affiliates, Sydney CND represented a younger, more vibrant alternative, willing to undertake riskier public demonstrations of anti-nuclear dissent. A prelude to the divisive direct action demonstrations of the Vietnam War years, Sydney CND demonstrated a key component of the relationship between youth, performance, and dissent as it operated in various urban environments. That it modelled itself on British CND was key to its appeal, and the group used the British model of marches, lively public demonstrations, and youthful spectacle to help enliven Sydney's other peace organisations in adopting bolder public approach. As anti-nuclear campaigns in Britain and North America had embraced large public demonstrations and experimented with direct action in the 1950s, Australian activists soon flirted with these energising ideas, and did so by mobilising themselves in and around the city and its suburbs.

Radial marches, urban expanses

The centrepieces of this new approach were the two 'radial marches' held at Easter in 1962 and 1963. Modelled on, and in solidarity with, British CND's Aldermaston marches, Sydney's peace groups aimed to utilise the momentum that had been developing in the years prior to stage a peace march with broad appeal and a timely sense of urgency. Organisers hoped to capitalise on widespread feelings of unease related to international Cold War tensions. The Bay of Pigs invasion of April 1961, the Berlin Crisis that ran for a large part of 1961, and the breakdown of the Geneva disarmament negotiations in late 1961 provided opportunities for the movement to attract new interest. More prescient for the Australian movement, perhaps, was inspiration from Britain and the United States, where British CND and its radical offshoot the Committee of 100, and the American groups SANE and Women Strike for Peace offered Australians an indication of what large-scale anti-nuclear protest might look like (see Irving 2016).

The Radial March for Sanity and Survival – as both events were formally known – used the familiar visual trope of a nuclear bomb's blast radius as a means to mobilise peace groups from the far reaches of the city's vast suburbs, and to engage with the imaginary spectacle of urban nuclear destruction by specifically situating themselves in its geography. In press releases promoting the event, the NSW Peace Committee emphasised that the two-day march "will take the form of radial columns of marchers converging on the City from the outer suburban perimeter of an area which would be devastated by an H. Bomb exploding over Sydney" (NSW Peace Committee 1962a). Participants appreciated the value of using the iconography of the city map in leaflets and posters, noting that this imagery "was bound to shock people to their senses, and make them realise how hopeless and horrifying a situation like that would be on a city like Sydney" (Anon. 1962). The perception of what Bishop and Clancey (2004) call "the city-as-target", for Sydneysiders, represented a key opportunity in connecting the geographic idea of their city with the Cold War trope of imagined nuclear destruction, something that sat at the core of peace movement advertising and propaganda in the early 1960s.

In the 1962 radial march, columns of marchers were organised from seven outer suburbs – representing the scale of destruction of an imagined nuclear attack – marching up to 25 miles over two days to converge for a large rally at the Trocadero, a historic dance hall in the city centre. The columns themselves, often numbering just several dozen each, gradually grew over the two days as they converged on the city centre, where 2,000 attended the final rally (*Peace Action* 1962, p. 3). Once in the centre of Sydney, however, there was no central square in which these converging columns of marchers could meet. Sydney lacked a large, central plaza akin to London's Trafalgar Square or New York's United Nations Plaza, public spaces that lent themselves to large political protest gatherings. The layout of Sydney's city centre, whose urban geography limited large gatherings outside official institutions such as Parliament House, had the effect of dampening possible impact and visibility that a centrally located demonstration might have had. The seven columns, too, faced restrictions imposed by the authorities. Marchers needed to walk single file en route to the city, and no signs, banners, or amplifiers were allowed (NSW Peace Committee 1962b). Police permits emphasised that a "street procession", rather than a "demonstration" was allowed (Gentle 1963). Friendly relations with police were essential, lest the march be forcefully disbanded, and although some participants "do not think the young people should dart amongst traffic with leaflets", no controversy resulted (Wilcox 1962). These restrictions succeeded, as they had throughout the 1950s, in dampening the peace movement's effectiveness at engaging in bold public demonstrations. But these barriers to public demonstration ought not to conceal the new ideas about urban space and geography that sat at the heart of Sydney peace activists' thinking. In connecting the imaginary spectacle of nuclear destruction with the politics of scale throughout Sydney's urban and suburban environments, activists engaged with new Cold War concepts of place and scale that their 1950s predecessors had failed to do.

In the wake of the 1962 radial march, *Outlook* surmised that the "organised walk" style of peace protest could prove popular. It was "flexible, informal, sociable", and well organised, and the publication reported "plenty of cars coasting along looking for those suffering from blisters or bunions; a leaflet marking milk bars and toilets." The active, physical aspect of the protest was motivating: "hundreds of people ... feel they've done something much more significant than merely attending a public meeting" (*Outlook* 1962c, p. 16). From its attendance, along with that at the previous year's Hiroshima Day events, the NSW Peace Council's secretary Geoff Anderson felt that he could observe "the beginnings of a real mass movement developing in this country" (NSW Peace Committee 1962c). The days of stale processions were over, Anderson argued; protest in Australia now involved new forms of expression and language. "We must learn to tune in to this vast mass of people who have a sentiment," he urged. Newer attendees at peace rallies "won't be regimented into a procession with a marshall, but a casual walk with singing and music, with freedom to drop out for a drink, appeals to them" (NSW Peace Committee 1962c). Anderson felt that if the movement were to grow, it was the involvement of young people that would help it find success, hinting at the explosion of youth and student activism that would dominate the anti-war movement in several years' time. A key aspect of urban social movements in western nations, the radicalisation of Australian youth and student politics in the late 1960s owed much to these early years of experimentation with public protest in Sydney (see Murphy 2015).

Youth protest and public dissent

Australia's growing numbers of politically aware and engaged youth stemmed from a rapidly expanding university student population, and also from a rise in secondary school enrolments. These changes were part of broader developments in education in Australia in the second half of the twentieth century that heralded a new middle class, one which welcomed, as Campbell (2007, p. 1) notes, "the emergence of reliable and accessible government high schools with clear pathways to well-paid white collar work and through the universities, teachers colleges and later, colleges of advanced education, into the professions." This coincided with relative political stability and a period of unprecedented economic expansion, marked especially by low unemployment and a rise in average earnings. What these changes implied for youth and student involvement in Sydney's peace movement were frequently spatial in nature: young people had access to white-collar employment, tertiary education, and transport that facilitated proximity to the urban spaces of protest eagerly targeted by anti-nuclear campaigns. Their involvement, largely through younger peace groups such as CND, was key in the experimental development of public protest through direct action, performance, and visual engagement with the symbolism and iconography of protest movements in Britain.

In the early 1960s, Australian youth attending the University of Sydney, Australia's oldest university, or its newer suburban counterpart the University of New South Wales, enjoyed access to social and economic conditions conducive to political activity, both on and off-campus. The peace committees, Anderson recognised, needed to appeal to this potential audience, as young activists' engagement with music, theatre, and performance was enlivening and an ideal boost for public demonstrations that aimed to attract passers-by amidst the confines of the city and its commercial spaces. The public, Anderson felt, "does not see us as a pack of Dismal Desmonds walking to our doom, but as happy people wanting to live." He envisaged "Ban the Bomb concerts with folksingers singing atomic blues, entertainments, choral, bands, etc ... Entertainments with admission by souvenir songbook, with bands and artists, audience participation" (NSW Peace Committee 1962c). Several June 1962 concerts fulfilled these ideas (NSW Peace Committee 1962d), but not all participants appreciated this new approach. For some, such a festive spirit did not belong at a sombre occasion, and detracted from the seriousness of the issue at hand. One activist remarked that "songs touched with comedy about a serious subject tend to impinge upon our intentions" (Anon. 1962). Others also emphasised that anti-nuclear protests were meant to impress upon bystanders the "hopeless and horrifying" prospect of an imaginary nuclear war, not the camaraderie and enjoyment of a youthful, singing parade (Anon. 1962). The contested nature of activism and its public face, as with all social movement activity, hints at the changing nature of activism in 1960s Australia. It also highlights just how vital public demonstrations in city streets had become by 1962, and concerns about the peace movement's public appeal indicate additional layers of a negotiated relationship between activists and the urban environment in which they operated.

Conclusion: 1963 and beyond

The radial march was held again at Easter in 1963; by this time, various national and international concerns again contributed to a perceived atmosphere of urgency. The announcement of the US Naval Communication Station at North West Cape in Western Australia – an American-operated satellite communications base that would formally open in 1967 – was a mobilising factor (Sydney CND 1963, p. 2; Barker and Ondaatje 2015). Additionally, the news that France would be holding atmospheric nuclear tests in the South Pacific from 1966 roused spirited opposition that would continue in Sydney intermittently – peaking in the early 1970s – until France halted its testing program in the mid-1990s (see Elliott 1997). These two issues were at the forefront of the 1963 radial march, with hundreds of marchers converging on the city from the same seven outer suburbs, combining to form a crowd of 3,000 at a final rally in the city's Hyde Park. Leaflets distributed before and during the march again emphasised the potential destruction that could be wrought on Sydney in the event of a nuclear attack. Marchers, it said, were "traversing an area which could be devastated by one nuclear bomb – a threat from which no city in the world is free" (*Peace Action* 1963, p. 3). This kind of international solidarity with an imagined urban nuclear devastation continued to be a prominent feature of local and regional anti-nuclear campaigning into the 1980s (see Schregel 2015).

Several months later, however, the Partial Test Ban Treaty was signed in Moscow by the United States, Britain, and the Soviet Union, effectively halting atmospheric nuclear testing and decimating the international anti-nuclear campaign's immediacy and sense of urgency. In November 1964, a National Service Scheme was announced, introducing conscription into the Army for eligible men, and as a consequence, the peace movement's attention was almost exclusively focused on Vietnam. CND groups morphed into anti-war committees, the Peace Committees re-routed their focus, and other, smaller groups followed suit. Just as in Britain, the particular moment of public marches utilising space and the igonography of imagined destruction for a popular anti-nuclear cause was gone.

What the exploration of urban and suburban protest in the early 1960s demonstrate is not only a key part of the Australian peace movement's heritage and its development of a more radical repertoire of dissent. By conceptualising and contesting the public spaces of protest, and how the peace movement ought to most effectively occupy those spaces, Sydney peace activists engaged in a negotiation of the "space, place, and scale" of the city (Miller 2000). Their evolving ideas about mobilising the public in these key years are of a unique historical moment, one which occurred at the confluence of an emerging youth culture, post-war affluence, and expanding suburbanisation. For the peace movement, making the best use of these factors meant engaging with the idea of Sydney as a city at risk, and the spaces of this risk, visualised with the aid of a geographically imagined landscape of nuclear destruction, formed a key part of these activists' urban politics.

References

Anderson, K. and Jacobs, J.M. (1999), "Geographies of Publicity and Privacy: Residential Activism in Sydney in the 1970s," *Environment and Planning A: Economy and Space* 31, no. 6, pp. 1017–1030.

Anon. (1962), Questionnaire responses, Box 2, Folder "Aldermaston Rally and March, Sydney, 28–29 April 1962," People for Nuclear Disarmament Records, MLMSS 5522, State Library of New South Wales, Sydney (hereafter PND Records).

Australian Broadcasting Corporation (1961), news footage, February 20, www.youtube.com/watch?v=ejjg71vRFgo (accessed 1 November 2017).

Australian Women's Weekly (1961), "Survival in a Nuclear Attack," September 27, p. 14.

Ball, D. (1980), *A Suitable Piece of Real Estate: American Installations in Australia*, Sydney: Hale & Iremonger.

Barker, A. and Ondaatje, M. (2015), *A Little America in Western Australia: The US Naval Communication Station at North West Cape and the Founding of Exmouth*, Crawley: UWA Publishing.

Bennett, G.C. (1987), *The New Abolitionists: The Story of Nuclear Free Zones*, Elgin: Brethren Press.

Bishop, R. and Clancey, G. (2004), "The City-As-Target, or Perpetuation and Death," in S. Graham, ed., *Cities, War, and Terrorism: Towards an Urban Geopolitics*, Cambridge: Blackwell, pp. 54–74.

Blackwood, M. et al. (1961), letter to the editor, *Australian Women's Weekly*, November 1, p. 30.

Boyer, P. (1994), *By the Bomb's Early Light: American Thought and Culture at the Dawn of the Atomic Age*, Chapel Hill: University of North Carolina Press.

Brown, T.S. (2013), "The Sixties in the City: Avant-gardes and Urban Rebels in New York, London, and West Berlin," *Journal of Social History* 46, no. 4, pp. 817–842.

Burtch, A. (2012), *Give Me Shelter: The Failure of Canada's Cold War Civil Defence*, Vancouver: UBC Press.

Cameron, J. (2014), "From the Grass Roots to the Summit: The Impact of US Suburban Protest on US Missile-Defence Policy, 1968–72," *International History Review* 36, no. 2, pp. 342–362.

Campbell, C. (2007), "The Middle Class and the Government High School: Private Interests and Public Institutions in Australian Education in the Late Twentieth Century, with Reference to the Case of Sydney," *History of Education Review* 36, no. 2, pp. 1–18.

Castells, M. (1983), *The City and the Grassroots: A Cross-Cultural Theory of Urban Social Movements*, London: Edward Arnold.

Cutter, S.L., Holcomb, H.B. and Shatin, D. (1986), "Spatial Patterns of Support for a Nuclear Weapons Freeze," *Professional Geographer* 38, no. 1, pp. 42–52.

Douglas, R. (2004), *Dealing with Demonstrations: The Law of Public Protest and Its Enforcement*, Leichhardt: Federation Press.

Elliott, L. (1997), "French Nuclear Testing in the Pacific: A Retrospective," *Environmental Politics* 6, no. 2, pp. 144–149.

Farish, M. (2003), "Disaster and Decentralization: American Cities and the Cold War," *Cultural Geographies* 10, no. 2, pp. 125–148.

Gentle, G.L. (1963), letter to G. Anderson, April 2, Box 2, Folder "Aldermaston Rally and March, Sydney, 20–21 April 1963," PND Records.

Gerster, R. (2014), "Exile on Uranium Street: The Australian Nuclear Blues," *Southerly* 74, no. 1, pp. 55–70.

Grant, M. (2010), *After the Bomb: Civil Defence and Nuclear War in Cold War Britain, 1945–68*, New York: Palgrave Macmillan.

Grant, M. (2016), "The Imaginative Landscape of Nuclear War in Britain, 1945–65," in M. Grant and B. Ziemann, eds., *Understanding the Imaginary War: Culture, Thought and Nuclear Conflict, 1945–90*, Manchester: Manchester University Press, pp. 92–115.

Gregory, D. and Urry J. (1985), eds., *Social Relations and Spatial Structures*, London: Macmillan.

Harvey, K. (2018), "How Far Left? Negotiating Radicalism in Australian Antinuclear Politics in the 1960s," in J. Piccini, E. Smith, and M. Worley, eds., *The Far Left in Australia Since 1945*, Oxford: Routledge, pp. 118–133.

Hobbs, H.H. (1994), *City Hall Goes Abroad. The Foreign Policy of Local Politics*, Thousand Oaks, CA: Sage.

Irving, N. (2016), "Answering the "International Call": Contextualising Sydney Anti-Nuclear and Anti-War Activism in the 1960s," *Journal of Australian Studies* 40, no. 3, pp. 291–301.

Irving, T. and Cahill, R. (2010), *Radical Sydney: Places, Portraits and Unruly Episodes*, Sydney: UNSW Press.

Lambert, M. (1961), letter to the editor, *Nepean Times*, September 21.

Lowe, D. (1999), *Menzies and the "Great World Struggle:" Australia's Cold War, 1948–1954*, Sydney: UNSW Press.

Matthews, M.E. (2012), *Duck and Cover: Civil Defense Images in Film and Television from the Cold War to 9/11*, Jefferson: McFarland.

May, E.T. (1999), *Homeward Bound: American Families in the Cold War Era*, rev. ed., New York: Basic Books.

Miller, B.A. (2000), *Geography and Social Movements: Comparing Antinuclear Activism in the Boston Area*, Minneapolis: University of Minnesota Press.

Mitchell, D. (2003), *The Right to the City: Social Justice and the Fight for Public Space*, New York: Guilford Press.

Murphy, J. (1993), *Harvest of Fear: A History of Australia's Vietnam War*, Sydney: Allen & Unwin.

Murphy, K. (2015), "'In the Backblocks of Capitalism:' Australian Student Activism in the Global 1960s," *Australian Historical Studies* 46, no. 2, pp. 252–268.

Nepean Times (1960), "Civil Defence Films at St. Marys," August 25.

Nepean Times (1961), "Babies Health Today—Atomic Aid Tomorrow," February 23.

NSW Civil Defence Organisation (1961), *Civil Defence: To the Citizens of Mosman*, pamphlet, Mitchell Library, State Library of NSW.

NSW Peace Committee (1960), "Minutes of Peace Meeting Held at BWIU Hall," October 8, Box 2, Folder "Executive Committee," PND Records.

NSW Peace Committee (1961), "Minutes of Meeting of Local Peace Group Representatives," March 4, Box 2, Folder "Peace Supporters, Peace Group Representatives," PND Records.

NSW Peace Committee (1962a), Press Release, March 20, Box 2, Folder "Aldermaston Rally and March, Sydney, 28–29 April 1962," PND Records.

NSW Peace Committee (1962b), Newport to Sydney Section leaflet, April, Box 2, Folder "Aldermaston Rally and March, Sydney, 28–29 April 1962," PND Records.

NSW Peace Committee (1962c), "Minutes of Interstate Meeting Held at Punt Road Methodist Church Hall," May 26, Box 2, Folder "National Representative Committee; Interstate Representative Committee," PND Records.

NSW Peace Committee (1962d), concert flyer, June, Box 2, Folder "Circulars, Notices, Press Releases, etc.," PND Records.

NSW Peace Council ([mid-1950s]), "Do You Know What the H-Bomb Will Do...," leaflet, Box 73, Folder "Notices, Circulars, Printed Material," PND Records.

O'Mara, M.P. (2006), "Uncovering the City in the Suburb: Cold War Politics, Scientific Elites, and High-Tech Spaces," in K.M. Kruse and T.J. Sugrue, eds., *The New Suburban History*, Chicago, IL: University of Chicago Press, pp. 57–79.

Oakes, G. (1994), *The Imaginary War: Civil Defense and American Cold War Culture*, New York: Oxford University Press.

Outlook (1962a), "Nuclear Protest: A Turning Point," editorial, 6, no. 3, May-June.

Outlook (1962b), "What Goes On," 6, no. 4, August.

Outlook (1962c), "Walk Not Run," 6, no. 3, May-June.

Peace Action (1961), "On August 5th, Australians Said: 'No More Hiroshimas,'" 2, no. 7, August.

Peace Action (1962), "Seven Ways to Sanity," May.

Peace Action (1963), "Sydney's Radial Walk and Rally," 4, no. 4, May.

Rigby, A. (1998), "A Peace Symbol's Origins," *Peace Review* 10, no. 3, pp. 476–477.

Scalmer, S. (2002), *Dissent Events: Protest, the Media, and the Political Gimmick in Australia*, Kensington: University of New South Wales Press.

Schildt, A. and Siegfried, D. eds. (2005), *European Cities, Youth and the Public Sphere in the Twentieth Century*, Aldershot: Ashgate.

Schregel, S. (2015), "Nuclear War and the City: Perspectives on Municipal Interventions in Defence (Great Britain, New Zealand, West Germany, USA, 1980–1985)," *Urban History* 42, no. 4, pp. 564–583.

Schregel, S. (2017), "Global Micropolitics: Towards a Transnational History of Grassroots Nuclear-Free Zones," in E. Conze, M. Klimke and J. Varon, eds., *Nuclear Threats, Nuclear Fear, and the Cold War of the 1980s*, New York: Cambridge University Press, pp. 206–226.

Sherratt, T. (2003), "Atomic Wonderland: Science and Progress in Twentieth Century Australia," Ph.D. thesis, Australian National University, Canberra.

Spearritt, P. (1999), *Sydney's Century: A History*, Sydney: UNSW Press.

Steinbach, J. (2002), "Nuclear Threats and Civil Defence in Australia, 1951–1957," *War & Society* 20, no. 2, pp. 91–106.

Summy, R. (1988), "The Australian Peace Council and the Anticommunist Milieu, 1949–1965," in C. Chatfield and P. van den Dungen, eds., *Peace Movements and Political Cultures*, Knoxville: University of Tennessee Press, pp. 233–264.

Sun (1946), "The Atom Bomb and Its Grim Lesson," editorial, October 29.

Sun-Herald (1960), "Gay Crowds See Festival," October 9.

Sydney CND (1963), *Newsletter*, no. 5, June.

Titterton, E.W. (1952), "One H-Bomb Could Destroy Any City," *Sunday Herald*, December 7.

Tribune (1954), "Shock for Australia in H-Bomb Effects," November 24.

Tribune (1960a), "'On the Beach' Spurs May 15 Peace Rally," May 4.

Tribune (1960b), "Hiroshima Day Rally Says: Act to Disarm," August 3.

Tribune (1961), "Sydney's 'Atomic Escape Practice' Misses the Point," February 15.

Weart, S. (1988), *Nuclear Fear: A History of Images*, Cambridge: Harvard University Press.

Wilcox, F.R. (1962), questionnaire response, Box 2, Folder 'Aldermaston Rally and March, Sydney, 28–29 April 1962', PND Records.

10 In the middle of the atomic Arena

Visible and invisible NATO sites in Verona during the fifties

Michela Morgante

INDEPENDENT SCHOLAR

Italy emerged from the Second World War with diminished international power but in a key geographical location between Eastern and Western blocs, a considerable asset for receiving economic support from USA and protection under the NATO shield. Accordingly, the Italian centrist leaders agreed that their communist opponents would be excluded from central government, thus taking advantage of the international context for their domestic policy objectives. Moreover, Cold War bipolarity was a crucial stabilising factor and modernisation of the country was pursued with large popular consent.

The influence of the USA on post-war Italy was epitomised by the Bilateral Infrastructural Agreement, signed in 1954, without any parliamentary involvement. The BIA, still classified today, was a short and rather vague understanding (Duke, 1989) ruling the presence of the Atlantic installations on 18 Italian sites. As far as we know, the key defence sites were focussed in the South and in the Tyrrhenian regions, intended to safeguard over 8,000 km of Italian coastline.

The defence of the southern Europe mainland was an equally delicate matter and Verona was to be its nerve centre in terms of technical and logistical support in time of emergency. This north-eastern medium-sized Italian city had hosted the Headquarters of Allied Land Forces Southern Europe (COMLANDSOUTH, codename: HALFSE) since 1951. The NATO command, led by Italian generals, was located in the heart of the city, in the historic Carli Palace. It was responsible for the strategic management of Italian troops and allies in the event of a conflict with Russia.

In this first stage of the Cold War, following the Warsaw Pact, the Italian north-eastern border was considered most at risk of invasion. When Austria's declaration of neutrality in 1955 opened a further potential gap in the Alps region, the USA government reaction was to activate a "combat-ready atomic-capable support force" (Fischer, 1958, p. 2). This was named the Southern European Task Force (SETAF) and was under the command of NATO COMLANDSOUTH, with its strategic headquarters in Verona and most of the troops stationed in nearby Vicenza. SETAF was supposed to be the prototype for other six similar units envisaged following the Korean War (Anderson, 1957, p. 10). Verona, therefore, with its NATO and US Forces settlements, was key to European defence, up until 1965 when all the Veneto installations were merged into the Vicenza base.

In the decade after 1951, the city was subtly, almost imperceptibly shaped by its new role. Based entirely on primary sources, this chapter analyses the manifold instances of such change. The coexistence between the Veronese and the American military is portrayed by looking at where they mixed in daily life – housing, schooling, recreational facilities. An account is given of the construction and the impact of these places on the local urban setting, including some restricted military areas. The nuclear risk and consumerism behind welfare and security pledged to Italy by the Atlantic alliance are closely examined in the second section, as additional factors in the renewal of the city.

Why was Verona chosen by NATO? Firstly, this was for geographical reasons, due to its position along the historical invasion routes, from Slovenia and the passes of Tarvisio and Brennero, which lead deep into the Italian economic core of the Po Valley. In the 1950s, Verona remained an important logistical and support point between Italian harbours and Continental Europe. Secondly, Verona, as an army stronghold until the end of 1800, boasted centuries of history as a military town. Throughout the city's fabric were pre-existing barracks and fortifications, which were still operating or easy to renovate. Moreover, nearby sites at Boscomantico and Villafranca had the potential to create two military airfields. Finally, from a political point of view, Verona was regarded as a safe city, firmly held by the national centrist party, the Democrazia Cristiana (DC). The Italian Communist Party was one of the largest in Europe, thus the Americans were perceived as challenged by "the Communists in front and the Communists behind." (Hessler, 1959, p. 29). However, the Church here was providing an adequate counterbalance; the Veronese mayor, Giorgio Zanotto, was a pragmatic Catholic reformist, who promoted his own vision of the town as a developing metropolis. From 1956 to 1964, he was the ideal representative of a citizenry that the local Community Relations Advisor of SETAF portrayed as "conservative, with roots dug way in the past, and very Catholic" (Stanghellini, 1960, p. 40).

Riding the same elevators, shopping in the same markets

The number of NATO and US personnel stationed in Verona, out of a population of around 200,000 in 1958 is not known, since they were not recorded in the national census and both propaganda and counter-propaganda were inclined to exaggerate the count. The figure of 1,000 for Verona and 4,000 for Vicenza was reliably reported by the local press (*L'Arena*, 1955A, p. 5); they probably increased over the years, so that in 1965, 2,000 military personnel could conceivably have been moved from the former leaving about 600 apartments on the Veronese market (*Il Gardello*, 1965). Instead, it should be noted that, out of error or falsehood, the local Communist Party announced that 8,000 "American occupiers" settled in the town in 1958 with their 2,500 family members (*Il Lavoratore*, 1958A).

The city almost doubled in population size from 1936 to 1961, a higher demographic increase than other Italian NATO cities: Verona + 43.8%, Naples + 36%, Livorno + 28% (Istituto Nazionale di Statistica, 1994). The threshold of 200,000 inhabitants was proudly announced as surpassed in June 1958 by the mayor, and promoted as an indicator of the city's prosperity. The Veronese population increased by 23% during the 1950s, as did the housing stock (*Edilizia veronese*, 1962, p. 70). In the post-war years, the house building sector was particularly intense, especially in 1956–1957 (*Edilizia veronese*, 1959, p. 59), when local elected representatives actually permitted a general deregulation, pending the urban plan's final approval.

The Americans had arrived in Verona at a particularly dynamic stage of development, a process that they did not steer but that was boosted by their investments, which they took advantage of. Except for a small number of enlisted men billeted in the local *caserma*, soldiers and non-commissioned officers were allowed to live "on the economy", that is in homes they had chosen and rented with their housing allowance. Officially there was no "Little America" in Verona. Blending in with the Veronese people and making small talk in the lobby with Italian neighbours (*L'Arena*, 1956) was indeed a key, and subtle, strategy against popular support for the Italian Communist Party.

Initial housing was temporarily provided by SETAF, but families had to find suitable alternative accommodation quickly, according to their needs and budget. The furniture was their own, integrated with second-hand pieces purchased from colleagues who returned home and newer pieces bought at the "post". The result was a kind of mismatched setting that did not necessarily configure with young brides' aspirations (McGraw, 2010). Since their posting in Verona typically lasted for only three years, NATO and SETAF members did not buy, but rented houses there, choosing from roughly 22,000 available (*Il Lavoratore*, 1960A). Their choices fell mainly within the new suburban districts, built by speculative entrepreneurs thanks to indiscriminate state subsidies for middle-class social housing in that period.

Despite the myth of friendship between the "ambassadors in uniform" and the people of the host countries, the US military did tend to create small *enclaves* with their colleagues' families in specific areas, as in many other overseas military installations (Alvah, 2007; Hawkins, 2005). Oddly, in Verona they did not choose the historic neighbourhood around the Passalacqua compound, Veronetta, probably because dwellings in this area did not meet modern standards. US officers preferred more distant and recently built neighbourhoods. Daily commuting across the ancient setting of downtown Verona, "along the Adige and stiched cross-tiles of San Fermo", was depicted by the 22-year-old officer and poet Charles Wright, who lived in the neighbourhood of Borgo Trento (Wright in Denham, 2009, p. 124).

Alternatively, some officers chose the Biondella neighbourhood to the east, a quiet hilly area with single-storey houses and four-family houses. However, coexistence with the Veronese people was not always idyllic: the newcomers did not go unnoticed due to the size of their cars, which were not only a luxury, but were enormous compared to the few owned by locals (one every 10 inhabitants in 1958). Some banal quarrels in the neighbourhood required intervention by the Military Police, a fact that was denounced by the local Communist press as an abuse of power (*Il Lavoratore*, 1959). The language barrier did not help "mutual understanding" and, as the Italian advisor to the SETAF Community Relations Division admitted, Verona and Vicenza were cities "among the least inclined towards novelty" and "not traditionally expansive." (Stanghellini, 1960) Life apparently ran more smoothly in the north-western part of the Veronese urban belt, in Chievo, S. Massimo or Borgo Milano; last but not least, it was the Borgo Trento district that, despite the official line, was to become their "Little America" (Morgante, 2010).

These areas were growing fast, sharing the same dull buildings, a lack of public spaces and green areas. In Borgo Trento, the interiors of newly built homes were unexpectedly well-equipped and finished, reflecting a wealthy individualistic community. This was the most prestigious district, the one most sought-after not only by the Americans but also by the local middle to high-class bourgeoisie. It doubled in size during the 1950s, with mostly 4–5 storey buildings. Despite its popularity, it represented the result of un-regulated development during delays of the local plan process, in short, a failure to control urban development through the planning system (*Urbanistica*, 1957A). The other area favoured by American families was the outer and even more popular Borgo Milano (Figure 10.1). The setting was made up of detached mid-rise buildings here too, rather different than those in contemporary American suburbs. Compared to the domestic customs of post-war America, the average size of Italian social housing units was also inadequate, so that two were often merged, especially when the family was expanding. In this context, the McGraws had to move from Borgo Trento to Borgo Milano (McGraw, 2010, p. 75), where prices were lower. The Gerards, similarly, set up home in Chievo, the best location for a serviceman stationed on the Boscomantico base, especially when raising a large family (Gerard, 2006, p. 157). SETAF officers in 1958 were likely to find a 150 sq.m apartment in a good residential area, for about 50,000 L. ($ 90) per month, equivalent to the monthly pay of a private first class. But even with four times the wages and a living quarters allowance granted to a second lieutenant, the income "was only enough to pay our food, rent and other living expenses" (Boyer D. Q., 2013, p. 167).

The local opposition party blamed the Americans for driving up the cost of living, viewing it as a result of the servility of Veronese representatives to NATO demands. The Communists argued that foreigners' financial means were incomparably higher than those of the locals and caused distorting effects on the housing market. In Verona, rents were higher due to "unfair

Figure 10.1 The neighbourhood of Borgo Milano.
Source: *Urbanistica* 1957.

competition" from people with stronger purchasing power, also exempt from both local taxes and freight rates (*Il Lavoratore*, 1958B). The Post Exchange was duty-free, which also fostered illegal trade (*New York Times*, 1958, p. 2). It was no use highlighting, as SETAF Comptroller's office did, that in 1956 in just one month more than 1 million dollars had been spent in the national economy (*SETAF Dispatch,* 1956A, p. 4).

It is not easy to ascertain whether the claims of either side were merely political rhetoric, but similar rumours also circulated in Vicenza. Supposing that the deployment of US forces did have an effect on the cost of living, then it was an equally distributed effect. In 1963, for example, rent in Verona does not seem to have been higher than those in another two NATO cities: 4,900 L. / sq.m per year on the outskirts and 6,600 in the city centre, as compared to 5,000/6,800 in Livorno and 6,500/7,500 in Naples (*Edilizia veronese,* 1963).

Schooling at the shoe factory and the Riverside

Urban growth required new community services, including schools. From the second half of the 1950s, the City of Verona had to face this need, more pressing in the growing new neighbourhoods, by granting substantial funding to these kinds of facilities. Of course, the Communist Party judged that the DC commitment to this matter was too weak. "Houses and Schools, Lower Taxation, Not Missiles" was the leftist slogan in 1960, merging the crusade for education with the petition for disarmament. If Americans paid the due usage fees for the bases – they claimed – Verona would get 7 billion L. annually, enough to build 100 new classrooms and other community services (*Il Lavoratore*, 1960B). In addition, the presence of American military households was definitely increasing the need for school buildings. The local clergy dealt with all this in an entrepreneurial way: they founded a Catholic private school tailored to NATO families' needs.

Unlike the scattered homes policy, the US approach to education in Verona was that of establishing a local branch of the Department Defense Dependents Schools (DoDDS) network. School was in fact an educational body not only devoted to children but to the community as a whole. As such, the school system was vital in making Verona a "miniature theater command" in the event of war, as Commander Harvey Fischer defined it, namely, a logistically self-sufficient base (Fischer, 1958, p. 6). The community relations programme that the Americans were trying to carry out along with the local school system was a key strategy (Stanghellini, 1960). Incidentally, this was an easy target, as education was firmly controlled by the DC party at the ministry level, as a result any kind of joint initiative in the field then was given formal approval. Cultural exchange events involving Italian teachers were closely covered by newspapers, which also promoted recurring celebrations of innovative American learning methods (*SETAF Dispatch*, 1959A). SETAF students were engaged in "know your neighbor" contests on Italian culture, with the Education Minister as chief of the jury. There were also annual visits to Italian museums (SETAF, 1962). Military wives often attended the opera at the renowned *Arena* with DoDDS teachers; other recreational activities and soccer in particular, the most popular sport in Italy, were strongly emphasised.

As with the number of US personnel and their families, it is not easy to determine how many foreign students there were in Verona at that time. Some 1,200 children were recorded in 1957 (*Time*, p. 23) but this number seems to pertain to SETAF Italian bases as a whole. It could instead be assumed, by many accounts, that there were about 200–300 pupils, from kindergarten to eighth grade. Schools for children of Atlantic servicemen were logically located in the western suburbs of Verona, near the American residential clusters. Younger children were enrolled in the new Marco Polo nursery school, operating from 1956 in Borgo Milano, with some 30 children (*L'Arena*, 1956). For the first elementary grades, as mentioned above, an American Catholic school was established in Borgo Trento, by the Catholic Chaplain, SETAF Cap. Edgar Pelletier and the Comboni Missionaries, which Cardinal Francis Spellmann, Archbishop of New York, visited in 1958 (*L'Arena*, 1958A).

For the early grades of secondary education, students could attend a school nicknamed the "shoe factory" (with reference to where it was housed) from the autumn of 1956 in Corso Milano (*SETAF Dispatch*, 1956B). To conclude formal education it was necessary to move to the on-base school at Vicenza; nevertheless the "shoe factory" also held extra-curricular classes in typing, maths and Italian. Local language learning was strongly recommended by the military authorities, it also contributed to better coordinated operations within the command, which, since 1959, had become a binational organisation (Daley, 1961, p. 9). The Riverside elementary-middle school was created by SETAF along the Adige riverbank by means of an informal agreement between the US military and Verona's mayor in December 1956. It was a small yet symbolic initiative and a perceivable sign of a forthcoming spread of the NATO settlement into the city's public land. This was perhaps the only new built-up area developed specifically for the Americans. The buildings could be clearly detected from the main road; they rose up in an area that should have been kept as open space to limit urban sprawl according to national planning practices. In fact, the Italian Urban Planning Act established that all border areas with no zoning designation were to be kept as rural areas. Moreover, this site was one of the very few municipally-owned estates in Verona, one that should have been spared from development to leave all options open for future spatial planning (*Urbanistica*, 1957b, p. 70).

Unsurprisingly, the construction of this innocuous looking building – a small but significant exception in the regulatory plan – was perceived as a sign of intrusiveness by the community. Moreover, this SETAF initiative was made public in the spring of 1957 (*Il Lavoratore*, 1957a, p. 1), at the time of the first revelations about the presence of atomic weapons in Verona, thus an age of considerable fear and hostility climate towards NATO. It was no coincidence that at the same time "anti-atomic self-protection" drills were introduced in Veronese schools as part of first-aid training (*Il Lavoratore*, 1960c). On the other hand, local communists and socialists were pressing the issue of the land-leasing fee, along with the lack of real debate within city council (*Il Lavoratore*, 1957b, p. 1). The DC counterpart defended its decision in name of the friendship between Italy and the United States of America and claimed that it offered wider economic benefits to the community. US military forces would pay a nominal fee of $1 for 12 years. But once displaced from Verona the US forces would return – according to DC representatives – some 50 million L. ($ 80,000) worth of properties, buildings later available for conversion into community facilities (City Council, 1957). SETAF also managed to impose their own conditions, such as exemptions from the contractual costs required by Italian law and the automatic renewal of the loan on its expiration date (Municipal Secretary, 1961).

The total area measured a couple of hectares and two very simple buildings were envisaged: the school (circa 150 square meters, with a laboratory, games room, kitchen and toilets), and the "activities building" (circa 400 square meters, with recreation areas, meeting room, canteen, offices). Works started in the spring of 1957 without a formal contract between the city authorities and NATO. The following summer the opening of the school, a single storey shed, 17.5m long and containing 8 classrooms, was proudly announced by the local pro-government newspaper (*L'Arena*, 1957). The Riverside School was led by the director Robert Generelli, who came to Verona as a military chaplain's assistant in 1956 and afterwards became manager of the US Army Youth Service and Morale, Welfare & Recreation (Sayo, 2004). The classes were sizeable; there were 35 pupils in grade 4 in 1963 (Verona American School FB Group). In itself though, "it was tiny and Kindergarten through eighth grade attended the same school together" (Brennan, 2007). It was not a prestigious building: its gambrel roof structure was quite similar to contemporary industrial sheds. At the rear, there was a large playground (460 square meters) with one field for football and another for baseball: the two national sports, symbolic of Italian-American fraternity.

Facilities for the morale of the troops

The area most profoundly shaped by the Americans in Verona was that of the Passalacqua *caserma*, about 25 hectares in the east of the historic city, with the barracks partly dating back to the sixteenth century, which were severely damaged in World War Two. Renovation aimed at creating "the post" produced a considerable change in previous planning decisions. In 1947, the Italian military agreed to return the area to the Municipality of Verona in order to permit its development as a residential expansion for 3,275 inhabitants. This action was judged as urgent given the critical housing shortages at that stage by the planner appointed for post-war reconstruction. However, shortly after, and rather inexplicably, the Italian military solicited for the repair of the barracks instead (Municipality of Verona, 1952, p. 14). This change took place shortly before NATO's LANDSOUTH activation in July

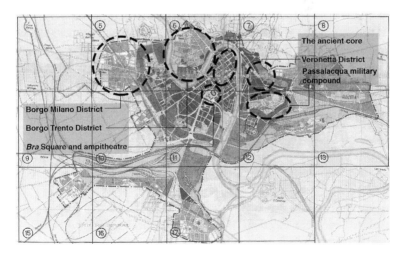

Figure 10.2 The Verona Reconstruction Plan (1948) showing the primary areas described in this chapter as settled or affected by US led development. This record shows the first envisaged transformation of the Passalacqua area into a residential area.

1951 and was a further case of NATO's apparent interference with adopted urban planning measures (Figure 10.2).

It is hard to believe that the veto on municipal plans was not connected with the new role to be played by Verona in the Atlantic defence framework. Passalacqua was a strategic installation, as it was close to the main road to Vicenza and the Porta Vescovo railway station. From 1954, the LAND-SOUTH command was approved by authorities with limited funds in the budget for construction ($ 1,208 assigned to new building + $ 2,440 to rehabilitation), but this increased exponentially between 1956 and 1958 (Military Budget Committee). Little is ascertainable about this early funding and even on the number of buildings extant on site at the moment of SETAF activation in October 1955. An aerial image taken at this time as a record for the Marshall Plan's implementation assessment shows the eastern half of the area blurred by the military censor.

Shortly thereafter, the local press gave an account of the first American Christmas celebration in Verona, which included carols and a large nativity scene on display at the Passalacqua entrance (*L'Arena*, 1955B). In the base's chapel, a Protestant Mass was celebrated every Sunday morning, while the Catholic service was officiated as well (*L'Arena*, 1956A). The following spring the first open festival was held, intended to demonstrate "Atlantic solidarity" and to provide the citizenry with a view of "the technical and organisational progress of the US Armed Forces". On this occasion, the weapons assigned to SETAF were shown (including nuclear-capable Honest John rockets) as well some model barracks (*L'Arena*, 1956B).

The bulk of the logistical and recreational facilities of the compound were mostly completed by the summer of 1957 (*L'Arena*, 1957). Appropriations for most of these works were allocated over the two fiscal years 1956–1957. At the hearings in May 1956, an amount of $4.77m was requested before the United States Congress, and this was part of the total $11.25m envisaged up to that time for three SETAF sites – the Vicenza and Verona bases, plus another classified installation in Italy, to date unidentified (U S Congress, 1956, pp. 185–188). This first tranche of grants referred to the facilities for military vehicles (parking and maintenance areas, petrol station), and others related to heavy engineering equipment, followed by funds for a runway, a hangar, a shed for signal facilities, combat training areas and a railroad junction. Other, less expensive, facilities were required for recreational uses: the officers and NCO messes (both open to personnel families), athletics fields, a broadcasting station for entertainment and a workshop for the soldiers' personal hobbies. All of the latter were considered vital in such an "isolated overseas location", for the morale and spirit of the troops (US Congress, 1956, p. 187).

In September 1956, the US military press put particular emphasis on a $16m saving from the SETAF budget for the Verona-Vicenza area, made by converting Italian buildings, by using local materials and shortening the building process (*Stars & Stripes*, 1956). To enable rapid progress with construction, existing funding was reallocated, including sums previously approved for Austria. In addition, soldiers were employed for construction work, instead of the Italian contractors expressly required by the terms agreed with Italy (Grathwol and Moorhus, 2009, p. 136). Yet, the employment of labour force from the city was regarded as crucial for consensus from the host community, and therefore was a key-issue in local propaganda. The main Verona newspaper repeatedly highlighted these works as an employment opportunity not to be missed. Indeed, more than 200 jobs had been promised (*L'Arena*, 1957). After all, the American army in Verona was exploiting several advantages of the local setting, not only the installations obtained from the Italian military, but also the building materials available in the area, mainly local natural stone. Efficiency, shortcuts in procedure and the efficacy of riding the wave of economic growth, were all heralded positively by the military and the conservative American press. Headlines like "SETAF Employs the Economy for the Sake of Economy" were rhetorical and a clear political message – according to the Western Bloc mantra, even governments should be able to seize the advantages of the free market to benefit their taxpayers, so as to produce shared wealth (Belmonte, 2008). However, this strategy was feasible only at the expense of the host countries. The United States was very much the dominant player, in that it had a free hand over local and national bureaucratic rules, contract law and urban planning regulations.

Figure 10.3 The model of the Passalacqua area.
Source: *SETAF Dispatch*, November 12 1956.

The master plan for Passalacqua was clearly defined in all its details by November 1956. A scale model with movable parts (Figure 10.3) was built by Italian military, as a sign of friendship, to show the present and future settlements. A movie theatre and a medical dispensary were already operational at that date – instead, the regular hospital, with its busy maternity ward, was located in the Vicenza base. (*L'Arena*, 1957). Soon afterwards a canteen, which would seat 600 personnel, was delivered to replace the provisional one (*SETAF Dispatch*, 1957). Construction works stopped then for a few months due to poor winter weather, but at the end of 1956 "SETAF Dispatch" announced plans for more facilities to be added to those already requested from the US Congress in May 1956, including a 168-seat snack bar, post exchange, barbers and laundry (*SETAF Dispatch*, 1956B). The "Class VI Post Exchange" (PX), established in May 1957, was a large retail store that employed 40 female clerks and provided any kind of "personal demand items", from clothing to furniture. All these duty-free goods came in from Germany and the United States (*Il Lavoratore*, 1957A).

In the summer of 1957, a swimming pool was opened, and under construction were the gym with its 6-lane bowling alley, some barracks for unmarried soldiers, a petrol station and two more workshops (total cost $230k). Not yet completed, but still scheduled, were the officers mess with bar and headquarters facilities for the command (*L'Arena*, 1957). The fully-automated laundry plant, with a huge dryer and ironing machines, that was operational by the spring of 1958, was a milestone of technological innovation introduced to Verona by Americans. This facility (cost included the PX $500,000) was at the service of all of the SETAF hospitals – and it employed 75 Italians, a conciliatory political move, as already mentioned (*SETAF Dispatch*, 1959B). This was the final provision in the reshaping of this area. Across the Adige River, the Joint Construction Agency rapidly constructed, some 60 buildings roughly similar to common industrial warehouses. $ 1.8 million worth of funding was spent between October 1955 and August 1957, not to mention works carried out by the army independently, such as the aforementioned swimming pool ($ 47,000).

The local newspaper commented on the imprint left on the townscape of Verona with a sense of admiration for "the perfect installations of the richest army of the world" (*L'Arena*, 1957). One-fifth of the expenditure needed for the Vicenza area, approximately $ 10 million to host 4,000 troops, was spent (Serafin, 2009–2010, p. 35). This was significant given that the resulting buildings were of limited architectural value.

Recreation and consumerism in the atomic Arena

Various recreational services were provided for at Passalacqua, which was conceived as a substantially self-sufficient compound also for leisure. Moreover, during their stay, the SETAF servicemen had the chance to enjoy Italy's largest lake, Garda, about 20 km away, to swim, sail and fish. Here two SETAF recreational centres were opened in the mid-50s, one for the officers in the village of Lazise, the other one in Cisano, for the lower-ranking personnel. This latter was in a private rented villa whose park provided the ideal venue for the popular American pastime of barbecuing (*L'Arena*, 1956C).

SETAF members were further provided with a recreational venue within Verona's ancient core, close to the main square in town. The *Bra* area was both a tourist attraction and a popular nightlife spot; it became a favourite for many American soldiers, including the memoir writer John Browning (2002). NATO personnel had become regulars at the Arena opera season and used to frequent the sidewalk cafes and the *ristorante Americano* on *Dietro Liston*. It was here where *Look* chose to portray life on overseas bases. "Off duty: boredom in Vicenza, beer in Verona and a break in Venice", according to the caption (*Look*, 1957, p. 140).

The *Arena Service Club* had been set up nearby, in a rented office in September 1956. It was the domain of the Verona Officers' Wives association (*SETAF Dispatch,* 1957B). The Club was located in a pompous oversized speculative building, erected in 1952–1953, not entirely in accordance with building permits, after a lengthy vacancy of the area due to uncertain provisions in monument protection regulations (Morgante, 2006, p. 145). Once again, the Americans proved to be rather ready to exploit such local

Figure 10.4 Interiors of the Arena Service Club.
Source: SETAF Dispatch, January 21 1957.

Figure 10.5 The gift of a TV to the Don Mazza Orphanage, 1957.

real estate market opportunities, provided by the gaps in regulations in an age of reconstruction frenzy as the post-war era.

Inside this condo, the Club was comprised of a grand ballroom with marble floors (Figure 10.4), a "gamesroom" with billiards, table tennis and shuffleboards, two other rooms for music practice and Italian language classes. Parties, exhibitions, concerts and screenings were often held, and documentary films from the United States Information Service on US culture and history were shown to the Veronese public (SETAF, 1962). However, one of the main features was the television inside the reading room (Figure 10.5). It should be noted that in a medium-sized city like Verona, where in 1955 only 1,000 families were subscribers to state television, new technologies were seen as fashionable features. Backing, in terms of progress and welfare, was the core of Italy's close alliance with NATO for military defence from the Soviet Bloc. In 1953 the *Mostra dell'Aldilà*, a touring government-initiated exhibit came to Verona and displayed the supposedly miserable everyday life of an *afterworld* – behind the Iron Curtain (*L'Arena*, 1953). The DC party wanted to promote a social model firmly linked to consumerism, especially through home-ownership oriented policies, a strategy that fuelled, among other things, the domestic appliance industries. Both the affluent Veronese society and NATO men in Verona lived according to this model.

The presence of 1958s Miss America in a shop in downtown Verona as a testimonial for Philco Atlantic Home Appliances, fitted this picture perfectly (*L'Arena*, 1958B). The modernisation of the domestic role of women was a clear message of rebuttal to the egalitarianism in Soviet society. There is, once again, political intent in the gifts provided by SETAF to a charity, consisting of knitting machines, washing machines and televisions. The recipients were the girls of the Don Mazza Orphanage, not far from the Passalacqua area, who received US contributions every year that were always conspicuously promoted in the media (SETAF photos series).

Even the revolutionary new atomic devices were part of the great dream of modern techno-consumerism, as described by General David Sarnoff, the American telecommunications tycoon, while on vacation in Italy in the summer of 1958. Nuclear energy, it was avowed, would allow everyone within 20 years to have colour televisions everywhere, simultaneous translation broadcasting across the globe, atomic batteries for homes and cars and video telephones. All of this would lead to general industrialisation, shorter working hours, and improved life expectancy (*L'Arena*, 1958C).

At the same time, a mainstream community like the Veronese witnessed the first organised anti-nuclear protest movements. The women's peace committee lead by Dora Russell, in parallel to her husband Bertrand in the Campaign for Nuclear Disarmament, passed through Verona on its way to Moscow (*Il Lavoratore*, 1958C). The local section of the Union of Women, a leftist organisation, was then able to collect more than 7,000 signatures in support of their cause (*Il Lavoratore*, 1958D). Moreover in 1958, a Veronese communist Member of Parliament, Silvio Ambrosini, advanced a bill to ban atomic missile launch pads from Italy (*Il Lavoratore*, 1958E). This was consistent with the party's official position, reiterated in April 1959 by its national leader, Palmiro Togliatti, at the first Regional Conference of Veneto Communists. On this occasion, the Secretary called on the government to strengthen the economy of the region instead of treating it like an "outpost of atomic destruction" (*L'Unità*, 1959, p. 8).

Nevertheless, both national and local communists, as well as DC party members, had long been fascinated by all things nuclear. Given the general concern for potential petrol rationing in the aftermath of the Suez crisis, several articles in *Il lavoratore*, the local Communist newspaper, celebrated in the late 1950s the new nuclear power plants in Russia, while sharply criticising the missile presence at the Verona-Vicenza base as a gesture of servility, though these issues were not entirely unrelated. The governing party had a more ambiguous and disjointed position. On April 29, 1954, it voted for international arms limitation; but in 1958 it signed agreements with the United States by accepting – the only state in Europe to do so – the hosting of medium-range nuclear-capable missiles on its bases as a means to raise Italy's status in international affairs (Nuti, 2011).

Even in the less troubled setting of local politics, the DC mayor repeatedly dodged questions on the nuclear threat raised by the opposition, referring to it as being out of local interest. Instead, the Veronese left was committed to this delicate issue, due to public sentiment in the post-Sputnik era. Since 1957, the rumoured presence of atomic weapons in Verona, in the face of the Soviet threat, had created apprehension not only for individuals' lives but also for the "local art treasures which belong to mankind as a whole" (*Il Lavoratore*, 1957C, p. 1). However, the local communists also dealt with the matter in a

250 *Michela Morgante*

Figure 10.6 Communist cartoon which satirised the Veronese DC party. The signs read: *vegetables, frozen foodvegetables, frozen food, missiles, steaks in cellophan, fresh white flowers* (DC symbol), Fanfani (a DC leader) speeches, super frauds. The party is claimed of supporting large retail chains at the expense of small shops.
Source: Il lavoratore, February 26 1959.

specious manner: at the opening of the first supermarkets in Verona, for instance, DC support for retail magnates and Atlantic defence provisions were gratuitously paired together, with a scathing tone (Figure 10.6).

A general climate of ideological confrontation might explain the campaign of paranoia launched against the construction of such banal infrastructure as a road underpass. The S. Zeno in Monte tunnel was a former World War Two air raid shelter, just 1 km from Passalacqua, which was to be refurbished for traffic to connect the developing eastern neighbourhoods to the city centre. This provision was resumed in December 1957 as part of the regulatory plan implementation. It was a straightforward facility to build, a tunnel of less than 300 meters and 12 m in diameter, costing 130 million L. ($ 200,000). However, the mayor was called upon by the city council to justify its usefulness and to provide adequate information on traffic (*Il Corriere del mattino,* 1957). In 1960, the local Communist Party announced, rather deceptively, that the true purpose of the works was to make a great atomic bomb shelter, as was already under way in some American cities (*Il Lavoratore*, 1960C) – this was not the case.

Whatever the risks, the American presence brought a touch of modern glamour to the quiet town, which was chosen as the set for several films and television programmes. SETAF's activation gave rise to releases by the Army Pictorial Center, including a 1957 episode of the popular *The Big Picture* television series largely dedicated to "the United States Army's only operational missile command in Europe" (*BP*, episode 378), and *Overseas Tactical Operations* (1962), set at Passalacqua and Boscomantico (*L'Arena*, 1962). In the same vein, in 1959, Peter Ustinov's play *Romanoff and Juliet* was staged on the base by SETAF's theatre group. It was a Cold War era parody of the Shakespearean drama, revolving around the love affair between two diplomats' children, an American and a Soviet (*SETAF Dispatch*, 1959C). As for Boscomantico, in the early 1960s, not only did director Elia Kazan visit the small airport with actress Susan Strasberg but another prominent Italian director, Michelangelo Antonioni, also filmed some of the key scenes of his *L'Eclisse* there. The location was the bar at the *Aeroclub*, where white and black African-American soldiers were shown drinking beer in a rarefied and almost dreamlike atmosphere. A few scenes afterwards a frame unexpectedly showed a full-page newspaper headline stating that "The Atomic Race" is underway out there. Furthermore, in the summer of 1961, Anthony Quinn was playing in the colossal *Barabba*, partially shot in Verona. Extras were recruited from among the children of SETAF personnel, including a young model based in town – Sharon Tate – one year before being portrayed in "Stars and Stripes" with her famous ride on a missile (Tate, 2014).

A modern outdated dystopia

The upbeat mood of the US military in Verona was abruptly interrupted in October 1962. Following the Cuban missile crisis, NATO personnel remained on full alert for several months and the American families had to stock up on supplies and medicines and be prepared for rapid evacuation (McGraw, 2010).

The military wives and Veronese community could never have imagined that at the same time the construction of a colossal underground structure was afoot, 30 km from Verona, in the Garda area. It was the secret bunker of COMLANDSOUTH, a glimpse of which was included in the programme of the National Defense College's "Cold War training curriculum" (Langford and Langford, 2011, p. 117). Designed between 1958 and 1960, "West Star" (or "Site A") started operating very late, in 1966, when SETAF command had already left Verona for Vicenza and shortly before Italy signed the Non-proliferation Treaty. Consisting of a vast space –13,000 square meters– excavated inside a low mountain, with about one-third dedicated to the operations rooms, a third for circulation and the remaining third taken up by sophisticated transmission equipment.

Until at least 1964, SETAF was responsible for the storage of nuclear warheads in the Verona area (NY Times 1964). Despite rumours to the contrary, the West Star bunker was never intended to contain missiles, it served as a wartime command headquarters both for LANDSOUTH and air defence forces. It was linked to a second minor bunker called *Back Yard*, about 50 miles away, in hills to the east of Verona. This underground network strongly evokes the underground military complex conceived by Mordecai Rosswald in *Level Seven*, a successful science fiction novel issued in 1960 in its Italian version by the Verona-based publisher Mondadori.

Access to Site A was through a tunnel more than 1 kilometre long. The underground space was comprised of about 100 rooms located on two levels. It could accommodate up to 500 civilians and military personnel, and was substantially equipped with facilities including a cafeteria, infirmary, bar, barber, smoking room, gym and dormitories. Rare interior photographs show a series of mostly undifferentiated spaces, carved out from four major tunnels, covered with barrel vaults in exposed concrete. It was equipped with standard office furnishings and shoddy lighting fixtures. There was no special care taken in the design, no 'high-tech' features, very few personalised touches by its users, except for the COMLANDSOUTH marble bathroom, huge geographic maps on the walls of the joint conference room and wallpaper with a landscape scene in the canteen. This type of mundane and unspectacular fit out, whilst typical, was not reflective of the high technological aesthetic of similar facilities imagined by contemporary cutting-edge design culture (Monteyne, 2011).

Being essentially a shelter for the chief command, safety conditions in case of nuclear, chemical or bacteriological attack were mandatory and guaranteed by means of devices such as watertight and blast-proof doors, air filters, water purification and decontamination showers. Constant air conditioning ensured 20–22°C temperatures, and spaces were soundproofed and isolated from radio interference to ensure the secrecy of communications. Until 2004, it was used four times a year for war simulations. This unreal space – described by NATO men to their wives as simply "a large cave" (Mc Graw, p. 62) – was experienced as part of a standard routine, which included two weeks in the field and four at the base. However, in the early 1960s, personnel working on Site A could be lucky enough to have a face-to-face encounter with General Bruce C. Clark, commander of all forces in Europe.

The threat of nuclear holocaust continued to stir up a mix of fear and fascination in the average perception of the Veronese. Even in the mid-1980s, the project of an innovative antinuclear shelter in a local condominium was advertised in the papers. In case of emergency, each of its 250 tenants would have just 3 m^3 available, 5 m below ground (*La Repubblica*, 1984). These kinds of underground anti-fallout shelters – either governmental, community or private – are one of the most explored subjects of English and American studies on nuclear era architecture. This is perhaps the more tangible feature shared between this case study of Verona and other cold war cities, as these bunkers all relied on the same rationale (secrecy, resilience, wiring, etc.). Besides that, the local planning decisions cannot be regarded as profoundly shaped by NATO defence measures. For decades

Verona maintained its dense structure and the potential vulnerability of downtown areas in case of nuclear war was never called into question in local debate. The main inner-city military area, Passalacqua, was inserted within the urban fabric as a "foreign body", a fact that went quite unnoticed in the compelling post-war urban renewal process, even though this central area remained utterly inaccessible to its inhabitants for a long time. Conversely, both in housing and leisure activities, the American community emulated the Veronese regarding choices of the most popular spots, thus sharing spaces – even in times of overwhelmingly dominant American mass culture – lead to a partially mutual hybridisation.

References

Anderson O., 1957, "Nuclear Army of Future is in Italy; Prototype for Atomic Support Commands", *Plattsburgh Press-Republican*, May 27.
Belmonte L. A., 2008, *Selling the American Way: U.S. Propaganda and the Cold War*, Philadelphia, PA: University of Pennsylvania Press.
Boyer D. Q., 2013, *Path Chosen. Life of a Lakota*. Bloomington, IN: Xlibris Corp.
BP, episode 378, *The Big Picture* series: https://www.youtube.com/watch?v=mpB5EJQVS-k.
Brennan J. K., 2007, *A Dance in the Woods*, Electronic Version. Albuquerque, NM: Casa de Snapdragon Publishing LLC.
Browning J., 2002, *Flint Hill*, Dorrance Publishing Co., Inc., Pittsburgh.
City Council, 1957, Session May 31, City Council reports collection, Verona City Archives.
Daley J. P. Major General, 1961, "Birth of a Binational Command," *Military Review*, no. 41, Jan.
Donna A., 2007, *Unofficial Ambassadors: American Military Families Overseas and the Cold War, 1946–1965*. New York: New York University Press.
Duke S., 1989, *United States Military Forces and installations in Europe*, Stockholm Peace Research Institute, Oxford University Press.
Edilizia veronese, 1959, no. 5, May.
Edilizia veronese, 1962, no. 7, July.
Edilizia veronese, 1963, no. 12, Dec., "Mercato immobiliare (quotazioni medie indicative)."
Fischer H.H. Major General, 1958, "SETAF Pilot Organization for the Future", *Army Information Digest*, vol. 13, no. 7, U.S. Department of the Army.
Gerard R. J., 2006, *The Road to Catoctin Mountain: A 20th Century Journey*. Bloomington, IN: Xlibris Corp.
Grathwol R., Moorhus D. M., 2009, *Bricks, Sand, and Marble: U.S. Army Corps of Engineers Construction in the Mediterranean 1942–52*, Center of Military History, Corps of Engineers, U.S. Army Washington, D.C.
Hawkins J. P., 2005, *Army of Hope, Army of Alienation: Culture and Contradiction in the American Army Communities of Cold War Germany*. Tuscaloosa, AL: University Alabama Press.
Hessler W. H., 1959, "'Honest John' in the Po Valley", *The Reporter*, vol. 20, Mar. 15.
Il Corriere del mattino, 1957, Sept. 25.

Il Gardello, 1965, Oct. 8., "Se ne va da Verona il comando U.S.A.".
Il Lavoratore, 1957A, "Verona bersaglio atomico," May 2.
Il Lavoratore, 1957B, "Servilismo ad oltranza," May 2.
Il Lavoratore, 1957C, May 16.
Il Lavoratore, 1958A, Feb. 20, "Gli americani a Verona hanno trovato l'America."
Il Lavoratore, 1958B, Dec. 11. "Vive preoccupazioni per il caro-vita a Verona."
Il Lavoratore, 1958C, June 26, "Una delegazione di donne inglesi ha portato un messaggio di pace a Verona."
Il Lavoratore, 1958D, Oct. 30, "Oltre 7.000 firme delle donne Veronesi contro i missile e le armi termonucleari."
Il Lavoratore, 1958E, June 19.
Il Lavoratore, 1959, Sept. 10, "Come al tempo delle SS."
Il Lavoratore, 1960A, Jan 8.
Il lavoratore, 1960B, Sept. 25, "Cosa significa il disarmo: case e scuole, meno tasse, non missili – Aumenta il costo della vita per la presenza delle truppe *USA*."
Il Lavoratore, 1960C, "Educazione atomica", May 22.
Il Lavoratore, 1960C, July 8, "La galleria di via N. Sauro."
Istituto Nazionale di Statistica, 1994, *Popolazione residente dei Comuni, Censimenti dal 1861 al 1991,* Istituto poligrafico e zecca dello Stato, Roma.
L'Arena, 1953, May 5 and May 9.
L'Arena, 1955A, Sept. 30.
L'Arena, 1955B, Dec. 24.
L'Arena, 1956A, Oct. 25.
L'Arena, 1956B, May 19 and May 20.
L'Arena, 1956C, Jan. 2.
L'Arena, 1957, Aug. 8.
L'Arena, 1958A, Oct. 3.
L'Arena, 1958B, Apr. 9 and Apr. 10.
L'Arena, 1958C, Aug. 5.
L'Arena, 1962, Nov. 9.
La Repubblica, 1984, "Diventa un condominio il rifugio antiatomico," *La Repubblica*, May 5.
Langford M., Langford J. W., 2011, *A Cold War Tourist and His Camera*, McGill-Queen's University Press, Montreal.
Look, 1957, Oct. 15.
L'Unità, 1959, Apr. 6, "Il discorso di Togliatti alla conferenza Veneta."
Malatesta L., Trevisan G., Pozza A., De Castro C. R., 2015, *Viaggio nelle basi segrete della NATO West Star e Back Yard*, Pietro Macchione Editore, Varese.
McGraw F., 2010, *Safe Landings: Memoirs of an Aviator's Wife*, Tate Publishing, Mustang.
Military Budget Committee, Reports Series on the Budget Estimates of Headquarters Allied Land Forces Southern Europe, NATO Archives Online.
Morgante M. (ed.), 2010, *Borgo Trento, un quartiere del Novecento fra memoria e futuro*, Fondazione Cattolica Assicurazioni, Verona.
Morgante M., 2006, *Il piano è redatto con giudizioso accorgimento, un po' azzardoso in qualche parte* in Vecchiato M. (ed.), *Verona. La Guerra e la Ricostruzione*, La Grafica, Verona.
Municipal Secretary, 1961, Internal Report Feb. 9, Verona City Archives.

Municipality of Verona, 1952, *Progetto di piano regolatore generale, Relazione tecnica*, Arena, Verona.
New York Times, 1958, Sept. 6, "Italian Smuggling Discovered."
Nuti L., 2011, *Italy's Nuclear Choices,* Unit on International Security and Cooperation (UNISCI) Discussion Papers, no. 25 (January).
Sayo J., 2004, "Sun Sets on US Military's 60-Year Stay in Verona, Italy," *Stars and Stripes Electronic Edition*, June 13.
Serafin M., 2009–2010, *Le basi americane a Vicenza. Dalla SETAF all'U.S. Army Africa*, dissertation, Università degli Studi di Padova, Facoltà di Scienze Politiche.
SETAF 1962, Community Relations Branch, *Report 21 April 1961*, 87th Congress, 2nd Session, Report No. 1549, pp. 116–118.
SETAF 1962, Community Relations Branch, *Report 21 April 1961*.
SETAF Dispatch, 1956A, no. 9, Nov. 19.
SETAF Dispatch, 1956B, no. 11, Dec. 3.
SETAF Dispatch, 1956B, no. 7, Nov. 5.
SETAF Dispatch, 1957A, no. 2, Jan. 14.
SETAF Dispatch, 1957B, no. 3, Jan. 21.
SETAF Dispatch, 1959A, no. 1, Jan. 5.
SETAF Dispatch, 1959B, no. 4, Jan 26.
SETAF Dispatch, 1959C, no. 10, Nov. 30.
SETAF Signal Photos series, Istituto don Mazza Archives, Verona.
Stanghellini E., 1960, "SETAF is a Friendly Word," *Army information Digest*, vol. 15, Aug., U.S. Department of the Army.
Stars & Stripes, 1956, Sept. 1, "SETAF Employs the Economy for the Sake of Economy."
Tate D., 2014, *Sharon Tate: Recollection*, Runningpress, Philadelphia, PA.
Time, 1957, vol. 69, 24 June, "Jeeps in the Marketplace."
U S Congress, 1956, *Military construction appropriations for 1957*, Eighty-fourth Congress, Second Session, May 14, pp. 185–188.
Urbanistica, 1957A, no. 22, July, "Verona aspetti del recente sviluppo edilizio."
Urbanistica, 1957B, no. 22, July, "Il Piano regolatore generale di Verona."
Verona American School FB Group, https://www.facebook.com/groups/VeronaAmericanSchool/photos/
Wright, "Driving to Passalacqua" (1960), in Denham R. D., 2009, *The Early Poetry of Charles Wright. A Companion 1960–90*, McFarland & Company, Jefferson and London.

11 Conceiving the atomic bomb threat between West and East

Mobilisation, representation and perception against the A-bomb in 1950s Red Bologna

Eloisa Betti

UNIVERSITY OF BOLOGNA

> If an atomic bomb of the same strength as the one dropped on Hiroshima should explode over Piazza Maggiore, the whole of our city would be reduced to a heap of rubble, no one would be saved from death. The irradiated city could not be crossed by any living being, without danger to his or her life and for that of others for several weeks.
>
> (*La Lotta*, 1950f)

The Italian Communist Party governed Bologna uninterruptedly between 1945 and 1989 (Baldissara 1994; Maccaferri, Pombeni 2013), while at national level Christian Democrats controlled the Italian government and the Parliament. In an era of "atomic urbanism" (Farish 2011) the city positioned itself ideologically in the Eastern bloc even if it belonged geographically and politically to the West. Bologna represented a peculiar case of "local communism" (LaPorte 2013; Wirshing 2013), elsewhere defined as "communisme municipal" (Bellanger 2013), a nuanced situation that did not reflect the typical bipartite cartography of the Cold War.

Bologna did not fit in the hegemonic geography and representation created by the Cold War. Considering the latter also as a spatial concept (Farish 2011), "Red Bologna", together with "Red Emilia", generated a geography of panic and paranoid political projections both at the national and international levels. The colour red was used by the Communists to visualise their own strongholds, while maps shown in rival propaganda often rendered the Emilia-Romagna region in a red to match the Soviet bloc, somehow suggesting that it could be a bridge for a possible invasion by the Communist East. The city and its region were actually labelled "Red", first by Communists themselves[1] (Togliatti 1946, Bertucelli 1999; Anderlini 1990) and later on by foreigners (Jaggi, Muller, Schmid 1977).

Other European cities were named "red", "rouge", "rote" by Communists militants themselves (Knotter 2011), who exhibited a shared sense of pride for their political identity (Hastings 1991). This was the case of "der rote Wedding" (Berlin), "Red Clydeside" (Glasgow), "Red Poplar" (London) and "la banlieue rouge" (Paris) in the inter-war period. In the Cold War era "little Moscows", as *Time* defined the Dutch rural village of Finsterwolde in 1950 (*Time* 1950), or "ville rouges", as *Life* addressed the French industrial town of Saint-Junien, could be found all over Europe as showed by comparative research concerning France, Belgium, the Netherlands, the UK and Italy itself (Knotter 2011; Lagrave 2004).

Although there were other Communist municipalities, Bologna was the biggest Communist-run Western city, located precisely in the country, Italy, with the most powerful Communist Party in the Western bloc (Duggan 1995). The city, then, threatened the established militarised geography of the Cold War, acting as a bridge between the two allegedly separate blocs. Bologna played an active role in the Global Cold War (Westad 2005), promoting international exchanges and interactions across the Iron Curtain (Autio-Sarasmo, Humphreys 2010; Autio-Sarasmo, Miklóssy 2011) as well popularising Soviet culture in the West. From the late 1940s, the city became the main stage of cultural and political cooperation with the Soviet Union and other communist countries such as East Germany, Romania, Vietnam and also China. Many official delegations from the East were given hospitality with festivals and other celebrations involving a large number of citizens, while Bolognese delegations visited the Soviet Union and other communist countries, sharing publicly their memories upon their return (Betti 2016).

Bologna became a threat not only for the Italian government but also for the United States, which explicitly required Italian authorities to take action to combat the Communist menace and "red cities" in particular (Marino 1991, 1995; Dogliani 2017; Brogi 2011). Anti-communist measures were particularly harsh in Bologna, leading to an escalation of political violence (Betti 2016, 2017). Bologna was a major arena for ideological rivalry. Communist militants fiercely criticised both the North Atlantic Treaty underwritten in 1949 and the invasion of Korea by the United States in 1950. The myth of the Soviet Union was intertwined with the "demonization" of United States policies (Kertzer 1996; Flores 1991; Aga-Rossi, Orsina 2004). The nuclear threat was perceived and represented by the Communists as closely related to the US warmongering, while the Christian Democrats blamed the Soviets for the very same reason (Brogi 2011; Dondi 1995).

Informed by the myth of a peaceful USSR (Schipperges 1980; Fincardi 2007), Communist propaganda (Novelli 2000) provided a peculiar aesthetic of the nuclear threat, specific to 1950s Bologna. Urban life in Bologna was influenced by an anxious representation of Cold War danger, especially that of nuclear weapons. Bolognese Communists created a Cold War scenario of urban disaster, informed by imaginative geography (Farish 2011). The Communists considered central Bologna as a possible target for a US nuclear attack. Dozens of articles on the effects of a possible A-bomb explosion along with drawings and various kinds of diagrams, montages, photos were published in the local Communist press to shock Left-wing Bolognese supporters into action. The perception of the atomic threat, galvanised through visual, oral and written sources, led the Bolognese to take part in dozens of rallies and demonstrations against the deployment of the A-bomb. Amongst them were many politicians, intellectuals, representatives of female associations, and trade unionists.

Left-wing women, in particular, played a major role in mobilising against the A-bomb, gendering the public narratives and iconography of the atomic threat. Bologna, in the Cold War period, was also an atypical case in regard to women's participation in the public and productive spheres. In 1950s Bologna, 80,000 women were registered with the Union of Italian women (UDI), some 63,000 were members of the Italian Communist Party (PCI) and 70,000 were part of the Leftist Italian General Confederation of Labour (CGIL) (Betti 2015; Furlan 1993). Some women unionists became General Secretaries in garment, textile and chemical unions during the 1940s and 1950s. A total of 95 female municipal councillors were appointed in 52 local municipalities of the provinces of Italian; in 1956, seven in the city of Bologna alone.[2] About 20,000 women worked in Bolognese factories (34.6% of the overall factory labour force), including male-dominated sectors like metalworking (Betti 2013). Women were, then, greatly involved in post-WWII politics, in the labour movement struggles and peace campaign, taking part in strikes, factory occupations, marches and rallies (Betti 2015).

Fighting the nuclear threat and campaigning for peace were effectively two parts of a common strategy for Italian as well as Bolognese Communists, who did not separate the agendas in their political discourse and activities (Andreucci 2005). Nevertheless, the widespread perception of the atomic menace led to groups other than Communists also assuming roles in the anti-nuclear movement, including Catholics (Cerrai 2011; Giacomini 1984). Left-wing activists who participated in rallies and demonstrations were often beaten up, put on trial and imprisoned during an escalation of political violence in Bologna in the early phase of the Cold War (Betti 2016). Due to their political activism, also Left-wing Bolognese women, as well as men, were the subject of various forms of repression (Betti 2017).

In this chapter, I will show the atypical nature of Bologna's Cold War political condition and how this manifested culturally, through an analysis of media, and socially, by examining the role of women in political activism. The chapter has been divided into three main parts. The first provides evidence of the high level of mobilisation against the A-bomb, looking at the various actors involved in a trans-local perspective, with a specific view of the role of women's activism and of the local case of Bologna. Here, I will also connect the urban level to the wider global context of the peace movement. The local account is based on archival documents and photographs as well as other written documents of specific events linked to the major actors in Italy and Bologna.

The second is devoted to the analysis of the multiple forms of representation of the nuclear threat. As already mentioned, iconography played a crucial role in the political strategy of the Italian Communist Party, the main political actor taken into consideration in the chapter. Posters, leaflets and the political press clearly testify to the existence of an intentional system of representation which required further exploration. Textual, visual and event driven propaganda will be investigated to understand how these were adapted and nuanced to the specific agendas of the Communists.

The third and final part examines the familial and gendered perception and memory of the nuclear threat through written accounts, reportage, memoirs, interviews to better understand the conventional social discourses that existed around the atomic threat and how these tallied or contrasted with views proffered by activism and forms of media representation.

Mobilising against the A-bomb: a gendered approach

> It is not an exaggeration to say that the great success of the collection of signatures for peace should be merited to the Italian democratic women; female "peace visitors", who in dozens of provinces went home by home introducing the text of the petition and the related form, women who arranged temporary tables for the collection of signatures in public markets and work places, women who organized parties, rallies, shows
> (UDI 1949)

In the early 1950s, the mobilisation against the A-bomb became one of the main goals of the "Partisans of Peace" who led the so-called "Cold War communist peace campaign" (Brogi 2011; Jenks 2003). The organisation was formally established during the "First World Congress of Peace" held in Paris in April 1949 and attended by more than 2,000 delegates from 72 countries (Giacomini 1984). The Peace Partisan movement was conceived by the Communist Information Bureau (popularly referred to as Cominform) to be "its main anti-American offensive in the West" (Brogi 2011). The Partisans of Peace were strongly attached to the Soviet bloc as well as Western Communist Parties on the other side of the Iron Curtain. However, the Partisans of Peace were also keen to involve liberals and Catholics with their campaigning in the West to expand the movement beyond the original Communist and Socialist milieu (Cerrai 2011).

The movement, which reached its peak between 1949 and 1956, was shaped by the reality of the A-bomb and the fear of the H-bomb. Concerns escalated after US President Henry Truman announced the decision to continue and intensify research and production of thermonuclear weapons in 1950 (Brogi 2011). A Permanent Committee of the World Peace Congress was formally set up after the Paris Congress. In its third meeting held in Stockholm in 1950, the Committee issued a Declaration, the so-called "Stockholm Appeal", which demanded "the outlawing of atomic weapons as instruments of intimidation and mass murder of peoples", and explicitly condemned the "government which first uses atomic weapons against any other country" as a war criminal (Wittner 1993).

Italy played a crucial role within the international movement of the Partisans of Peace. The Italian branch was the most active participant amongst the Western European states (Cerrai 2011). Unsurprisingly, Pietro Nenni, leader of the Italian Socialist Party, belonged to the Permanent Committee of the World Peace Congress (Giacomini 1984). Moreover, the crucial role of Italy was recognised by the international movement that selected the Italian city of Genoa to host the second World Congress of Peace. After the opposition from the Italian government, who did not grant the necessary permissions for the organisation of the event, the Congress eventually took place in Warsaw (Giacomini 1984).

Nevertheless, thanks to the organisational skills of the Italian Communist Party and through close cooperation with the Socialist Party, the Partisans of Peace became widespread all over Italy, and successfully managed to involve some Catholics and liberal democrats in their activities (Cerrai 2011). Due to the strength of the Partisans of Peace, Italian mobilisation against the H-bomb was the largest in Western Europe: more than 19,000 local committees were actively involved in the collection of signatories for the 1950 Stockholm Appeal. According to the estimate of the movement, more than 16 million Italians signed the Appeal (Giacomini 1984), 1/3 of the Italian population (47,398,489 in 1951).[3] Left-wing trade unions (Italian General Confederation of Labour) as well as women's (Union of Italian Women) and youth (Italian Communist Youth Federation) associations combined to strategically recruit more and more Italians to the peace movement (Cerrai 2011).

"Red Emilia", and its capital Bologna, were the frontline in the mobilisation for peace during the early Cold War. In 1950, the Mayor of Bologna, Giuseppe Dozza (Lama 2007), took part in the third meeting of the Permanent Committee in Stockholm. After returning to Italy, he played a key role, alongside the neorealist novelist Renata Viganò and the Communist MP Emilio Sereni, in the First Peace Congress of the Bologna area, organised by the local Committee of the Partisans of Peace. Several intellectuals, politicians, leaders of the unions and women's association took part in the Congress, which was very well attended (*La Lotta* 1950a). The Bolognese Committee also strongly supported the petition for the Stockholm Appeal. According to estimates, about 500,000 signatures were collected in the Bologna area alone (*La Lotta* 1950e; *L'Unità* 1950a), around 2/3 of the overall Bolognese population (763.907) signed the petition. In Italy, Bologna was second only to Genoa (Giacomini 1984).

The peace movement and the campaign against the H-bomb was strongly opposed. The Italian government, through police forces and prefects, tried to hamper the peace campaign in many ways, as they were seen as a threat to the Cold War order (Marino 1995, Dogliani 2017). Repression by police forces during the anti-communist escalation of the early 1950s was particularly brutal in 'Red' cities (Betti 2017; Bertucelli 2017). Between 1948 and 1954 in Bologna, there were 13,935 trials for law and order offences, 7,531 of which resulted in guilty verdicts.[4] According to the "Committee of Democratic Solidarity" *(l'Unità* 1948a), whose aim was to provide legal aid and concrete support to activists who were put on trial, many experienced imprisonment and other forms of repression (Ponzani 2004). Dozens of Bolognese were put on trial and sentenced to prison for taking part in peace demonstrations, against the Atlantic Treaty and the use of the H-bomb.

Demonstrations and protests increased in the city, after the press reported the declaration of the US President Henry Truman and other members of the US Congress to be in favour of the production (and use) of the A-bomb (*La Lotta* 1950b and 1950c). US politicians were labelled "enemies of humanity" by *L'Unità* (1950b). A number of female activists were arrested during those demonstrations; a woman was imprisoned for 10 days, simply because she was handing out flyers against the US General Matthew Ridgway, involved in the Korean War at the time (Betti 2017).

Women had a strategic role in the mobilisation for peace and against the atomic threat (Michetti, Repetto, Viviani 1984), in particular the Union of Italian Women (UDI), the largest women's association in post-war Italy affiliated to the International Women's Democratic Federation (WIDF) (Pojman 2013). The latter played a major role in the foundation of the Partisans of Peace (Giacomini 1984, Cerrai 2011). The UDI's mobilisation for peace and against the atomic bomb preceded the establishment of the international movement, as the association first launched their collection of signatures in 1947 during the so-called "peace week" (Cerrai 2011). In 1948, conferences labelled "Assise della pace" were promoted by left-wing and UDI women (Gabrielli 2005; Michetti, Repetto, Viviani 1984) at national (Rome) as well as at local level (*Noi Donne* 1948) in cities like Bologna (*La Lotta* 1948). Additionally, UDI national and local congresses showed an enduring commitment to peace, attached to the association's anti-fascist legacy (Gabrielli 2005). They condemned the use of nuclear weapons and emphasised the reality of the atomic threat, [5] being on the frontline collecting signatures for the "Stockholm Appeal" and taking part in several demonstrations against the deployment of the A-bomb in the early 1950s (*La Lotta* 1950c; *L'Unità* 1950b). UDI women also managed to engage high-profile liberal and Catholic women, a significant number of whom signed the "Stockholm Appeal" (Cerrai 2011).

The strength of the UDI in the Bologna area, and its commitment to peace, led to a huge mobilisation against the H-bomb (Betti 2017). Photographs taken by both amateur and professional photographers, now preserved in the UDI historical archive of Bologna, are crucial to understanding the extent to which the campaign for peace and against the A-bomb was highly gendered and particularly visible in the urban space of Bologna. Photographs show how public squares and streets became the core spaces of peace mobilisation mostly led by women (Cerrai 2011), but also which kind of flags, dresses and gadgets were used to create a sense of belonging among peace activists as well as to symbolically mark out the spaces themselves.[6]

Figure 11.1 shows how in 1949 peace marches were very well attended by women and children, thanks to the involvement of the UDI and the Communist Youth Association "Pioneers". Streets, squares and other public spaces, like parks, were peacefully invaded by activists between 1949 and 1950.[7] Women were also photographed in rallies under the slogan "friends of peace, friends of the USSR", reminding us of the atypical political alliance

Figure 11.1 A peace march in Bologna in 1949. The procession was formed predominantly of women and children.
Source: UDI Archive of Bologna.

Figure 11.2 Bolognese citizens crowd a square in a rally for peace during 'peace week'.
Source: UDI Archive of Bologna.

of Bologna with the Soviets. Figure 11.2 shows thousands of Bolognese taking part in a rally during for the so-called "peace week".[8] The word "peace", created with light bulbs was placed on the town hall of the Municipality of Bologna (Figure 11.3), which also testified to the allied support of the Communist-led local authorities. Socialist and Communist municipalities were united in an association known as "Democratic Municipalities" (Gaspari 2006), which was strongly involved in the campaign for peace and against

Representation and perception in Red Bologna 263

Figure 11.3 The word 'peace' (pace) in lightbulbs, with an illuminated globe symbol mounted above the entrance to Bologna's town hall.
Source: UDI Archive of Bologna.

the atomic bomb. Several left-wing local authorities in the Bologna area passed motion in favour of the "Stockholm Appeal" (*La Lotta* 1950b).

Peace mobilisation in the Italian context was shaped by women's activism, which anticipated the foundation of the Partisans of Peace as an international movement. Bologna was at the frontline in the mobilisation thanks to the huge activities of UDI women. As the quote at the opening of this section shows, women were able to transcend the private/public divide, directly bringing the Cold War peace campaign inside Italian and Bolognese homes. The atomic menace represented a constitutive part of the Communist peace discourse and political agenda, producing an entire set of propaganda materials, as the following section will show.

Representing the nuclear threat

> The population would be overwhelmed by the ruins; the survivors would be more unfortunate than the dead: living corpses with blinded eyes,

blood dripping from their noses, their mouths, their ears; they would vomit their own guts, they would die within a few days. In the surviving population there would appear a terrible epidemic: injections, treatments would serve no purpose: on treated limbs gangrene would appear.

(*La Lotta* 1950f)

The representation of the nuclear menace was intertwined with various forms of media promoted at different levels by the Partisans of Peace and their supporting organisations. Between 1949 and 1959, several international, national and local periodicals were created (Cerrai 2011; Giacomini 1984). *Le partisans de la Paix* (*La paix* from 1951) was founded during the Paris Congress in 1949. It was translated into several languages, including Italian with the title *La pace* (1951). In Bologna, a local magazine *Bollettino della pace* (1951) was established by the Bolognese Committee of the Partisans of Peace.

The Italian Communist Party had a crucial role in the production of media that promoted peace and was against the atomic threat (De Giuseppe 2000). Both agendas were included in the publications devoted to Communist activists such as *Il quaderno dell'attivista*, issued fortnightly by the "National Commission for Propaganda" of the Italian Communist Party (Flores 1976). At a local level, *Il propagandista*, edited by the "Press and propaganda" commission of the PCI of Bologna, clearly included "Peace" among its key topics[9]. The so-called "propagandista" (a communist activist involved in propaganda activities) was a key figure for the PCI in promoting party ideology as well as its commitment to peace and anti-nuclear stance. According to the Bolognese publication, the "propagandista" should always promote, and engage, in political discussion, defending the party's ideology and enhance its policies in the office, in the workplace, on the metro and in other public places.[10]

Each week dozens of articles that dealt with the commitment of the people of Bologna for peace were published in *La Lotta,* the newspaper of the Bologna branch of the Italian Communist Party – this amounted to thousands between 1950 and 1956. The magazine is particularly interesting in terms of its iconography, as it published drawings, several realised by famous Italian Communist artists (Duran 2014) such as Renato Guttuso and Aldo Borgonzoni, alongside articles with a mixture of photographs and captions, aimed at visually linking the nuclear threat to other political events such as the Korean War, the North Atlantic Treaty, and the European Defence Community (Figure 11.4).

Due to the fact that the Communist militants in Bologna saw themselves as closely linked to the Communist bloc (Betti 2016), they frequently reacted strongly to what they considered to be "American imperialism" and the "warmongering" of the US (Aga-Rossi, Orsina 2004). They fiercely criticised both the North Atlantic Treaty underwritten in 1949, considering it to be an act of belligerence against the Communists (*La Lotta* 1950g), and

Figure 11.4 Extract from *La Lotta* newspaper showing a range of references to the global aspect of the struggle for peace and disarmament.
Source: La lotta, 21 December 1951.

later on the invasion of Korea by the US. Using drawings and photographs to blame US warfare and the Christian Democrats for their allegiance, the pages of *La Lotta* declared that "Americans in Korea [are] worse than the SS" (*La Lotta* 1950). Contrastingly, the USSR together with the figure of Stalin were associated with peace.

Even President Truman's actions were seen as harbouring criminal intentions. The "Truman war" was visually depicted as a struggle between a man personifying the province of Bologna and invading US troops (*La Lotta* 1951b). Figure 11.5 depicts a US soldier subtly, but deliberately rendered wearing a uniform akin to that of the Nazis who occupied the Bologna area during the Second World War and destroyed Bolognese infrastructure. The reference to Marzabotto, the village where the bloodiest Nazi massacre took place, was a clear symbol of war and atrocity for the Bolognese and, as such, was aimed to shock and sway Communist activists. The Korean War was deeply associated by the Bolognese Communists to the nuclear threat. Both articles and drawings were used to create montages, containing explicit references to the Stockholm Appeal. Several pages of *La Lotta* that presented such ideas carried the footer "Sign for the Stockholm Appeal" (*La Lotta* 1950h), itself associated with notions of peace and security.

266 *Eloisa Betti*

Figure 11.5 Illustration from *La Lotta* newspaper suggesting the extents to which US military incursion affected the Bologna city region.
Source: La lotta, 3 June 1950.

The projected effects of a possible nuclear attack in downtown Bologna on the part of the United States were also published in *La Lotta,* in the form of a map featuring impossible perfect rings suggesting the degree of atomic destruction (Figure 11.6). The graphic representation was not realistic, but a clear example of imaginative geography (Farish 2011). An abstracted depiction of the city, showing its boundaries and principal monuments was deployed. Concentric circles were used to emphasise the potential damage that a detonation would create. Each demarcated zone was accompanied by a caption that described the destruction specific to that area. This representation is one illustration of how various parties with alternate agendas depicted different geographies of atomic urbanism. The extended captions explained what would happen if an A-bomb, like the one dropped on Hiroshima, was detonated over Bologna. Within the prose an explicit comparison with the WWII bombings by the Nazi was forwarded, and the consequences for people's health of an A-bomb explosion were stated in a very dramatic way – as in the quote at the beginning of this section.

Along the same lines, a drawing depicting a man staring at the sky with his eyes wide open and a hand on his mouth was published on a full page spread in *La Lotta* (1950d), together with the slogan "Sign against the Atomic bomb" referring to the Stockholm Appeal (Figure 11.7), preceded by the sentence "when you see the bombs falling then it will be too late". Another drawing portraying a skeleton with an atomic bomb

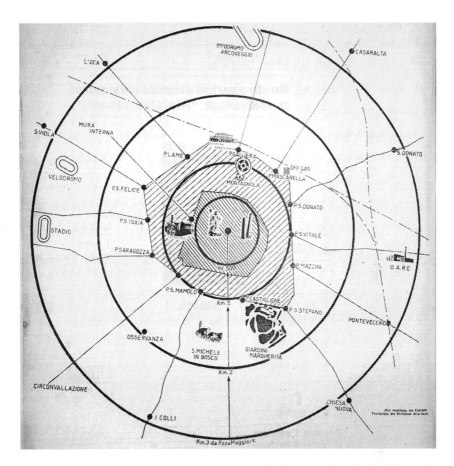

Figure 11.6 Extract from *La Lotta* newspaper showing the imagined effects of a nuclear strike on central Bologna and the sites within a series of concentric zones 1 kilometre apart from each other.
Source: La lotta, 19 June 1950.

in his hand, clearly symbolising death, surrounded by human skulls was realised by the Communist painter Aldo Borgonzoni (Figure 11.8) with the caption beneath that read: "War and death come from America" (*La Lotta* 1951a).

In 1955, another article was devoted to the largest nuclear test that occurred on March 7 in the Nevada desert, "Atomic experiment threatened human kind" titled *La Lotta* (1955), which published a photograph of the explosion. The article recalled the need to sign the "Vienna Appeal", adopted on January 19, 1955 by the World Council of Peace specifically "against preparation for Atomic War" (*La pace* 1955). Bolognese archives preserve copies of the appeal, showing a white dove symbolising peace. According to an estimate made by the Partisans of Peace, approximately 10 million Italians signed this second appeal (Giacomini 1984).

Figure 11.7 Extract from *La Lotta*. 'When you see the bombs fall it will be too late'. 'Sign against the atomic bomb'.
Source: *La Lotta*, 3 June 1950.

Figure 11.8 Illustration by Aldo Borgonzoni from *La Lotta* newspaper. 'War and death come from America'.
Source: *La Lotta,* 2 February 1951.

Similar representations of the nuclear threat were promoted by the Communists through posters, leaflets and what was called a "wall newspaper" – a special newspaper edition published as a poster and affixed in public spaces (Dondi 1995; Andreucci 2005; Gundle 2000). The fight for peace and against the atomic threat was made highly visible in the urban domain, thanks to the wall newspapers posted across the city. For this very reason, on several occasions, wall newspapers were confiscated and their editors put on trial (Betti 2016).

In addition to wall newspapers, other events contributed to popularise the Communist representation of the atomic threat throughout the city. The *Festa dell'Unità* was an annual event promoted by Communists all over Italy, successfully providing a blend of food, dancing and politics. During those Festivals, exhibitions and large parades took place (Tonelli 2012, Dondi 2004; Baravelli 2005). In the early 1950s, "peace" as well as the "atomic threat" were over-represented in these events, which created a new urban setting for the representation of anti-nuclear and peace activism in many Italian cities.

Again, the Communist newspaper *La Lotta* provides concrete visual evidence of *Festa dell'Unità* hosting anti-nuclear and peace parades. In 1949, women and men dressed in traditional costumes paraded with a giant rainbow flag symbolising peace (*La Lotta* 1949). In 1950, a parade where women and men dressed completely in white took place under the slogan "peace" (*La Lotta* 1950). The dove of peace was particularly popular in such parades. In 1954, a "peace dove" model was installed over a huge ball with the sign "peace"; while at the entrance of the *Festa dell'Unità* a slogan mentioned the H-bomb and European Defence Community, as shown in Figure 11.9 (*La Lotta* 1954c). The Communists were, in fact, against the European Defence Community, considering it an instrument of war. In *La Lotta* the European

Figure 11.9 A model of a globe with the word 'peace' and a white dove. This was mounted above the entrance to the Bolognese *Festa dell'Unita* in 1954.
Source: *La Lotta*, 3 September 1954.

Defence Community was often associated to the US warmongering and the atomic threat itself.

Photographs from a private archive of a former trade unionist, representing the 1954 Mayday Festival, show the presence of the Partisans of Peace movement amongst trade union celebrations, confirming the close allegiance between left-wing organisations. A large space within the Festival was devoted to an exhibition realised by the Bolognese Committee of the Partisans of Peace. The atomic threat was portrayed along with other symbols of the "Cold War Communist peace campaign", such as Stalin (who died a year earlier) and the Rosenberg family. Julius and Ethel Rosenberg were executed in New York in 1953, after have been found guilty of spying for the Soviets. During the trial, a worldwide campaign under the slogan "save the Rosenbergs" was launched and supported especially by Communist organisations. Bologna was again at the frontline as propaganda materials, including several articles published in *La Lotta*, testify. At the centre of the 1954 exhibition of the Bolognese Committee of the Partisans of Peace, a statue of Julius and Ethel Rosenberg, with their two children, was erected (Figure 11.10).[11]

The Bologna case shows to what extent the geography of the atomic age was truly global (Farish 2011), how the local and the global were intertwined spatially and symbolically in the everyday life of urban communities. An interconnected atomic world was created, drawing on the visual memories of the Hiroshima and Nagasaki disasters to stress the global scale of

Figure 11.10 Photograph showing a statue of the Rosenberg family at the centre of the 1954 exhibition of the Bolognese Committee of the Partisans of Peace. The Rosenbergs were executed in the United States after being found guilty of spying for the Soviets.

Source: Author's scan from private collection.

the atomic menace produced by 1950s nuclear testing. Alternative to other Western case studies, due to the atypical political constitution of Bologna, the dominant imaginative geography in the city was reversed. Due to the strength of the Communists, Bolognese citizens were exposed to a very specific imaginary of urban disaster, associated with the warmongering and nuclear activities of the Unites States. In the cultural Cold War, informed by the atomic threat of the early 1950s, urban spaces were both a target of the bomb and a site of propaganda against the bomb. Posters, wall journals, exhibitions, rallies, parade and marches created specific urban histories and shaped the urban spaces themselves.

Perceiving and remembering the atomic fear

> From what I have learned from the newspapers I think that before the serious danger of the H-bomb all the people and the major powers must agree to ban such destructive weapons. Personally, in the interests of my family, I am against all weapons of mass destruction.
> *La Lotta*, 1954b.

The aforementioned forms of political propaganda and the representation of the atomic threat had an impact on the individual perception of the Bolognese, which can be investigated thanks to different sources. Contemporary interviews and written accounts, collected and reproduced by *La Lotta* in the spring of 1954, are useful to understand the shared perception of the atomic fear among Communists militants and other Bolognese citizens. Concepts such as "home", "family", "body" emerged in the voices of Bolognese talking about the H-bomb.

In 1954, after the US nuclear test at Bikini Atoll, renewed campaigning against the H-bomb took place globally as well as in Italy (De Giuseppe 2000), leading in January 1955 to the adoption of the previously mentioned "Vienna Appeal". Among the interviewees there were female and male workers and even nuns and priests. The priest of San Venanzio di Galliera declared "I do not believe that there are men who intend to use this bomb, but it is right that the governments be committed not to deploy it, that they should think more about peace than war." (*La Lotta* 1954a). The nun, alternatively, honed in on the fear of the nuclear menace, saying that "the atomic bomb is something very serious, it is very frightening" (Ibid.).

In addition to individual voices, *La Lotta* also collected the declaration of the 25 Municipal councils of the Bologna province. These declarations clearly referenced "the population's anxiety" and the need to avoid "the catastrophe of humankind", which could have been caused by an H-bomb explosion (*La Lotta* 1954a). Several additional voices, even from members of other political parties were collected to show how mobilisation against the H-bomb went beyond the traditional Socialist and Communist milieu. In the Bolognese village of Galliera a local Committee for peace, outside of any party political affiliation, was created (*La Lotta* 1954b). At the same time the possible alliance between Communists and Catholics, envisioned by the PCI leader Palmiro Togliatti, was also sought by the Bolognese

> I am sure that an agreement between the Communists and the Catholics, as the Right Honourable Togliatti has said, is achievable because in the face of such frightening destruction there can be no honest person who for political reasons can refuse to support it.
>
> (*La Lotta* 1954b)

Almost 30 years later, a collection of memoirs, published by the Bologna PCI for its sixtieth anniversary, suggest that the vivid perception of the atomic threat became less relevant in the preceding years. The collection, *Comunisti* (1983), contained more than 80 short memoirs, which focused on specific aspects of the life and political activism of the Bologna Communists. Three short memoirs written by Communist women focus on peace-related topics, including the atomic threat. Ada Patelli, born in 1911 and member of the PCI of Bologna from 1944, described her activism in the collection of signatures for "peace" in the municipalities of the Bolognese Apennines. Ada was both a member the Communist party as well as of the Union of Italian Women (UDI) and so she took a trip, together with other female activists, to a small municipality in the Bolognese hills to collect signatures. The area was still underdeveloped in the 1950s, so travelling around was not so easy, especially for women. Although she did not provide further details, we can imagine that Ada and her friends were involved in the collection of signatories for the "Stockholm Appeal" or the "Berlin Appeal". Ada's memoirs reveal the difficulties she encountered in persuading women to sign the appeal, due to the hostility of the local priest and the legacy of the fascist regime. The atomic threat was not explicitly mentioned in Ada's account, possibly suggesting that peace was the prevailing agenda amongst Italian Communists in the final years of the Cold War, while the atomic threat had gradually waned (Patelli 1983).

Tina Anconelli, born in 1925 and member of the Bologna PCI from 1945, was a member of the Union of Italian Women, in charge of the Imola branch. She recollected the severe consequences of her activism for peace and her involvement in the Partisans of Peace in the early 1950s. Tina was arrested twice, and imprisoned for several days, simply for distributing leaflets against the Korean War and later on against the European Defence Community (Anconelli 1983). Dina Stanzani, born in 1918 and member of the Bologna PCI since 1945 as well as of the UDI, remembered her involvement for peace and against the atomic bomb in the post-war years. She recalled women's involvement against the atomic bomb, before the establishment of the Partisans of Peace. In 1947, the Union of Italian Women launched a national petition against the A-bomb, with 6 million signatures collected across the country and brought to the United Nations in 1948. Dina recalls that more than 7,500 women activists were involved in the collection of signatures in the Bologna area alone, collecting 150,000 signatures altogether. The manufacture of the so-called "peace flag" had been crucial, according to Dina, in order to involve more women and families in the mobilisation against the atomic bomb and for peace (Stanzani 1983).

Since the end of the bipolar political order, the collective memory of Cold War-related events has diminished (Romero 2009) in Italy as well as in the Italian Cold War city *par excellence*, Bologna. The city, as well as the Emilia-Romagna region, is no longer "Red", the collapse of the Communist Party after 1989 led to a complex process of renegotiation of the city's Communist identity (Capelli 2013; Sante Cruciani 2014). Very few Cold War events were remembered in Bologna after 1989 and very few signs of 'Red Bologna' remained, despite an enduring communist sub-culture (Capelli 2013; Baccetti 2003).

Mobilisation for peace, however, is still present in the individual memory of some Bolognese and certain objects that were preserved still remind us of such mobilisation. Among them, the so-called "peace flag", stitched together by women mainly belonging to the Union of Italian Women, like Dina, is one of the most significant. The flag was made thanks to the collective efforts of working-class women, who contributed patches of fabric they had lying around in their own houses. For this reason, the 1950s "peace flag" looked like a colourful patchwork of multifarious fabrics (Gabrielli 2005; Ombra 2005). Several flags sewn in the early 1950s during the massive peace mobilisation described above have been preserved by women's associations, exhibited in public spaces and sometimes adopted as a symbol to oppose conflict more generally.

In the Bologna province, the "peace flag" became a tool for creating an educational project for high school students: "the peace game". Students were invited by the women's associations (including the UDI) to re-create a "peace blanket" with small piece of fabric of their choice. Each of them could contain also specific slogan, within the wider concept of "peace." Hundreds of students took part in the project and a final exhibition with the blankets was held, hosted by Municipality of Bologna.[12] Although most of the "peace flags" were made in the early 1950s during the Cold War peace campaign, the Cold War and the atomic threat have disappeared in the collective memory as well as in the public discourse related to the subject itself. In post-Cold War Era the "peace flag" became a symbol of female anti-war activism, with little ideological connotation.[13]

The Cold War atomic threat sometimes emerges in the individual memory of Bolognese, born in the 1920s and 1930s, who remember the huge campaigns of the early 1950s and the forms of repression experienced at that time. The memories collected in the new millennium are deeply informed by the diverse political situation, showing a generalised collective amnesia regarding Cold War imaginary and topoi, as emerged by recent interviews I realised between 2013 and 2014 (Betti 2016).

The Communist activist Albertina, however, clearly recollected her participation in the 1949 Paris World Congress of Peace as a pivotal event in her life, thanks to which she could understand what solidarity meant as well as the value of human cultural diversity through encounters with people from all over the world. In her personal archive, Albertina still kept several mementoes of the Paris event, such as her registration card, booklet as well as photographs.[14] Nothing was mentioned regarding the atomic menace, nor was any object she preserved from that period related to the nuclear threat.

The 1950s nuclear threat seems to have disappeared from the collective as well as the individual memories of the Bolognese, while the 1980s atomic menace created by Chernobyl disaster is still alive. It is feasible that the vivid memory of the mobilisation that occurred in the 1980s (Liberatore 1996; Giugni 1999), which led to the closure of all nuclear power plants in Italy, has obscured the memory of the previous campaigns, due to the different political landscapes but also to the passing of several key anti-bomb protagonists. Moreover, unlike other Western countries, there were no atomic shelters in the Italian cities, except for the Soratte Bunker (Rome), converted into an atomic shelter in 1967 (completed only in 1972) for the protection of the Italian government and the President of the Republic.[15]. The age of atomic urbanism did not leave any material traces in the urban fabric of Italian cities, including Bologna. There are no ruins of civil defence or atomic shelters, while there is still a strong urban memory of WWII anti-aircraft shelters, which have been preserved, studied and are used in guided tours (Paticchia, Brunelli 2015).

Conclusion

Red Bologna is a peculiar and relevant case study for understanding alternative imaginative geographies and their impact in the representations and perceptions of the atomic urbanism. Being a Communist city in the West, Bologna represents a misaligned Cold War history, able to challenge the Global Cold War order due to its international relevance as a Red city and its global relations. Bologna created its own Cold War imaginary in the 1950s, informed by the atomic menace and disaster scenarios. The role of periodicals in promoting dramatic representation of atomic disasters has clearly emerged by the analysis of the Communist weekly journal *La Lotta,* which created a counter narrative and iconography of the atomic menace. The United States was identified as the enemy, responsible for endangering Bolognese communities with nuclear testing and the potential deployment of the H-bomb.

This study of Bologna has also shown the extent to which mobilisation against the atomic menace was gendered, unveiling the central role of women in organising pro-peace events and wide-ranging political activities against the atomic threat, especially collection of signatures for the international appeals. Both the iconography and memory of the mobilisation were gendered, as photographic sources and collection of memoirs and interviews have illustrated. Urban spaces were the sites of a possible nuclear explosion but also the places of activism. Cold War cities, like Bologna, were peacefully invaded by peace activists with their own symbols, creating a special urban setting, again highly gendered. The local, the national and the global level were truly intertwined not only in the representation but also in the mobilisation against the nuclear menace. Italy, and Bologna along with it, preceded the international movement precisely thanks to female activism. Women were also able to transcend the public/private divide, bringing anti-atomic campaign in people's homes.

The absence of urban traces of the atomic age, along with the vanishing identity of Bologna as a Red city, were probably responsible for the generalised amnesia of the 1950s representation and mobilisations against the atomic threat, which this chapter has tried in some way to recover.

Notes

1 The concept was introduced by the Communist Party leader Palmiro Togliatti in his 1946 speech in Reggio Emilia (Togliatti 1946).
2 Union of Italian Women Archive (Bologna), b. 2 "Anni '50", f. "VI Congresso provinciale UDI 1959", UDI Bologna, *Atti del 6° Congresso provinciale dell'Unione donne italiane di Bologna 2-3 maggio 1959*.
3 As the only estimate of the signatures come from the movement itself, they could have been manipulated for propaganda purposes. We do not have any counter evidence, though, to prove such manipulation. In comparative terms the estimate is viable in placing Italy and Bologna at the top of such collections of signatories.
4 Gramsci Foundation, Emilia-Romagna, Italian Communist Party Archive, Federation of Bologna, Series "Commissioni, Sezioni di lavoro e dipartimenti", Sub-series "Lavoro di massa", f. 1953–1955, "Documentazione sugli arbitri, gli illegalismi, le rappresaglie…", July 1954.
5 Union of Italian Women National Archive (Rome), cronological section, b.15, f. 160, 162, 163, 164.
6 Union of Italian Women Archive (Bologna), photographic section.
7 *Ibidem.*
8 *Ibidem.*
9 Gramsci Foundation, Emilia-Romagna, Italian Communist Party Archive, Federation of Bologna, Series "Commissioni, Sezioni di lavoro e dipartimenti", Sub-series "Stampa e propaganda", f. "Bollettini della stampa e propaganda".
10 *Ibidem.*
11 Arcangelo Caparrini private archive.
12 *Il Gioco della Pace in mostra* http://www.comune.bologna.it/quartieresantostefano/notizie/151:39136/ [Accessed 8 August 2018].
13 *Ibidem.*
14 Albertina Bitelli private archive.
15 Soratte Bunker webpage: http://www.bunkersoratte.it/ [Accessed 8 August 2018].

References

Aga-Rossi E., Orsina G. (2004), *L'immagine dell'America nella stampa comunista italiana*, 1945–53, in P. Craveri, G. Quagliarello (eds.), *L'Antiamericanismo in Italian e in Europa nel secondo dopoguerra*, Soveria Mannelli, Rubettino. pp.35–49.

Agosti A. (1999), *Storia del Partito comunista italiano 1921–1991*, Rome, Laterza.
Anconelli T. (1983), *Contro la Guerra contro la Ced* in *Comunisti: i militanti bolognesi del PCI raccontano* Rome, Editori Riuniti, pp. 201–202.
Anderlini F. (1990), *Terra rossa. Comunismo ideale e socialdemocrazia reale. Il Pci in Emilia-Romagna*, Bologna, Coop Il Nove.
Andreucci, F. (2005). *Falce e martello: Identità e linguaggi dei comunisti italiani fra stalinismo e guerra* fredda, Bologna, Bononia University Press.
Autio-Sarasmo S., Humphreys B. (eds.) (2010), *Winter Kept Us Warm: Cold War Interactions Reconsidered*, Helsinki, Aleksanteri Institute.
Autio-Sarasmo S., Miklóssy K. (eds.) (2011), *Reassessing Cold War Europe,* London and New York, Routledge.
Baccetti C. (2003), *After PCI: Post-communist and Neo-Communist Parties of the Italian Left After 1989*, in J. Botella, L. Ramiro (eds.), *The Crisis of Communism and Party Change. The Evolution of West European Communist and Post-Communist Parties.* 71-88. Barcellona, Institut de Ciències Politiques i Socials ICPS.
Baldissara L. (1994), *Per una città più bella e più grande. Il governo municipale di Bologna negli anni della ricostruzione (1945–1956)*, Bologna, Il Mulino.
Baravelli A. (2005), *Fare Festa. Bologna e il Festival dell'Unità (1945–2005),* Bologna, Manifesta Press.
Bellanger E. (2013), *Le "communisme municipal" ou le réformisme officieux en banlieue rouge*, in E. Bellanger, J. Mischi (eds.), *Les territoires du communisme. Elus locaux, politiques publiques et sociabilités militantes*, Paris: Armand Colin, pp. 27–52.
Bertucelli L. (1999), *L'invenzione dell'Emilia rossa. La memoria della guerra e la costruzione di un'identità regionale (1943–1960)*, in L. Paggi (ed.), *Le memorie della Repubblica*. Florence, La Nuova Italia, pp.12–28.
Betti E. (2016), *Bologna in the Cold War. Perspectives of Memories from a Communist City in the West*, in K. Pizzi, M. Hietala (eds.), *Cold War Cities: History, Culture and Memory*, Bern, Peter Lang, pp. 171–201.
Betti E. (2017), *Gendering Political Violence in Early Cold War Italy. The Case of Bologna*, in P. Casanellas, A.S. Ferreira (eds.), *Violência política no século XX. Um balanço.* Lisbon: Instituto de História Contemporânea da Universidade NOVA de Lisboa, pp. 673–683.
Betti E. (2015), *Making Working Women Visible in 1950s Italian Labour Conflict. The Case of the Ducati Factory*, in K.H. Nordberg, H. Roll-Hansen, E. Sandmo, H. Sandvik (eds.), *Myndighet Og Medborgerskap. Festskrift til Gro Hagemann pa 70–arsdagen* 3.september 2015, Oslo, Novus, pp. 311–322.
Boarelli M. (2007), *La fabbrica del passato. Autobiografie di militanti comunisti (1945–1956),* Milan, Feltrinelli.
Bollettino della pace (1951), n.3, 1 February.
Brogi, A. (2011), *Confronting America: The Cold War between the United States and the Communists in France and Italy.* Chapel Hill, NC: University of North Carolina Press.
Burtch, A. (2012), *Give Me Shelter: The Failure of Canada's Nuclear Civil Defence Program, 1945–1963*, Toronto, UBC Press.
Capelli C. (2013), *Il filo spezzato: il 1989 e la memoria collettiva dell' "Emilia rossa,"* in M. Carrattieri, C. De Maria (eds.), *La crisi dei partiti in Emilia-Romagna negli anni '70/'80*, "E-Review Dossier", 1, pp. 55–84.
Cerrai S. (2011), *I partigiani della pace in Italia. Tra Utopia e sogni egemomico*, Padua, libreriauniversitaria.it edizioni.

Comunisti: i militanti bolognesi del PCI raccontano (1983) Rome, Editori Riuniti.
Cruciani S. (2014), *Il modello emiliano dall'Italia repubblicana all'unione europea*, in C. De Maria (ed.), *Il modello emiliano" nella storia d'Italia. Tra culture politiche e pratiche di governo locale*, Bologna, Bradypus.
De Giuseppe M. (2000), *Gli italiani e la questione atomica negli anni cinquanta*, "Ricerche di storia politica", n.1.
Dogliani P., Matard-Bonucci M.A. (eds.), (2017), *Democrazia insicura: violenze, repressioni e Stato di diritto nella storia della Repubblica (1945–1995)*, Rome, Donzelli.
Dondi M. (1995), *La propaganda politica dal '46 alla legge truffa: temi della comunicazione pubblica nel confronto fra Pci e Dc*, in A. Mignemi (ed.), *Propaganda politica e mezzi di comunicazione di massa fra fascismo e democrazia*, Turin, Abele, pp. 185–197.
Dondi M. (2004), *Le Feste dell'Unità: rito laico tra politica e tradizioni popolari*, in A. De Bernardi, A. Preti, F. Tarozzi (eds.), *Il Pci in Emilia Romagna. Propaganda, sociabilità, identità dalla ricostruzione al miracolo economico*, Bologna, Clueb, pp. 119–137.
Duggan C. (1995), *Italy in the Cold War Years and the Legacy of Fascism*, in C. Duggan, C. Wagstaff (eds.), *Italy in the Cold War. Politics, Culture and Society, 1948–1958*. Oxford, Berg Publishers, pp.64–79.
Duggan C., Wagstaff C. (eds.) (1995), *Italy in the Cold War. Politics, Culture and Society, 1948–1958*. Oxford: Berg Publishers.
Duran A.R. (2014), *Painting, Politics and the New Front of Cold War Italy*. Farnham, Surrey, UK: Ashgate.
Farish M. (2010), *The Contours of America's Cold War*, Chicago, University of Minnesota Press.
Farish M., Monteyne D. (2015), *Introduction: Histories of Cold War Cities*, "Urban History", 42, pp.543–546.
Fincardi M. (2007), *C'era una volta il mondo nuovo. La metafora sovietica nello sviluppo emiliano*, Rome, Carocci.
Flores M. (1976), *Il Quaderno dell'attivista: ideologia, organizzazione e propaganda del Pci degli anni Cinquanta*, Milan, Mazzotta.
Flores M. (1991) *Il mito dell'Urss nel secondo dopoguerra*, in P.P. D'Attorre (ed.), *Nemici per la pelle. Sogno americano e mito sovietico nell'Italia contemporanea*, Milan, Franco Angeli, pp. 491–507.
Furlan P. (1988), *Gli anni della ricostruzione*, in L. Arbizzani (ed.), *Il sindacato nel bolognese. Le Camere del lavoro di Bologna dal 1893 al 1960*, Rome, Ediesse.
Gabrielli P. (2005), *La pace e la mimosa. L'Unione donne italiane e la costruzione politica della memoria (1944–1955)*, Rome, Donzelli.
Gaspari O. (2006), *Dalla Lega dei comuni socialisti a Lega autonomie. Novant'anni di riformismo per la democrazia e lo sviluppo delle comunità locali*, Rome, Edizione Alisei.
Giacomini R. (1984), *I partigiani della pace. Il movimento pacifista in Italia e nel mondo degli anni della prima Guerra fredda*, Milan, Vangelista.
Giugni M. (1999), *Mobilitazioni su ambiente, pace e nucleare*, "Quaderni di Sociologia" [online], 21, DOI: 10.4000/qds.1400.
Grant M. and Ziemann B. (2016), *Understanding the Imaginary War: Culture, Thought and Nuclear Conflict, 1945–90*, Manchester, Manchester University Press.

Gundle S. (2000), *Between Hollywood and Moscow. The Italian Communists and the Challenge of Mass Culture, 1943–91*, Durham and London, Duke University Press.

Hastings M. (1991), *Halluin la Rouge 1919–1939. Aspects d'un communisme identitaire*, Lille, PU SEPTENTRION.

Jaggi M., Muller R., Schmid E. (1977), *Red Bologna*, London, Writers and Readers.

Jenks J. (2003) *Fight Against Peace? Britain and the Partisans of Peace, 1948–1951*, in M.F. Hopkins, M.D. Kandiah, G. Staerck (eds.), *Cold War Britain, 1945–1964*. Cold War History Series. London, Palgrave Macmillan, pp.82–97.

Kertzer D.I. (1996), *Politics and Symbols: The Italian Communist Party and the Fall of Communism*, New Haven and London, Yale University.

Knotter A. (2011). *"Little Moscows" in Western Europe: The Ecology of Small Place Communism*, "International Review of Social History", 56, pp. 475–510.

l'Unità (1948), "L'insediamento del Comitato per la solidarietà democratica," 30 luglio 1948.

l'Unità (1950a), "Mentre gli americani aggrediscono la Corea, Bologna intensifica la campagna antiatomica," Cronaca di Bologna, 4 July 1950.

l'Unità (1950b), "Un appello dell'UDI alle donne cattoliche," Cronaca di Bologna, 5 July 1950.

La Lotta (1948), "8 marzo. Tutte le donne vogliono la pace," 5 March 1948.

La Lotta (1949), 9 September.

La Lotta (1950a), 3 March 1950.

La Lotta (1950b), "Esigiamo la proibizione dell'arma atomica" by G. Dozza, 25 March 1950.

La Lotta (1950c), "Contro l'arma atomica il 'NO' unanime dei bolognesi," 3 June 1950.

La Lotta (1950d), "Firma contro la bomba atomica," 3 June 1950.

La Lotta (1950e), "Centinaia di migliaia di firme raccolti dai comitati della pace," 23 June 1950.

La Lotta (1950f), "Se l'atomica cadesse in Piazza maggiore Bologna diverrebbe un immense cimitero," 19 June 1950.

La Lotta (1950g), "Il popolo Bolognese leva la sua protesta contro I criminali di Truman," 30 June 1950.

La Lotta (1950h), 30 June 1950.

La Lotta (1950i), 28 July 1950.

La Lotta (1951a), "Uno schizzo del compagno pittore Aldo Borgonzoni: 'La Guerra e la morte vengono dall'America," 2 February 1951.

La Lotta (1951b), "Così per la Guerra d Truman si vorrebbe trasformare la nostra provincia," 21 December 1951.

La Lotta (1954a), "Bolognesi di ogni condizione condannano le armi di sterminio," 23 April 1954.

La Lotta (1954b), "A Galliera un largo Comitato cittadino guiderà la Lotta di quella popolazione," 7 maggio 1954.

La Lotta (1954c), 3 September1954 [caption p.1]

La Lotta (1955), "Gli esperimenti atomici minacciano l'umanità."

La pace (1955), "Appello ai popoli contro la preparazione della guerra atomica," a.5, February 1955.

Lagrave R.M. (2004), *Le marteau contre la faucille. Introduction*, "Etudes rurales", n. 171/172, Special issue *Les "Petites Russies" des campagnes francaises*.

Lama L. (2007), *Giuseppe Dozza. Storia di un sindaco comunista*, Reggio Emilia, Aliberti Editore.
LaPorte N. (2013), *Introduction: Local Communisms Within a Global Movement*, "Twentieth Century Communism", 5, 2013, pp. 1-12.
Lazar M. (1992), *Maisons rouges: les partis communistes français et italien de la Libération à nos jours*, Paris, Aubier.
Liberatore A. (1996), *Policy Responses to Chernobyl in Italy, France and Germany: A Comparative Analysis*, "Review of European Community & International Environmental Law", 5, pp. 211–217.
Light J. (2003), *From Warfare to Welfare: Defense Intellectuals and Urban Problems in Cold War America*, Baltimore, Johns Hopkins University Press.
Maccaferri M., Pombeni P. (2013), *I partiti politici durante la "Prima Repubblica"* in A. Varni (ed.), *Bologna in Età contemporanea (1915–2000)*, Storia di Bologna, Vol. 4, Bologna, Bononia University.
Marino G.C. (1991), *Autoritratto del PCI staliniano: 1946–1953*, Rome, Editori riuniti.
Marino G.C. (1991), *Guerra fredda e conflitto sociale in Italia: 1947–1953*, Rome, Salvatore Sciascia.
Marino G.C. (1995), *La repubblica della forza. Mario Scelba e le passioni del suo tempo*, Milan, Franco Angeli.
Michetti M., Repetto M., Viviani L. (1984), *Udi: laboratorio di politica delle donne*, Rome, Cooperativa libera stampa, Rome.
Noi Donne (1948), "Messaggere di Pace," 15 February 1948.
Novelli E. (2000), *C'era una volta il Pci: autobiografia di un partito attraverso le immagini della sua propaganda*, Rome, Editori riuniti.
Ombra M. (ed.) (2005), *Donne manifeste: l'UDI attraverso i suoi manifesti 1944–2004*, Milan, Il saggiatore.
Patacchia V., Brunelli M. (eds.) (2015), *Memorie sotterranee. I rifugi antiaerei a Bologna tra ricerca, tutela e valorizzazione*, Bologna, Istituto per i Beni artistici culturali e naturali della Regione Emilia-Romagna.
Patelli A. (1983), *Duecento firme per la pace a Camugnano* in *Comunisti: i militanti bolognesi del PCI raccontano* Rome, Editori Riuniti, pp. 199–200.
Pojmann W. (2013), *Italian Women and International Cold War Politics 1944–1968*, New York, Fordham University Press.
Ponzani M. (2004), *I processi ai partigiani nell'Italia repubblicana. L'attività di Solidarietà democratica 1945–1959*, "Italia contemporanea", 237.
Rodano M. (2010), *Memorie di una che c'era: una storia dell'UDI*, Milan, Il Saggiatore.
Romero F. (2009), *Storia della guerra fredda. L'ultimo conflitto in Europa*, Turin, Einaudi.
Schipperges M. (1980), *Il mito sovietico nella stampa comunista* in P.P. D'Attorre (ed.), *La Ricostruzione in Emilia-Romagna*, Parma, Pratiche.
Stanzani D. (1983) *Auguri di pace ricambiati con il carcere* in *Comunisti: i militanti bolognesi del PCI raccontano*, Rome, Editori Riuniti, pp. 203–204.
Time (1950), "The Netherlands: Little Moscow," 4 December.
Togliatti P. (1946), *Ceto medio e Emilia rossa. Discorso pronunciato a Reggio Emilia*, Rome, Stabilimento UESISA.
Tonelli A. (2012), *Falce e Tortello. Storia politica e sociale delle Feste dell'Unità (1945–2011)*, Rome, Laterza.
UDI (1949), *Sotto le bandiere della pace – Due anni di attività dell'Unione donne italiane dal secondo al terzo congresso nazionale*, Rome, Edizioni Noi Donne.

Westad, O.A. (2005) *The Global Cold War: Third World Interventions and the Making of Our Times*, Cambridge, Cambridge University Press.

Wirsching A. (2013), *Comparing Local Communisms*, "Twentieth Century Communism", 5, 2013.

Wittner, L.S. (1993), *One World or None. A History of the World Nuclear Disarmament Movement through 1953,* Stanford, Stanford University Press.

Wittner, L.S. (1997), *Resisting the Bomb: A History of the World Nuclear Disarmament Movement, 1954–1970*, Stanford, Stanford University Press.

12 Making a 'free world' city
Urban space and social order in Cold War Bangkok[1]

Matthew Phillips

ABERYSTWYTH UNIVERSITY

Between two of Bangkok's most important economic arteries, Silom and Surawong roads, lies Patpong, one of the most visited red light districts in the world. Those who make the trip do so mostly at night. There they are met with bustling street markets. Vendors hawking tourist keepsakes, and crude invitations to sex shows. Few would notice the buildings above the neon lights: unlit windowless façades that give little away of what goes on behind blackened concrete walls, to say nothing of a more illustrious past.

Today the commodity is sex, but this was not always the case. Patpong was once a major centre of international business and commerce. IBM, Shell, and American President Line were all based there along with some of the finest restaurants and hotels in Bangkok. The library of the United States Information Service (USIS) was located there, as was the covert CIA operation, Air America, and the equally as obscure Rockefeller IBEC center. Pristine white walls protected modern air-conditioned interiors that by 1965 were consuming more electricity than Korat, a city in Thailand's rural northeast (Grossman 2009, 145). The transformation of Patpong into the centre of Thailand's sex trade took place during the latter stages of Vietnam War when American military soldiers flooded into the city for 'Rest and Recreation' (R&R): code word for days of sex, drugs and booze (Steinfatt 2002, 299).

Patpong was built by the son of Luang Patpongpanich, the head of a Chinese migrant family from Hainan who had been honoured with a Thai name from King Rama VI (r. 1911–1925) at the turn of the twentieth century. Thailand declared war on the United States in 1942, when son Udom was studying at the University of Minnesota. As a result, he was invited to join the Office of Strategic Services (OSS), precursor to the CIA. His clandestine training complete and his US contacts intact, Udom never joined an invasion force. Instead, he returned to Thailand at the end of the war to begin work on creating one of the most valuable pieces of real estate in the country (Steinfatt 2002, 298). Bought for $2,400 in 1946, by 1965 Patpong road was worth $5 million (Backman 2005; Grossman 2009, 145).

How Udom attracted such riches remains largely obscured from public view. It is nonetheless possible to incorporate his story into a broader account about how Bangkok was impacted by the Cold War. Specifically, Udom's story sheds light upon how a newly established nationalist elite was able to consolidate its position by building alliances with key American actors. Recent scholarship has established that the Cold War experience in Asia was distinct, shaped as much by local priorities as those of the superpowers (Westad 2007; Wallerstein 2010; James and Leake 2015). In this light, nationalist elites are noted for their ability to extract security, aid, technical expertise and ideological support to bolster their respective positions. In Thailand, there are countless examples of how men like Udom were able to channel resources gained through the Thai–US exchange to support their own internal interests.

As a sphere of relative peace, Thailand was viewed by US policymakers as a bastion against communism (Fineman 1997; Glassman 2005). While Bangkok might have been peripheral to events in Korea, Laos and Vietnam, its status as a friend and ally to the capitalist world meant the United States emerged as the undisputed hegemon in the country. Generally, therefore, American officials in Thailand were focussed on shoring up political support while pulling the country into the US-led global economy. This placed Americans in a central position at a critical time in the development of Thailand's social order. There is already a lively discussion on the cultural impact of the Cold War in Thailand (Thak 2007; Wasana 2009; Harrison 2010; Phillips 2016; Ford 2017). This chapter extends that discussion by foccusing on how the Thai–US alliance led to an innovation in high society lifestyle practices that part reinforced, part reinvented, elite notions of distinction (Bourdieu 2009 [1979]).

Henri Lefebvre argues that social space is the outcome of a sequence of operations that reflect social relations (Lefebvre 2007 [1974]). Similarly, contestation over Bangkok's urban landscape during the Cold War directly reflected shifting alliances within Thailand's national elite. Twenty-first century scholarship has shown how design and construction in Bangkok played a fundamental role in shaping class-consciousness and reflecting power dynamics (Peleggi 2002; Wong 2006, Prakitnonthakan 2009; King 2011; Noobanjong 2013; Chua 2014; Ünaldi 2016). Others have established that through the Cold War, the Thai monarchy was able to strengthen its position by linking up with the Sino-Thai capital and aligning itself with American foreign policy (Gray 1991, 1992; Connors 2007 [2003]; Handley 2006). This chapter focusses on how changing approaches to the production of external and internal space, expedited by a trans-Pacific configuration of finance, ideology and culture, served to unify Thailand's enlarged nationalist elite and while reinforcing a strict social hierarchy.

No Hilton here

Annabel Jane Wharton (2001) has described the city as a palimpsest: a contested site, the surface of which 'inevitably involves the erasure of one set of social relations by another, the superimposition of an altered structure of authority over an older system' (Wharton 2001, 1). It was in this manner that Hilton Hotels exported US power and ideology during the Cold War. As Conrad Hilton himself explained, Hilton Hotels were built to provide 'spaces that might act as a challenge – not to the peoples who have so cordially welcomed us into their midst – but to the way of life preached by the Communist world'. The buildings were seen as a gift to the city; dazzling monuments of American modernity that provided a new standard in comfort and hospitality, but which also spelt out 'friendship between nations' (Wharton 2001, 8). Their presence would reorient the city, both towards the architectural forms displayed and to the primacy of American mastery of the material world. They also provided a unique space within which US policymakers, business leaders and cultural producers could imagine their relationship to the host country as one of unambiguous harmony.

John W. Houser, vice-president of Hilton Hotels International, proposed an 'outstanding' hotel for Bangkok to the Thai government in December 1955. The new building, he claimed, would serve the people of Bangkok by providing a showcase for Thai products and skills as well as a space to meet, inform and entertain, creating 'greater understanding between people thus brought together.' Yet, while being a 'service to the community' it would not necessarily be 'profitable from the investors standpoint.' Financing of the building, its furnishing and equipment, he made clear, must come from either the government, private investors or a mixture of the two (Houser 1955). The value, in other words, spoke for itself.

To back up his claims, Houser directed the Thai government towards the recently completed Istanbul Hilton. Opened in June 1955, the Istanbul Hilton had 244 rooms and was serviced by air-conditioning systems and elevators imported from the United States. Constructed with reinforced concrete, the building towered above its surroundings. Uniform lines of windows and balconies ran across the façade, making it unlike anything the city had ever seen. It was, 'the first elite, large-scale modern structure in the city', characterized by a 'lucid rationality', that drew 'commercial, entertainment and government elites' into a space inscribed with the presence of anti-Communist America' (Wharton 2001, 35).

Architecturally dominated by signs of US efficiency, the Istanbul Hilton also sought to reflect the local surroundings through sensory and visual additions. Teakwood screens referenced vernacular Ottoman architecture, while Turkish carpets 'relieved the severity of the marble floor.' On the 'Bosphorus Terrace', Turkish hostesses wore bright uniforms inspired by traditional costumes as they prepared and served coffee in the local manner, while lavishly draped curtains in the 'Tulip Room' referenced a sultan's tent. Such signifiers were 'deployed as a sign of the Other within a dominant aesthetic of American Modernity' (Ibid., 26). In the context of the Cold War, they also dramatized the harmonious nature of the Turkish–US exchange; presenting material evidence that local cultural forms could exist in harmony with US-informed modernity.

Bangkok was clearly an ideal site for such a hotel. In the early years of the Cold War, with much of Asia engaged in winning independence from the colonial powers, Bangkok was a site of relative stability. Meanwhile, the post-war rise in air transport made Bangkok's Don Mueang airport an important regional hub. In the late 1940s, a number of United Nations agencies made the city their regional headquarters and by the middle of the 1950s Bangkok was already a centre of diplomacy and economic activity. In considering the Hilton offer, the Thai government acknowledged that Thailand was now playing host to tourists, businessmen and international congresses throughout the year and lacked a first-class hotel. A new Hilton was certain to draw more international visitors. Yet, in the end, the cost of the project was deemed too great and more importantly, the government was already in the process of constructing The Erawan, its own first-class hotel, completed in late 1956 (Phibun 1955).

Urban hearts and minds

From 1951 to 1957, US economic aid to Thailand amounted to roughly $222 million (Pike 2011, 195). This was used primarily to reinforce the power of Bangkok over the rest of the kingdom (Tambiah 2013 [1973]). New roads, like the Friendship Highway [*Thanon Mitraphap*] to the north, connected the capital with the interior, making it easier for Bangkok-based elites to both access and control large swathes of the country. Further improvements to ports and railways meant raw materials could be extracted and transported more easily, supporting both US strategic interests and stimulating Thai economic growth, which averaged 5% over the same period. A postwar surge in opium production, particularly over the Thai border in the Shan State of Burma provided substantial private revenue streams, boosting the personal power of a relatively small group of government officials and well-connected elites (McCoy 2003 [1972], 183–192).

Most notable was Police Chief Phao Sriyanond, whose *Soi Ratchakru* group built a business empire that included fishing, timber and sugar, as well as gambling, prostitution and narcotics (Thak 2007, Pike, 2011). Phao was also responsible for the state-owned Syndicate of Thai Hotels and Tourist Enterprise, which was set up within the Ministry of Finance and was responsible for the Erawan (Ünaldi 2016). A good deal of Phao's personal power came from his relationship with the CIA, who had worked with Phao to arm Chinese Nationalist (KMT) troops based in northern Burma following the Communist victory in China and helped set up the Border Patrol Police (BPP) in 1951. This specialized police unit was charged with protecting security in the border areas and, by 1953, had over 4,000 armed personnel (Cooper 1995, 153). Through a covert front company named Sea Supply Corporation, the CIA provided arms, communication equipment and transportation to Phao, helping him build a virtual monopoly of Burmese opium exports and placing him at the centre of Bangkok's political and economic life. This included collecting protection money from Bangkok's wealthiest Chinese businessmen and forcing himself onto the boards of more than 20 corporations (McCoy 2003 [1972], 184; Skinner 1958, 192–194).

These strongman tactics were successful in connecting innovative, government-backed, revenue streams with Thailand's existing capitalist class, but they also provoked discord, and placed the United States in a vulnerable situation in supporting what increasingly looked like a mafia state. From 1955 in particular, an increasingly hostile Thai press frequently referred to Phao as a gangster-like figure. In part recognition of these vulnerabilities, US officials looked to shore up popular support among Bangkok's urban community, in the hope that this would have a related impact on the rest of the country. In planning documents drafted by United States Information Service (USIS)[2] officials in Bangkok during 1952, it was explained that, 'the capital city of Thailand exerts great influence over the entire kingdom' and that 'Bangkok thinking' influenced the nation psychologically. Of particular importance were the 1,500 or so opinion leaders, comprised of writers, journalists, lawyers, doctors, scholars, "liberals" and members of Thai royalty. Similar dynamics applied to the Chinese community. At around 3 million persons, and making up around 16.6% of the population, they dominated the economic life of the nation (US Embassy Bangkok 1952).

William Donovan, a key architect of the OSS, took a position as ambassador to Thailand in September 1953, where he remained for just under a year. A champion of people-to-people propaganda, Donovan contributed to an aggressive expansion of psychological operations in the capital. Through 1953, Viet Minh successes, combined with a perceived increase in Chinese propaganda activity, convinced US officials that Southeast Asia was a major front in the battle between the 'free' world and the communist one. As one of the few countries where US officials could operate freely, Thailand hosted a raft of projects intended to shore up support. These included dissemination of United States Information Agency (USIA) news wire material, mobile movie units and information centres. Under Donovan, there was also a special program of indoctrination, whereby key individuals in specific Thai government ministries were indoctrinated over a three-month period to associate Communism with a threat to 'Buddhism, the kingship, private property rights, and the government bureaucracy.' Each individual was then encouraged to 'conduct a similar indoctrination to his own civil service, there being about 1,000 in each of the ministries' (USIA 1954).

While the impact of such schemes is yet to be fully understood, the release of documents such as those published through the CIA Freedom of Information reading room clearly raise questions about the extent of US involvement. Beyond doubt, however, was the clear wish of US officials in Bangkok to actively align Thai society with an American centre. 'American prestige in Southeast Asia and particularly in Thailand', Donovan explained in May 1954, 'will depend upon American cultural penetration that identifies Thai interests, ideals, and institutions with those of the United States' (Donovan

1954). While continuing to support the current administration, the US ultimately needed to secure a harmonious and long-standing commitment from across all of Thailand's elite constituencies that at least appeared free from strongman tactics. Over the years, he cautioned, the good will generated by aid programs and military assistance,

> will be lost unless the Thai leaders of tomorrow and a substantial part of the Thai people grow in the belief that the common ground they share with America and Americans is far greater and more enduring than the inevitable differences that will arise to shake their mutual confidence in each other.
>
> (Ibid)

Celestial capital

Despite the freedom granted to Americans in Bangkok during the 1950s, the dissemination of US proclamation of 'harmony' could only take place within the existing political and cultural landscape. Americans also had to necessarily demonstrate awareness of a society that remained cautious, even hostile, to overbearing international influence. Such feelings had their roots in the nineteenth century, when Siam (renamed Thailand in 1939), was integrated into the economies of imperial Europe. This had left the city with three largely distinct quarters or spheres: a foreign quarter, where westerners resided; a more diverse economic quarter dominated by Chinese migrants; and the existing site of royal authority on Rattanakosin. At this time, European residents lived and worked under separate rules and regulations in the southern part of the city – a consequence of unequal treaties signed with Siam in the mid-nineteenth century. With European trade the dominant engine of urban life, Thai monarchs were forced to employ enormous resources to ensure the palace remained the spiritual and cultural centre of gravity.

Particularly through the reign of King Chulalongkorn (r. 1868–1910), members of the royal family invested significant sums in emulating the fashions, tastes and lifestyles of the imperial monarchies of the Europe, successfully presenting themselves to westerners as at least 'partially' or 'imperfectly' civilized. At the same time, the project to become relevant globally reinforced distinction between modernized royals and their less civilized subjects. New building projects, often powered by the royal wives and relatives, many with Chinese ancestry, reorganized the urban space to reassert the unity and centrality of Thailand's royals over a rapidly changing society. As Maurizio Peleggi has explained, these plans showed little interest in creating space for the public (Peleggi 2002). Rather, the emphasis was on the construction of self-contained worlds that imitated the supposed best of Europe. The most notable such addition was the construction of Suan Dusit, a sprawling area of gardens and dwellings that sat within wrought-iron fences and high walls, and that upon completion became the new home of both the king, his harem and his princes.

Absolute monarchy was replaced by a constitutional regime in 1932 and efforts soon began to occupy key sites of royal power and open up new, publically owned national space (Wong 2006). From 1938, the first government of Phibun Songkhram redesigned Rattanakosin to privilege the common Thai population, subvert physical markers of royal authority, undermine the influence of Chinese capitalists, and curb the superior status of European residents. A large part of Suan Dusit was turned into a public zoo and Ratchadamnoen Avenue, a wide road built at the turn of the century for royal processions was converted into a national thoroughfare, complete with mid-point monument to the constitution.

The Second World War removed much of the ideological vigour from Phibun's assertive chauvinistic nationalist ideology. Thailand entered into an alliance with the Japanese in 1942 and the population suffered from Allied bombings and a collapse of imported goods that severely undermined living standards. While Great Britain sought to punish Thailand for its declaration of war, the United States was more interested developing a close partnership with the Thai elite. From very early during these post-war years, the American media portrayed Thailand as a special location where US military officials, businessmen and civilians, were made to feel welcome, and where (unlike elsewhere in Southeast Asia) they could operate without fear of anti-colonial, anti-white, violence (Phillips 2016).

Adversarial space and competing modernities

Through the 1950s, Thai and American social worlds in Bangkok remained largely separate. Much of the small American community who resided in the capital rented properties in the new suburbs on the eastern fringe of the city, particularly in Bang Kapi (Askew 2002, 240). US propaganda, on the other hand, was generally disseminated through sites explicitly associated with the US state, such as the United States Information Centre. Opening up sites of effortless 'exchange' was more difficult, and often involved going on the ideological offensive. This approach was typified by the US exploitation of Bangkok's annual constitution fair, an event that dated back to the pre-war era of mass nationalism but which during the early Cold War became a site for competing nations to attempt to capture Thai imaginations. US officials were rattled in 1953 when the Soviet Union invested heavily in a lavish display that showcased technological achievement. Through 1954, under Donovan's watch, the USIA launched a cultural counter-offensive, coordinating one of its most extravagant contributions to the fair to date. An elaborate Macy's fashion show featured the latest US styles and synthetic fabrics. Nearby displays featured a host of US-made consumer goods while a huge Cinerama screen, exported to Asia for the first time and put on free of charge for the citizens of Bangkok, demonstrated the technical brilliance of American industry and cinematography (Phillips 2016).

In 1956, jazz clarinetist Benny Goodman began a world tour at the fair, playing for two weeks in the American pavilion and kicking off his first night with an additional one-hour jam session with King Bhumibol Adulyadej (Von Eschen 2004, 46). Each night 10,000 members of the Bangkok public reportedly attended two performances. Many took time to pass through the USIS booth, which included a display entitled 'Century of Friendship' and a series of three large painted panels presenting an idealized relationship between the two countries.

A total of 800,000 people were estimated to have visited the 1956 fair and of that 90,000 passed through the USIS exhibit (US Embassy Bangkok 1956). Yet, for most, the displays of American technology and cultural pre-eminence offered little more than a glimpse into a possible future for the city's population. Even if some of the products on show were available to purchase, few could ever hope to afford them. This created a glaring disjuncture between the aims of the message and the reality of daily life. On top of this, investment in the fair was clearly adversarial. Through the 1930s, the fair had been a stage, upon which the Thai state sought to convince Bangkok society about its own modernising efforts, yet by the early 1950s it had become as much a site for rival nations to present their respective public relations messages.

Success at the fairs, therefore, did not necessarily translate into outright support for American involvement amongst Thai high society. Between 1955 and 1957, support for Phibun and Phao's strongman regime collapsed dramatically, and American aid became implicated in supporting an immoral and corrupt administration. National elections in February 1957, widely seen as rigged in favour of Phibun's own political party, instigated a political crisis that lasted through to September. Increasingly, anger towards the government was matched with hostility towards the United States, and in particular to the Sea Supply Corporation, which many now took for granted was a CIA front. On 16 September, a crowd of angry protesters and military officers marched on the Sea Supply office, resulting in a standoff described by Daniel Fineman as 'the single most frightening clash between Americans and Thais in the history of relations between the two countries' (Fineman 1997, 242).

Yet, while anger at US political institutions peaked, it did not necessarily translate into hostility towards Americans themselves, nor the ideology that they espoused. While small in number, the academics, officials and businessmen who travelled to Thailand during this time invariably came with a belief that the American development, underpinned by mass consumerism and liberal democracy, represented the best model available. Later referred to as 'modernization theory' these ideas were largely developed as a direct rejection of the Marxist view of history, which judged revolution as the primary means to implement change. Yet, its proponents also tended

Figure 12.1 A surviving example of a design created by Jacqueline Ayer that draws from Thai motifs adapted to reflect modern tastes and fabrics. Date unknown.
Source: ©Jacqueline Ayer Estate.

to discount *all* routes to modernity that deviated from the US model. This meant that existing notions of vernacular Thai modernity, including those espoused by the nationalists post-1932, were discounted as a poor imitation, while only that which could be seen as a reflection of a pre-modern way of life could successfully be viewed as authentically Thai (Figure 12.1).

Crucially, these assumptions were not limited to academics or policymakers. They dominated Thai–American interactions, including commentary on Thai attempts to embrace modern living. Carol Hollinger, who came to Thailand in the 1960s to teach English at Chulalongkorn University, was typical in noting how throughout the country, 'the use of the products of the West has meant the borrowing of the worst and gaudiest of our civilization.' This meant that the 'natural love of exciting colour and poetic design, historically Thai, defects to the West and is reborn in absurd houses of pastel stucco and squirming rococo design' (Hollinger 2000 [1965], 47). Paralleling US-derived development theories, idealized Thai interiors were invariably those that demonstrated a fusion of 'modern' elements imported from the west and local ones borrowed from the Thai past.

Unsurprisingly, those who best achieved harmony were American. Most notable was Jim Thompson, who in the aftermath of the Second World War set up a company selling Thai silk to Americans. His approach was to modernise the Thai product in order to create something that could be marketed in high-end boutique stores abroad as well as to visiting Americans. But he was also interested in introducing new designs, based on traditional Thai motifs. In the mid-1950s, Nelson Rockefeller became interested in sponsoring a company in Bangkok that would do a similar thing for with a variety traditional Thai crafts, and upon Jim Thompson's recommendation hired a young designer from New York named Jacqueline Ayer. Her company, Design-Thai, was supported through Rockerfeller's International Basic Economy Corporation (IBEC) and was therefore located in the IBEC centre on Patpong Road at the zenith of the streets status as a locus of commercial and clandestine activity in the capital. While there she produced new textiles that borrowed traditional Thai designs and adapted them for fifties tastes. As *Bangkok World* explained in 1962, Design-Thai's mission was to 'prove that in the decorative field the Thais have been great in the past and appear currently to have been lost in the surge of western imitation' (*Bangkok World* 1962). Like Jim Thompson's silk, the designs were soon being sold in up-market outlets across the world yet, as Ayer herself remarked, the one group who did not appreciate them were high society Thais. Despite their production for a high-end clientele, the locally produced cotton prints were 'too much a reminder of their "country bumpkin" cousins' (Ayer, n.d.).

Consumerism and urban development in post 'revolution' Thailand

Both Prime Minister Phibun Songkhram and Police Chief Phao were forced to flee Thailand following a coup in September 1957. Army Chief Sarit Thanarat, who led the revolt, had secured his own route to power by both exploiting Phao's collapsing popularity and gaining partial control over the opium trade (McCoy 2003 [1972], 188–191). A year later, in October 1958, Sarit consolidated his position by launching a 'revolution' [*patiwat*] in which he revoked support for electoral politics and imposed a strictly authoritarian style of governance. At the same time, Sarit rehabilitated the monarchy. Unlike his predecessor, who had sought to limit royal jurisdiction over the symbolic capital of the country's past, Sarit exploited the monarchy for his own ends (Thak 2007). Conveniently, American officials had long been convinced that King Bhumibol and Queen Sirikit provided valuable assets in their fight against Communism and supported the move with enthusiasm.

While presenting his regime as nativist in character, Sarit embraced US ideology in manner unlike any previous Thai leader. Newspapers, television and radio were all strictly controlled under his watch, and messages fell in line behind a strongly pro-US, anti-communist propaganda. US-informed developmentalism was also employed to shape government policy and state ideology, and a new Thai word, *kan phatthana* [development], was coined to explain the changed way of thinking. The associated reliance on technocratic support, learnt in one way or another from the United States, necessarily reinforced the role of foreigners [*farang*] within Thai society. As a result, access to American knowledge, ideology and culture, became a precursor to social advancement, and for the first time, new spaces began to emerge that actively facilitated the process.

Throughout the 1960s, the Thai economy grew rapidly. With assistance from the World Bank, the Thai government set up a host of agencies that together drafted Thailand's first economic plan. The aim was to demonstrate that the government would now refrain from interfering with private enterprise, paving the way for Chinese capitalists to operate more freely and encouraging investment from abroad (Pasuk 1980, 446). The change in policy also allowed for powerful Thai royalists to form closer bonds with new money. Under Sarit, the relationship between Chinese capitalists and the Thai state was thus transformed from one of competition to one that aspired towards harmony (Girling 1981, 81).

By the 1960s hundreds of Thais had received an education in the United States, including influential managers, planners and economists. Schooled in American theories of development, these technocrats now played a vital role in shaping the economy, opening it up to foreign trade and encouraging the development of public infrastructure for private enterprise. Greater access to the Thai interior helped make sure that agricultural products, primarily rice, remained Thailand's main legal export. At the same time, consumer goods were largely imported or assembled locally by foreign or joint foreign-Thai enterprises. This helped feed what was already a principally urban-based consumer culture and further detached Bangkok life from that of the rest of the country.

For the generation who lived through the period, the changes were dizzying. Across the city canals were filled in, roads were extended, and large new buildings were constructed. City planners struggled to get a grip. In line with Sarit's reliance on American advisors and foreign-trained Thai technocrats, the government made several attempts to develop a city plan. The most notable was drawn up by US consultant team Litchfield and Associates in 1960 but was never fully implemented. Rather, the construction frenzy made plans unworkable (Pasuk 1980, 455). While restructured under Sarit (largely to remove overt signs of direct involvement), the narcotics trade continued to flood the Thai capital with illicit funds, adding to a property boom already fuelled by increased foreign investment and demand for hotel and office space. By one account, the income from narcotics funded over half of all new construction in Bangkok during the period, overseen by Sarit who also took a controlling influence in the Sino-Thai owned Thai Farmers Bank (Pike 2011, 197).

Writing in 1964, a local journalist lamented the extraordinary speed with which the city was changing. 'One only has to look about when sitting in cars caught in the inextricable mass of traffic in morning and evening rush hours', he explained, 'to notice the new offices, shops, flats, houses, government buildings and the like springing up all over the place.' All, he made clear, with an 'apparent lack of any overall plan, of zoning, urbanization, or protection and creation of amenities' (Bangkok World 1964).

At the same time, between 1947 and 1976, the population of Bangkok rose from roughly half a million to 4.5 million, including large numbers of people from the least developed part of the country in the rural Northeast, known locally as Isan.[3] This contributed to a sense amongst Bangkok high society that modernising forces, for all their benefits, were not without problems. Through the 1960s, Bangkok developed one of the worst traffic problems in the world, while the speed with which buildings were being put up raised fears about the quality of the construction. The rising number of residents placed a continual strain on services, and led to an increase in petty crime. The dream of a city that catered for mass nationalist consumption rapidly retreated from view, only to be replaced with a distinct elite-orientated world, not so dissimilar from that of the early twentieth century. The difference was that the rationale for these spaces was the ever-greater presence of international arrivals who, unlike the imperialists before them, took great pleasure in embracing local culture, even if, in truth, they were merely replicating the Orientalist assumptions of a previous generation (Klein 2003).

At the same time, the renewed role of the monarchy in national life helped temper urban sensitivities about embracing Thai styles, particularly as the royal family itself began to embrace a traditional Thai crafts. In 1960, Queen Sirikit designed an entirely new wardrobe using Jim Thompson's Thai silk, and the Bangkok press increasingly sought to educate the urban consumers about new Thai styles that displayed national traditions, but which were also adored by foreigners. One such example appeared in the Thai language newspaper, *Prachathipatai* in January 1966. Introducing readers to a high-end art and craft shop run by the Department of Industrial Promotion, and named *Narayana Phand*. The shop sold a range of cultural products from hand carved dining tables, to models of the ceremonial royal barges and dolls in Thai dress. The article explained that the popularity of such goods internationally was furnishing Thai cultural products with a new kind of value that had replaced what was once meant by the term 'national' product.

> Around two decades ago, we, who are middle-aged, may have heard of a shop called 'Thai Industries' [*Thai Utsahakam*] and perhaps remember that this shop played a significant role in alleviating our suffering during World War II. In those days when imported goods were barred from the Thai market, it was 'Thai Industries' that mitigated the trouble of the Thai people by providing consumables that were needed in the circumstances. Nowadays, the brand 'Thai Industries' has long closed down, only *Narayana Phand* is a replacement.
>
> (Prachathipatai 1966a)

In a world where imported goods flooded a distinct urban market, foreign interest in traditional Thai culture made high-end gift shops important sites of a newly imagined nation; a place where the spending power of wealthy foreign and local consumers played an important role in protecting examples of Thai culture. Anti-imperialist tropes of economic sovereignty and protections were thus replaced with a cosmopolitan elite inversion. Sites that presented such harmony proliferated, establishing spaces within which a trans-Pacific elite could roam freely, interacting in spaces that felt both relaxed and free, but where entrance was in truth highly regulated through a regime organized by money, knowledge and power.

Sites of intersection: the emergence of free world space

Sarit's commitment to the United States in the Cold War clearly left Thailand vulnerable to shifts in global power. With limited ability to shape the political or economic environment globally, the mobilization of cultural resources became vital to securing ongoing US interests. The setting up of the Tourist Organization of Thailand at the start of Sarit's self-styled 'revolution' was just such a public relations intervention. Tourist literature intended for Americans presented the Thai people as friendly, peaceful and inherently conservative: all narratives taken directly from the pages of American publications such as *Life* magazine, and often supported by academic literature.[4] At the same time, tourism provided a valuable opportunity to reinforce the role of the Thai monarchy. Having been maligned for a generation, tourist attractions that placed performances of Thai high culture centre-stage, such as the Royal Barge Ceremony, affirmed the monarchy as an inherently valuable asset to the nation (Phillips 2017). They also provided opportunities for the newly forged Sino-Thai high society to mix with American visitors, reinforcing the impression that by sharing consumption practices, they existed within the same sphere.

One of the first examples of this came in December 1960. Lieutenant-General Chalermchai Charuvastr, who headed up the Tourist Organization, held a press conference at the Erawan Hotel in which he announced the introduction of a metered taxi service. He also revealed that the Erawan would be given a more Thai atmosphere: the chambermaids would wear Thai-style outfits in the traditional style and Thai food would be added to the menu. Thai-style entertainment, intended to rekindle the romance of old Siam, would also be introduced (*Bangkok Post* 1960). This trend was soon repeated across the city. The Sbai-thong restaurant, for example, offered diners the opportunity to 're-live the splendor of Siam in the olden days', with 'meticulously prepared Siamese dishes and classical dances as enjoyed by the nobility centuries ago' (*Bangkok Post* 1965a). From May 31 1966 The Oriental, a colonial-style hotel that sat on the banks of the Chao Phraya River, sought to expand its 'Thai-inspired' offering. Introducing a nightly Thai food festival that featured Thai décor in the Jim Thompson designed

Riverside Supper club, the hotel boasted Thai popular music and dancing, along with Thai shirts for all male guests and a 'Thai welcome for the Ladies' (*Bangkok Post* 1966a).

The new emphasis on 'Thai-style' internal space saw an increased interest in interior design. Now, the various Design-Thai outlets became sites of elite consumption where a well-networked high society, including Americans, Thai and Sino-Thai elites could purchase the latest Thai-inspired fabrics and furnishings. With a main store on Ploenchit Road, further outlets included a boutique in the Rama Hotel Arcade, a spacious four-room factory store on Sukhumvit Soi 69 and a further outlet at Pattaya Beach (*Bangkok Post* 1966b). In July 1965, a week-long grand sale began at the Sukhumvit Road outlet and was, from early in the morning, crowded with of what the *Bangkok Post* described as 'society ladies.' Amongst those identified was Mrs Graham Martin (wife of the US Ambassador) and Mrs Taylor (wife of the General Manager of Caltex petroleum), two daughters of former Prime Minister Phibun Songkhram and a host of television celebrities (*Bangkok Post* 1965b). At about the same time, the *Bangkok Post* also began a weekly feature called 'Home', intended to reflect a growing interest in design that reflected 'harmony' between local and imported styles (*Bangkok Post* 1966c). The first edition featured the penthouse of an American resident on Patpong Road.

In September 1965 a new hotel opened on Ratchadamnoen Avenue that sought to capitalise on the new trends. The Majestic occupied one of the modernist buildings built during the first Prime Ministership of Phibun Songkhram [1938–1944]. It had 75 guest bedrooms, a spacious lobby, a 200-seat restaurant, a 100-seat tearoom and an arcade of gift shops. There was also a bar that had a glass wall, providing a full view of a tropical garden and the swimming pool. The renovation of the hotel was reported to have cost 6 million baht (roughly $290,000 at the time), and was overseen by Chareon Chirangboonkul, whose company, Siphya Construction Company had been established only a year earlier (*Bangkok Post* 1965c). Replicating the zeitgeist of the time, the *Bangkok Post* claimed that the opening of the hotel marked 'the first time in the history of interior decoration in Bangkok [that] all modern international styles have been revolutionized to fit in with Thai design' (Ibid). As with the approach of Design-Thai, these 'revolutionary' designs were largely created by borrowing from the Thai past [*samai boran*] and, in this case, after visits to the National Museum. Bedroom chairs were created that resembled elephant howdahs, while chairs in the dining room were a 'modernized version' of those used by Rama I (r. 1782–1809) when he went into war (Ibid).

Over the following year, Chareon's new company would also have a hand in the construction of the Siam Intercontinental. Thai partners were to finance the majority of the project, but unlike the Hilton proposal, Pan American Airlines (of which Intercontinental was part) committed to putting up a portion of the cash. A new Thai company named Bangkok Intercontinental Hotels Company (BIHC) was formed in 1958, and after winning the support of Sarit and General Chalermchai (who became the head of the Bangkok-based company), the project progressed smoothly. At the outset, the Thai government agreed to put up just under half million dollars, while Pan American put up $740,000. The rest of the capital was to come from local Thai interests, made up predominantly of Sino-Thai business leaders. Further financing came with loans, provided by the Import Export Bank of Washington along with a raft of local banks (Garnett 1959).

After some debate about location, the new hotel was built on land that belonged to Sra Pathum Palace. This placed it on the periphery of the existing royal centre and bestowed the project with the prestige and authority of royal sanction. In fact, the hotel was initially labelled the Hotel Royal Bangkok, in deference to the esteemed nature of the location (Ibid). Crucially, it was also in touching distance of the new developments linked to both the post-war commercial and suburban centres sprouting up to the east, as well as to the more historic centres of commerce to the south and south-west (King 2011, 88–94). Positioned between these existing spheres, the new development was an open sign that the era of competing interests, locally and nationally, had for now at least come to an end (Figure 12.2).

Designed by the American architect Joseph P. Salerno, a student of Frank Lloyd Wright, the hotel was built to reflect harmony. Touted from the beginning as Southeast Asia's newest and most luxurious hotel, the design was intended to harness the best of the west, while paying homage to its Asian location (Ünaldi 2016, 149). As the *Bangkok Post* explained, the single story structure embraced both 'the latest in American design' and 'the most traditional Thai style' (Razak 1965). Upon opening in 1966, a special supplement explained that 'nothing has been overlooked in the meticulous attention to detail and complete utilization of the most charming aspects of Thai artwork.' Promising 'functional beauty', every room was intended to provide

Figure 12.2 Siam Intercontinental: Bangkok 1967.
Source: ©Roger Wollstadt.

a blend of the exquisitely modern and the beautifully refined. Soft, Thai-inspired lighting and furnishings made spaces such as the Pagoda Gift Shop sites of idealized consumption in the new guise. In sites of entertainment, such as the Naga bar (named after the Buddhist serpent god), business deals and international politics could be discussed in an interior that was 'decorated with large, locally hand-carved wooden screens of Naga figures which are legendary in the art of Thailand' (*Bangkok Post* 1966d). A victory for US hegemony, the Siam Intercontinental also represented a proclamation of friendship between the Thai and American nations. The opening ceremony, presided over by HRH The Princess Mother, was billed as an event 'typical of the change in relations between countries all across the world, and even more markedly in this immediate area' (ibid).

In the Thai press, Siam Intercontinental was seen as more than just a masterpiece in design: it signalled the expansion of opportunity for the capital's urban elite. Covering a total of 71 *rai* (113,600 sq. meters), the hotel recentred the city, intersecting with old and new locales of power. Long a signal of elite standing, the era of the automobile, also a key symbol of the American era, was ushered in with plenty of space for guests to park their cars. It also provided access to a road that connected to the building to the rest of the city and to Don Mueang airport to the north. Upon its completion it would become a new focus for commerce and finance (*Prachathipatai* 1966b).

Eventually the iconic building, which Serhat Ünaldi describes as *kan phatthana* in built form, became 'an institution for tourists and Bangkokonians alike.' The unique blend of American modern and Thai-inspired embellishment became ubiquitous. BIHC, which later changed its name to Siam Piwat, continued to develop the area for decades, creating a far clearer centre to Bangkok and securing the fabulous wealth of General Chalermchai and his family (Gluckman 2016). Siam was therefore the place where the intersecting interests of monarchy, military and Sino-Thai business families, facilitated by the American presence moved from possibility to reality (Ünaldi 2016, 13). The hotel thus became a popular site for society weddings, the ultimate practice in elite network consolidation (Ünaldi 2016, 131).

A disarming embrace?

By the mid-1960s, as the Vietnam War was heating up, Bangkok could be navigated by high society Thais, Sino-Thais and international guests alike through a series of interior spaces that proclaimed harmony. The new system of licensed taxis connected hotel lobbies with boutiques, restaurants, nightclubs and tourist attractions: social spaces framed by, and determined through, the alignment of Thai and US interests. These sites, product of violent military engagements on the periphery, had a profound impact on the capital. Thailand's construction boom became oriented towards the creation of service-focussed industries featuring an imagined harmony between an imported 'modernity' and Thai traditions at the heart of the transaction.

International buffet dinners, cheery chambermaids and smiling airhostesses – all played a role in adding a Thai inflection to international tastes.

Thailand's status as a long-term ally to the United States, combined with the willingness post-1958 to accept the pre-eminence of western knowledge and power, placed a premium on access to trans-Pacific relationships as the defining feature of the Thai elite. Over time, however, urban space became increasingly insular: a particular brand of cosmopolitanism that privileged harmony and friendship through a mutual embrace of Thai culture in its most exoticised and commodified form. Willingness to embrace an orientialist vision of their own culture explains in part how Bangkok's high society came to accept the increasingly visible sex trade that dominates portions of downtown. At the same time, retreat into ever more exotic and unattainable space helps explain how they insulated themselves from the less attractive outcomes of Cold War relationships. The influx of American GI's, travelling into the city for days of debauchery, offended the new air of distinction. Air conditioning being the ultimate signifier of elite consumption, the presence of these new Americans expedited the elite retreat from the street to the cool interiors of the hotel lobby, the shopping mall and the luxury automobile.

Udom managed brilliantly even as the centre of gravity shifted north towards the Siam Intercontinental. American corporate and government headquarters having expeditiously relocated, Patpong became defined by what happened at street level. By 2005, a decade after Udom's death, the land managed by the family corporation was worth an estimated $100 million, providing an annual rental income of $3 million (Backman 2005, 204). Having expertly satisfied regional and geopolitical demands in the aftermath of World War II, Udom in the end built a fortune by placing the dark side of development on show for the world to see.

Notes

1 This paper was first presented at the Asia Research Institute seminar at the National University of Singapore in October 2016. I thank all those who attended for useful feedback, much of which has been considered and implemented. Many thanks to Brian Curtin, Thiti Jamkajornkeiat, James Warren and Christine Gray for reading earlier versions and making helpful comments. Please note that in line with convention in Thai studies, Thai authors are referenced by the first name rather than the last name.
2 Note that the United States Information Agency (USIA) was the name given to the organisation in 1953. It had previously been referred to abroad as the United States Information Service (USIS), and this continued to be the name applied to buildings such as the USIS library.
3 Note that the most significant influx of migrants arriving in Bangkok began in 1957, following a major drought in the Isan region. See Thak Chaloemtiarana, *Thailand: The Politics of Despotic Paternalism*, Ithaca: Southeast Asia Program Publications, Southeast Asia Program, Cornell University, 2007, pp. 74–75.
4 There are countless such examples, but a particularly clear example can be found with the work of William J. Klausner, much of which was produced for official and as well as use.

References

Askew, M. (2002) *Bangkok: Place, Practice and Representation*, London, Routledge.
Ayer, J. (nd) Unpublished diary, consulted with permission from the Jacqueline Ayer Estate.
Backman, M. (2005) *Inside Knowledge: Streetwise in Asia*, Basingstoke and New York, Palgrave Macmillan.
Bangkok Post (1960) "Erawan Hotel a 'Thai Treat' for Tourists Expanding," December 20, 1960.
Bangkok Post (1965a) "Dine at Sbai-thong," July 24, 1965.
Bangkok Post (1965b) "Design Thai Sales Attract Huge Crowds," July 7, 1965.
Bangkok Post (1965c) "Special Supplement on the Opening of the Majestic Hotel," September 24, 1965.
Bangkok Post (1966a) "Thai Food Festival at the Riverside Supper Club," May 31, 1966.
Bangkok Post (1966b) "It's the Season for Fun and Design Thai has the Clothes," May 4, 1966.
Bangkok Post (1966c) "Homes: The First in a Weekly Series on Modern Living," May 28, 1966.
Bangkok Post (1966d) "The Siam Intercontinental, Bangkok Post Supplement."
Bangkok World (1962) "From West to East: A Renaissance in Thai Design," Bangkok World Sunday Magazine, Bangkok, March 11, 1962.
Bourdieu, P. (2008) [1979] *Distinction: A Social Critique of the Judgement of Taste*, Abington, Routledge.
Caldwell, J.C. (1968) *Massage Girl and Other Sketches of Thailand*, New York: The John Day Company.
Chatri Prakitnonthakan (2009), *Silapa-sathapattayakam Khana ratsadon sanyalakthan kan mueng nai choeng udom kan* [The Art and Architecture of the People's Party: Symbols of Political Ideology] Bangkok, Matichon.
Chua, L. (2014) "Honey, I shrunk the Nation-State: The Scales of Global History in the Thai Nationalist Theme Park," In: Lasansky, D.M. (ed.), *Archi.Pop: Mediating Architecture in Popular Culture*, London and New York, Bloomsbury Academic, 75–88.
Connors, M.K. (2007) [2003] *Democracy and National Identity in Thailand*, Copenhagen, NIAS Press.
Cooper, D.F. (1995) *Thailand: Dictatorship or Democracy*, London, Minerva Press.
Donovan, W. (1954) US Ambassador to Thailand to State Department "Renewal of the Educational Exchange (Fulbright) Agreement with Thailand," May 26 1954 (US NA): Decimal File: 511.923/5–2654.
Fineman, D. (1997) *A Special Relationship: The United States and Military Government in Thailand, 1947–1958*, Honolulu, University of Hawai'i Press.
Ford, E. (2017) *Cold War Monks: Buddhism and America's Secret Strategy in Southeast Asia*, New Haven, Yale University Press.
Garnett, G. (1959) Pan American World Airline Systems to Ministry of Economic Affairs, Government of Thailand, June 23, 1959.
Girling, J. L.S. (1981) *Thailand: Society and Politics*, Ithaca and London, Cornell University Press.
Glassman, J. (2005) "On the Borders of Southeast Asia: Cold War Geography and the Construction of the Other," *Political Geography*, Vol. 24, No. 7, 784–807.

Gluckman, R. (2016) The Queen of Siam: Chadatip Chutrakul Aims to Energize A District in Bangkok, *Forbes Asia*, April 2016: https://www.forbes.com/sites/forbesasia/2016/04/06/the-queen-of-siam-chadatip-chutrakul-aims-to-energize-a-district-in-bangkok/#78b95f0b60a0 [Accessed November 5, 2016].

Gray, C.E. (1991) "Hegemonic Images: Language and Silence in the Royal Thai Polity," *Man*, New Series, Vol. 26, No. 1, 43–65.

Gray, C.E. (1992) "Royal Words and Their Unroyal Consequences," *Cultural Anthropology*, Vol. 7, No. 4, 448–463.

Grossman, N. (2009) *Chronicle of Thailand: Headline News Since 1946*, Bangkok, Bangkok Post.

Handley, P. (2006) *The King Never Smiles: A Biography of Thailand's Bhumibol Adulyadej*, New Haven, Yale University Press.

Harrison, R.V. (2010) "The Man with the Golden Gauntlets: Mit Chaibancha's Insi thorng and the Hybridization of Red and Yellow Perils in Thai Cold War Action Cinema," In: *Cultures at War: The Cold War and Cultural Expression in Southeast Asia*. 22–36. Ithaca, Southeast Asia Program Publications, Southeast Asia Program, Cornell University.

Hollinger, C. (2000) *Mai Pen Rai Means Never Mind: An American Housewife's Honest Love Affair with the Irrepressible People of Thailand*, Bangkok, Asia Books.

Houser, J.W. (1955) "Vice President of Hilton International to Pote Sarisin, Thai Ambassador to the United Stets," *National Archives of Thailand*, Vol. 3, SR 0201.30/4, pp. 6–8.

James, L. and Leake, E. (2015) *Decolonization and the Cold War: Negotiating Independence*. London and New York, Bloomsbury.

King, R. (2011) *Reading Bangkok*, Singapore, NUS Press.

Klein, C. (2003) *Cold War Orientalism: Asia in the Middlebrow Imagination 1945–1961* Berkley, University of California Press.

Koompong Noobanjong (2013) *The Aesthetics of Power: Architecture, Modernity and Identity from Siam to Thailand*. Bangkok, White Lotus.

Lefebvre, H. (1991) *The Production of Space*, Oxford, Blackwell Publishing.

McCoy, A.W. (2003) *The Politics of Heroin: CIA Complicity in the Global Drug Trade*, Chicago, Lawrence Hill Books.

Peleggi, M. (2002) *Lords of Things: The Fashioning of the Siamese Monarchy's Modern Image*, Honolulu, University of Hawai'i Press.

Phibun Songkram (1955) to Adjunct General Department of the Cabinet for Political Affairs, "*Sanoe kho rap dam noen kichakan rongraem*," 16 April 1955.

Phillips, M. (2016) *Thailand in the Cold War*, New York and London, Routledge.

Phillips, M. (2017) "Ancient Past, Modern Ceremony: Thailand's Royal Barge Procession in Historical Context," In: Lambert, P., Weiler, B. (eds.), *How the Past was Used: Historical Cultures c. 750–2000*. 44–61. Oxford, Oxford University Press *published for* The British Academy.

Phongpaichit, P. (1980) "The Open Economy and Its Friends: The 'Development' of Thailand," *Pacific Affairs*, Vol. 53, No. 3. 440–460.

Pike, F. (2011) *Empires at War: A Short History of Modern Southeast Asia Since World War II* London, I. B. Tauris (googlebooks version).

Prachathipatai (1966a) *Narayana Phand: sun silapa-hattakan* [*Narai Phan*: centre of art-handicrafts], January 6, 1966.

Prachathipatai (1966b) *khayai rongraem Sian Inter 404 hong* [Siam Inter to expand by 404 rooms), January 23, 1966.

Razak, A. (1965) "US Designs Meet Thai Traditions in New Hotel," *Bangkok Post*, October 5, 1965.

Skinner, W.G. (1958) *Leadership and Power in the Chinese Community in Thailand*, Ithaca, Cornell University Press.

Steinfatt, T. (2002) *Working at the Bar: Sex Work and Health Communication in Thailand*, Westport, Ablex Publishing.

Tambiah, S.J. (2013) [1973] "The Galactic Polity in Southeast Asia," *HAU: Journal of Ethnographic Theory*, Vol. 3, No. 3, pp. 503–534.

Thak Chaloemtiarana (2007) *Thailand: The Politics of Despotic Paternalism*, Ithaca, Southeast Asia Program Publications, Southeast Asia Program, Cornell University.

Ünaldi, S. (2016) *Working towards the Monarchy: The Politics of Space in Downtown Bangkok*, Honolulu, University of Hawai'i Press.

US Embassy, Bangkok (1952) "Revised Draft of the Country Plan for Thailand," November 5, 1952.

US Embassy, Bangkok, (1956) "Special Events and Projects: The Thai Fair Program" (US NA) Decimal File 511.92/6–3053.

USIA (1954) "Briefing on USIA Program in Thailand," Washington, 12 May 1954. https://www.cia.gov/library/readingroom/docs/CIA-RDP61S00750A00010 0140216-5.pdf [Accessed 12 January 2017].

Von Eschen, P. (2004) *Satchmo Blows Up the World: Jazz Ambassadors Play the Cold War*, London and Cambridge, MA, Harvard University Press.

Wallerstein, I. (2010) "What Cold War in Asia? An Interpretative Essay," In: Yangwen Z., Liu H., Szonyi M. (eds.), *The Cold War in Asia: The Battle for Hearts and Minds*. Leiden and Boston, Brill, 15–24.

Wasana Wongsurawat (2009) "From Yaowaraj to Plabplachai: The Thai State and Ethnic Chinese in Thailand during the Cold War," In: Vu, T., Wongsurawat, W. (eds.), *Dynamics of the Cold War in Asia: Ideology, Identity, and Culture*, New York, Palgrave Macmillan.

Westad, O.A. (2007) [2005] *The Global Cold War: Third World Interventions and the Making of Our Times*, Cambridge, Cambridge University Press.

Wharton, A.J. (2001) Annabel Jane, *Building the Cold War: Hilton Hotels and Modern Architecture*, Chicago and London, University of Chicago Press.

Wong, K.F. (2006) *Visions of a Nation: Public Monuments in Twentieth-Century Thailand*, Bangkok, White Lotus.

Cold War telecommunication and urban vulnerability
Underground exchange and microwave tower in Manchester

Martin Dodge and Richard Brook

UNIVERSITY OF MANCHESTER; MANCHESTER SCHOOL OF ARCHITECTURE

Introduction

Protecting national telecommunications in the event of an atomic bomb attack was seen as vital for the continuity of core government operations. In Britain, the General Post Office (GPO) was the major national organisation, part of central government, a quasi-military uniformed service that had monopoly responsibility over the provision of all inland telecommunications. Early in the Cold War period, the GPO was charged by Whitehall with providing sufficient telecommunication, with resilient infrastructure and hardened facilities, for the nation to fight and survive an atomic war.

Despite the fiscal austerity of the late 1940s and early 1950s, expenditure related to atomic warfare was prioritised in terms of capital, resources and personnel. This included significant investment in new telecommunications infrastructure apparently justifiable for the overriding demands of national defence. Much of the new hardware was also purposefully designed to be dual-purpose, in that it operated fully for civilian traffic in peace time but could be switched in emergencies to grant access only to privileged priority users.

In the late 1940s and through the 1950s, this involved reconfiguring the major cable routes through vulnerable city centres and building underground exchanges. Peter Laurie in his book *Beneath the City Streets* (1970, p. 149) explained '[t]he exchanges and the organisations they are to serve must be housed in well-protected places, because they are the ganglia of the thermonuclear bomb resistant brain. If they are damaged, the government creature is blind, deaf and dumb'. By the 1960s, the approach was based more on the technology of microwave transmission and involved blast-resistant concrete towers built primarily away from population centres.

302 *Martin Dodge and Richard Brook*

Figure VE 3.1 View along main tunnel carved out as a rough egg-shaped space. Workers busy putting up concrete lining, 1956.
Source: BT Archives, ref. TCB417/e20845.

Figure VE 3.2 Part of the main tunnel fully lined with concrete, connection to side tunnel evident, 1956.
Source: BT Archives, ref. TCB417/e20847.

Construction of the 'Guardian' underground telephone exchange

Deep under the centre of Manchester is a Cold War era bunker, conceived as part of the national strategy to secure telecommunications from the atomic threat. It comprises a large network of reinforced concrete tunnels and housed a major telephone exchange. Within the GPO, it was known as Scheme 567 and given the code name 'Guardian'. It was one of three similar city centre installations constructed under secretive conditions in the latter half of the 1950s. Companion underground telephone were in central London (codenamed 'Kingsway') and in the middle of Birmingham (the 'Anchor' exchange). It seems that similar deep underground telecommunications facilities in Bristol and Glasgow were initially planned during this period but never constructed.

Figure VE 3.3 Upper floor of main tunnel being fitted out with racking to support telecommunication equipment, 1957.
Source: BT Archives, ref. TCB417/e21959.

No official maps of the full tunnel extents or original engineering plans for the 'Guardian' underground exchange have been released into the public domain but the general position of the core bunker and location of surface connections surface has been determined from other sources (such as plans deposited for the Parliamentary act authorising construction, Post Office Works Act 1958). However, a substantial number of original photographs of the construction of the tunnels and the fit-out of the exchange were taken by the GPO in the 1950s. These are available from BT Archives and we reproduce a selection here.

Figure VE 3.4 Completion of the shaft head no. 6 at George Street which would provide the main access to the exchange, March 1960.
Source: BT Archives, ref. TCB417/e24701.

After several years of tunnelling, the major below ground construction work on the bunker was completed by the end of 1957 and the central underground spaces were fitted out with extensive telecommunications and supporting electrical equipment. As an operational exchange it came to life on 7 December 1958, when the first traffic was received at 8 am. Completion of the associated surface structures took several more years.

Cold War telecommunication 305

Figure VE 3.5 Completing the utilitarian surface building and ventilation tower at George Street, December 1956.
Source: BT Archives, ref. TCB417/e25446.

306 *Martin Dodge and Richard Brook*

Operational spaces in the main tunnels of the 'Guardian' underground telephone exchange. The selection of photograph is illustrative of the scale of machinery needed to create an operative 'bubble world' in a bunker after the Bomb drops.

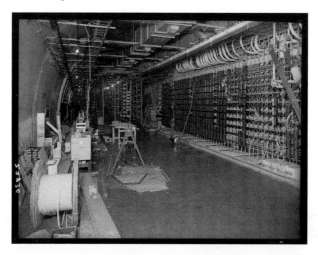

Figure VE 3.6 Completing the installation of the telecommunication switching equipment, 1959.
Source: BT Archives, ref. TCB417/e22939.

The defining characteristic of the bunker space – its strength from subterranean position and concrete construction – however were not effective. Even before construction was complete in 1958, advances in nuclear weapons yield and the development of intercontinental ballistic missiles meant that the tunnel design and its depth underground would not have protected the telecommunications equipment and personnel working within.

Figure VE 3.7 View of banks of batteries to provide emergency backup power before the diesel generators start-up, 1959.
Source: BT Archives, ref. TCB417/e24227.

The 'Guardian' exchange purportedly cost just over £2 million to construct, with £1.6 million expended on the tunnelling works, a significant sum in the time of austerity in 1950s Britain. The planning and design of the tunnels was by Sir William Halcrow & Partners, who were leading engineering consultants of the time; the main civil engineering contractor was Edmund Nuttall, Sons & Co. Ltd.

Figure VE 3.8 Plant room with equipment for ventilation and cooling, 1959.
Source: BT Archives, ref. TCB417/e22938.

The main part of the 'Guardian' telephone exchange is situated around 34 metres below ground, (actual tunnel depths were deemed a key secret during construction and remain imprecise estimates) and is comprised of a set of interconnected equipment tunnels under the Chinatown area of Manchester city centre. Smaller cable tunnels, just over 2 metres in diameter, that double as emergency escape routes, extend out to two vertical shafts in Salford and Ardwick.

Figure VE 3.9 The multiple backup electricity generators and ancillary equipment. This took up almost as much space as the primary telecommunication equipment, 1959.
Source: BT Archives, ref. TCB417/e24044.

Certain sections of the tunnels were built large enough to contain upper and lower level working spaces. On the main level of the exchange, in the largest and longest tunnel (known as A.T.8, 'apparatus tunnel'), were GPO engineers who worked to maintain the analogue telecommunications switching equipment. It could handle traffic from 1,488 incoming circuits. It was here that the familiar perforated vertical face of the main distribution frame (MDF) and the repeated racks of electromechanical switches, icons of mid-century communications, were situated. The lower level A.T.8 was partitioned into series of rooms that provided limited support for the welfare of a small cadre of around 35 engineers required to keep it functioning (first aid station, dining room, maintenance office, toilets and a cloakroom).

Figure VE 3.10 Rest room for engineering staff, 1959.
Source: BT Archives, ref. TCB417/e24232.

The other large interconnected tunnels A.T. 3, 5 and 7 contained subsidiary equipment necessary for the functioning of the underground exchange, along with supporting work spaces. The provision of electricity was elemental to bunker operations in an emergency and the power plant was of the "no-break" type incorporating a motor-alternator which, in the event of a mains failure, was run from a 240-volt battery until the prime-mover supply took over. Three 279kW diesel-alternator sets were to provide a standby power supply in the event of mains failure. Connected to A.T.5 was a smaller side tunnel that served as the fuel store for the standby generators. A.T.3 contained electrical switchgear in the upper half and banks of lead acid batteries for emergency power in the lower floor. Lastly A.T.7 contained a substantial workshop area and technical stores, along with equipment associated with ventilation for the tunnel complex.

Figure VE 3.11 Basic ablution facilities, 1959.
Source: BT Archives, ref. TCB417/e24228.

By the early 1970s, the tunnels were fast becoming an outdated product of the early Cold War and obsolete for even basic telecommunications operations due to advances in digital exchange technology. Permanent staffing underground ended at some point in the late 1980s and shifted largely into a care and maintenance approach, with necessity to keep pumps working to prevent flooding. The bunker continues to operate as an infrastructure space, housing fibre-optic cables. In terms of the extant surface presence of the 'Guardian' exchange, there is a scattering of anonymous buildings and a couple of entrances located in commercial premises, which provide secure access via deep lift shafts.[1]

Cold War telecommunication 311

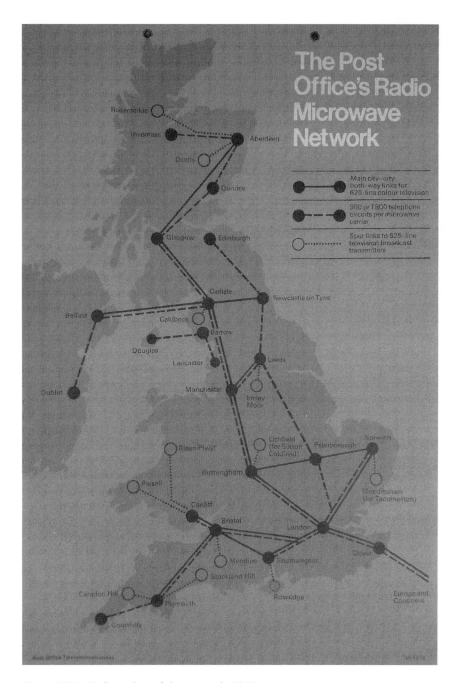

Figure VE 3.12 Overview of the network, 1969
Source: BT Archives, ref. TCC297/Tele_Ed_16.

Concrete microwave tower

One A later part of Cold War telecommunications provision was a network of high bandwidth links using super high frequency radio beams running along a chain of antennas in line-of-sight of each other. This became the national-scale microwave relay network by the late 1960s, also known as 'backbone'. (It built upon experiments in the 1930s and advances in technology made during the Second World War.) Engineering a secure, high capacity telecommunication network running through airwaves was appealing to central government and the military. The transmission towers could be widely spaced and were significantly cheaper compared to burrowing wholly new cable routes that bypassed vulnerable cities. Furthermore, evidence from the aftermath of atomic bombs in Japan in 1945 showed that steel lattice towers and concrete chimneys were some of the most blast-resistant structures.

The slender microwave towers were positioned prominently in the landscape for unobstructed line-of-sight transmission. As dishes and horns evolved, so did the towers. The GPO engineering, coupled with the cost-conscious, practical approach of the Ministry of Public Building and Works resulted in the design and development that might be described as the 'form follows function'. The bare concrete structures suggested efficiency and an unadorned functionality emblematic of modernist ideals.

The growing demand of consumers for long distance telephony and in particular the distribution of television services in the post-war decades was a power economic rationale and also a convenient cover story for defence infrastructure that could be hidden-in-plain sight. The chain of microwave towers was deemed vital for defence communications but in planning enquiries and public reports was described as fulfilling primarily civil needs.

The first major inter-city microwave link was from London to Birmingham which opened in 1949 initially to relay television signals to the Sutton Coldfield BBC television transmitter. Increased bandwidth and higher definition for future channels following the Television Act of 1954 further contributed to a convincing business case for a national microwave network.

Cold War telecommunication 313

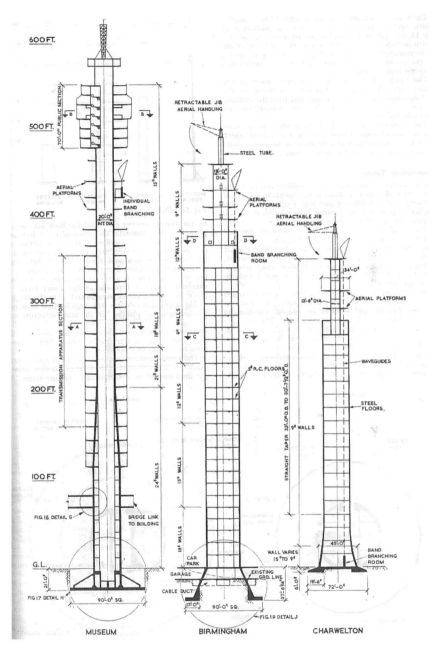

Figure VE 3.13 Design of three different types of concrete tower used on the microwave network. The Heaton Park tower is based on the 'Charwelton' design.

Source: Creasy L.R., Adams H.C. & Silhan, S.G. (1965) 'Radio towers' in *The Structural Engineer*, Vol. 43, No. 10, October 1965, pp. 323–336.

Figure VE 3.14 View of the Heaton Park microwave tower from within the park, 1966.
Source: BT Archives, ref. TCB417/e33223.

Cold War telecommunication 315

Figure VE 3.15 Close-up view of the horn reflectors on the Heaton Park microwave tower, 1966.
Source: BT Archives, ref. TCB417/e33232.

The standout architectural structures of the microwave network were the large cylindrical towers built from exposed reinforced concrete in the 1960s. Around ten were constructed, the most celebrated was the Post Office Tower in central London, opened by Prime Minister Harold Wilson in October 1965. Interestingly the companion structure for Birmingham city centre was to follow the same approach as London but a 'totally unexpected change in the design of the tower proposed for aesthetic reasons' meant that it was built square in section with distinctive corner profiles. Whilst of the same construction method, the city centre towers and those in less urban settings had alternative appearance. Seven rural and suburban concrete towers were built to a design named 'Charwelton' – its form approved by the Royal Fine Arts Commission. They were sited at Stokenchurch (Buckinghamshire), Charwelton (Northamptonshire), Pye Green (Staffordshire), Sutton Common (Cheshire), Tinshill (Leeds), Wotton-under-Edge (Gloucestershire) and Heaton Park (Manchester).

Manchester's microwave tower joined an existing police radio antenna on an elevated position in Heaton Park, a couple of miles north of the city centre in 1965. At the head of the tower, four cantilevered discs provided the platforms for a mixture of aerial types. Six dishes and five horn antennas served three relay points, Horwich, Sutton Common and Windy Hill; these connected to Carlisle, Birmingham and Leeds respectively. The most visually striking was undoubtedly the horns. The largest of these paraboloid aerials were 8m tall and made from ribbed aluminium, their shape and scale offered greater capacity for receiving multiple wavelengths. The inverted pyramid forms had chamfered faces over the horns' apertures made from a pearlescent Hypalon-coated terylene sheet. To the observer, these horn antennas provided a distinctive futuristic silhouette for the microwave tower.

The microwave transmission network was superseded by fibre-optic cables from the 1990s and towers became largely redundant. The large horns were subsequently taken down from the Heaton Park tower although the structure remains useful for hosting mobile phone antennas and other local line-of-sight communication links.

Note

1 Notes on captions: Figures VE 3.1–3.5, Construction of the 'Guardian' underground telephone exchange. Figures VE 3.6–3.11, Operational spaces in the main tunnels of the 'Guardian' underground telephone exchange. The selection of photograph is illustrative of the scale of machinery needed to create an operative 'bubble world' in a bunker after the Bomb drops. Figures VE 3.12–3.15, Design of the microwave towers. (All photographs courtesy of BT Archives, unless otherwise noted.)

Index

Note: **Bold** page numbers refer to tables, *Italic* page numbers refer to figures and page numbers followed by "n" refer to end notes.

Aarhus 201, *202*, 203–204, 214
Abakumov, Y. 126
active defence 23–28
After the Bomb: Civil Defence and Nuclear War in Britain, 1945–68 (Grant) 6
Air Raid Precautions Act (1937) 130
Albertina 274
Aman, Anders 144
ambiguity 190–191
Amemiya, Shōichi 111n22
Anconelli, T. 272
Anti-Aircraft Command (AAC) 67–68
Anti-Aircraft Operations Room (AAOR) 68, 69, *69*
anxious urbanism 185, 186
Applebaum, Anne 141
architectural design 16–17
Arena Service Club *247*, 247–248
Arman, H. 80, 83, 84, 91
armed camp 126–127
Armed Services Procurement Act of 1947 42
Asada Takashi 99, 104, 105
atomic-resistant building 16–17
atomic urbanism 2, 4, 8, 9, *10*, 11, **11**, 12, 14, 16–19, 21, 23, 25, 28–30, 104, 197–198
Atomic Weapons Requirements Study (1959) 155
Atomic Weapons Research Establishment 74, 226
Augur, T. 13, 15
Australia 219–226, 230–232; *see also* Sydney
Ayer, J. *289*, 290
Aylen, J. 23

Ballistic Missile Early Warning System (BMEWS) 189
Bangkok 3, 26, 281–282; Chinese capitalists 286–287; consumerism and urban development 290–293; Hilton 282–284; sites of intersection 293–296; Thai–American interactions 287–290, 296–297; Urban 284–286
Bangkok Intercontinental Hotels Company (BIHC) 294–296
Bangkok Post 294, 295
Bangkok World 290
Beauregard, R.A. 185
Bell Laboratories 47, 52n11, 52n16
Beneath the City Streets (Laurie) 301
Bennett, David 137
Berlin Blockade 7, **8**, 55, 66
Bettis Atomic Laboratory 38
Bierut, Bolesław 132, 134, 139
Big Picture, The 182, 184–185, 250–251
Bilateral Infrastructural Agreement (BIA) 236
Bishop, R. 229
Bologna 256–259, 274–275; A-bomb 257–263; *Festa dell'Unità* 269; *La Lotta* 264–267, *265–268*, 269–272, 274; left-wing 257, 258, 260, 261, 263, 270; nuclear threat 263–274; Partisans of Peace 259, 260, 263, 264, 267, 270, *270*, 272; peace mobilisation 257–265, 267, 269–275; UDI 258, 259, 261, 263, 272, 273
Bolsheviks 128, 134, 142
Borgo Milano 239, *240*, 241
Borgonzoni, Aldo 264, 267, *268*
Borgo Trento 238, 239, 241
Borough 37, 43–46, 48–50
Brabrand Boligforening 204

318 *Index*

Bristol 55–56; Anti-Aircraft Command 67–68; atomic attack 69–70; Civil Defence Corps 61–65; Civil Defence Region 7 60–61, 71; Civil Defence Service 56–57; defence of 65–66; Eastwood Park and Falfield House 57–59; Germany, territorial gains 56; Gun Defended Area 68–69; hydrogen bomb attack 72–74; Royal Observer Corps 66, 67, *67*; Sub-Control 70–72
British Nuclear Culture (Hogg) 6
Browning, John 247
Brown, K.: *Contemporary British History* 6; *Plutopia: Nuclear Families, Atomic Cities, and the Great Soviet and American Plutonium Disasters* 7
Building in the Atomic Age (Hansen) 16
Burnett, Alan D. 125
Burtch, A. 224

Cahill, R. 221
Campaign for Nuclear Disarmament (CND) 72, 226, 228, 229, 231, 232
Camp Century 25, 182, *183*, 184, *184*, 185, 188, 190–198
Camp Fistclench 191, 192, 194
Can we plan for the atomic bomb? (Clayton) 15–16
Central Committee of the League of Communists of Macedonia (Muličkoski) 179, *179*
Chmielewski, Jan 130
Chongqing 117, *121*
Churchill 38, 40, 42–50
Ciborowski, Adolf 169, 174, 177
Cities of Knowledge: Cold War Science and the Search for the Next Silicon Valley (Pugh O'Mara) 6–7
City Trade Centre (Popovski) 179, *179*
civil defence 19–23, 55–65, 70–74, 125, 130, 187, 201, *202*, 203, 204, 207, 213, 214, 219, 223–225, 274
Civil Defence Corps (CDC) 61–65, 70, 74
Civil Defence Organisation *see* New South Wales (NSW) Civil Defence Organisation
Civil Defence Region 7 (CDR7) 60–61, 71
Civil Defence Service (CDS) 56–57, 62
Clancey, G. 229
"Class VI Post Exchange" (PX) 246
Clayton, P. 15, 29
Cocroft, W. 5
Colgan, W. 198

COMLANDSOUTH 236, 251, 252
Communist Party of Australia (CPA) 225, 228
Comunisti (1983) 272
Contemporary British History (Hogg and Brown) 6
Czechoslovakia 125, 126, 129, 139, 143, 169, 176, 177
Czernin, Ottokar 128

Danish authorities 25, 182, 185, 189–192, 194, 196, 204–205
Danish welfare state 201, 213–214; Aarhus 201, *202*, 203–204; Glostrup 207, 211, *211–213*, 213; housing and sheltering 204; schoolchildren and refugees 204, 207
de-concentration and dispersal planning 12–16
Defense Production Administration's (DPA) 46, 47
demilitarisation 148–149, 162, 164n7
Democratic Municipalities 262
Democrazia Cristiana (DC) 4, 237, 240–242, 248–250, *250*
Denmark 26, 177, 188–192, 201, 203, 204
Department Defense Dependents Schools (DoDDS) network 241
Department of Architecture of the Estonian (DAE) 85–86
Distant Early Warning (DEW) Line 189
Don Mazza Orphanage 248
Donovan, W. 285, 287
Dozza, Giuseppe 260

Economic Study of the Pittsburgh Region (ESPR) 39, 40, 42
Edgerton, D. 6, 125
Effects of Atomic Weapons, The (1950) 17
Eisenhower, D. D. 19
Elder, John 47
Electric Labyrinth 107–109
Engel, J.: *Local Consequences of the Global Cold War* 7
ensemble, urban 26, 77, 78, 82, 83, 87, 90–92, 94, 94n1
Entangled Geographies: Empire and Technopolitics in the Global Cold War (Hecht) 6
Estonia 25, 77–81, 83–87, 89–94, 148–155, 157–160, 162, 163, 164n7
Estonian National Museum 151, *151*, 158, 159, *160*, 162

European Defence Community 264, 269, 272
Exhibition Hall 99, 103, 104
EXPO '70 97, 98, 108, 109, 110n2

Farish, M. 185–187
Fighter Command 67, 68
Flowers Hill 60, 61
From Warfare to Welfare: Defense Intellectuals and Urban Problems in Cold War America (Light) 6
Futurama II 197

Gamera vs. Jiger 97, 98, 108–110
Gellerup Plan 203, 204, *205, 206*
Gellerupplanen 204
General Post Office (GPO) 301–303, 309
Generelli, Robert 242
Glostrup 207, 211, *211–213*, 213
Goldzamt, Edmund 128, 132, 142
Gomulka 133, 139, 144
Goodman, Benny 288
government challenge 18–19
Grant, M. 6, 22, 214
Greenland ice cap 182–185; atomic urbanism 197–198; Camp Century 182, *183*, 184, *184*, 185, 188, 190–198; militarisation and modernisation 188–190; nuclear annihilation 185–188; nuclear power 190–192
Gun Defended Area (GDA) 68–69
Guttuso, Renato 264

Hamilton, Ian 125
haniwa 99, 103
Hansen, H. C. 191
Hansen, R. 16
H-bomb 72, 73, 259–261, 269, 271, 274
Headquarters of Allied Land Forces Southern Europe (HALFSE) 236
Hecht, G.: *Entangled Geographies: Empire and Technopolitics in the Global Cold War* 6
Hennessy, P. 6
Herald, S. W. 49
Hersey, John: *Hiroshima* 222
hiatus 137–138
Hietala, M. 4, 5, 28
Hilberseimer, L. 14
Hilton, Conrad 283
Hilton Hotels 283–284, 294
Hiroshima **8**, 9, 16, 98–109, 112n29, 222, 230, 256, 266, 270; Peace Memorial Park 99, 100, 102, 104, 106, 109, 113n51
Hiroshima (Hersey) 222
Hogg, J.: *Contemporary British History* 6
'Homo Sovieticus' *127*, 128
Hotel Royal Bangkok 295
Houser, John W. 283
Howson, F. Henry 138
Hutcheson, J. A. 43, 45–47
hydrogen bomb attack 72–74

ice cap 182, 185, 196, 198; *see also* Greenland ice cap
Il quaderno dell'attivista 264
Industrial Civil Defence Service (ICDS) 62–64
Institute of Town Planning and Architecture of Skopje (ITPA) 169
International Basic Economy Corporation (IBEC) 281, 290
International Women's Democratic Federation (WIDF) 261
Interstate and Defence Highways Act (1956) 18–19
Irving, N. 226
Irving, T. 221
Isozaki, Arata 105–109
Istanbul Hilton 283
Italian Communist Party 237, 238, 256, 258, 260, 264, 272
Italy 108, 176, 236, 237, 241, 242, 244–246, 248, 249, 251, 257–261, 269, 271, 273–275; *see also* Bologna; Verona
Itō Teiji 106

Jakubec, Ivan 125
Japanese Urban Space 106–108, 112n36
Jōmon 100–102, 111n14, 111n19, 111n20

Kahn, H. 12
Kalm, M. 81
Karafantis, L. 43
Kardelj, Edvard 177
Kawazoe Noboru 100, 102, 103, 105
Kennedy, J. F. 5
Khrushchev 132, 138–140, 144, 177
Kiev 137, 138
Kivi, R. L. 158
Kohlrausch, Martin 130
Kohtla-Järve 77, 78, *79, 80*, 84, *88, 89*, 89–91, *90*, 92, *92*, 93, 94
Konstantinovski, G. 179
Kubo, M. 5, 27

Lamb, Z. 14
Lapp, R. 15, 21; *Must We Hide?* 9, 14
Laurie, P.: *Beneath the City Streets* 301
Le Corbusier 103, 104, 106, 117, 169, *203*
Lefebvre, H. 282
Lengiproshacht 77, 90, 91
Lengorstroyproyekt 77, 86, 90, 91
Leningrad metro 77, 83–85, 87, 89, 90, 92, 126, 137, 155
Leslie, S. 43
Light, J.: *From Warfare to Welfare: Defense Intellectuals and Urban Problems in Cold War America* 6
Liupanshui 116, *118*, 119, *119, 120*, 121, *121*
Local Consequences of the Global Cold War (Engel) 7
Lorentz, Stanislaw 135
Lowe, D. 228
Lowry, Ira S. 40–41
L'Unità 261
Lu Yujun 117

Maas, Winy 162
Macedonia 168, 169, 179
Manchester *21*, 301–303; microwave towers 309–310, *311–313*; telephone exchange 301–303, *304–309*
Manhattan Project 5, 9, 10, 15
Mao Zedong 116, 117
Martin, R. 27
May, E. T. 227
Meigas, V. 80
Menzies, Robert 223
Metabolist architectural movement 105, 106
Michta, Andrew A. 130
microwave towers 309–310, *311–313*
Mihailovic, Josif 168
Miller, B. A. 227
Mitchell, D. 220
Monroeville Doctrine 37–38, 42, 50–51
Monteith, A. C. 45, 47
Monteyne, D. 15, 185, 187, 214
Mordvinov, A. 83, 92
Moscow metro 126, 127, 130, 133, 134, 136, 137, 139, 142, 144; *see also* Warsaw Pact and metro
Mostra dell'Aldilà 248
Mozingo, L. 43, 47
Muličkoski, P. 179
Museum of Contemporary Art 177, *178*

Must We Hide? (Lapp) 9, 14
Mutually Assured Destruction 74

Nagasaki 9, 16, 98, 222, 270
NAM 170–172, 178
National Fallout Shelter Program 187
NATO 23, 136, 139, 143, 144, 153, 188, 190, 191, 197, 236–243, 247, 248, 251, 252
New South Wales (NSW): Civil Defence Organisation 71, 223–227, 229–231; Peace Committee 227, 229
'Ninth Project' *173*, 174
Nørrevangsskolen 206, 207, *208*
North America *3*, 6, 8, 12, 14, 19, 148, 219, 220, 223, 228
nuclear annihilation 185–188
nuclear power 190–192
Nucleus Force 68, 69

Office of Strategic Services (OSS) 281, 285
Okamoto, Tarō 100–101, 111n14
On The Beach (Shute) 222
Operation Sunshine 195
Ōtani Sachio 99

La pace (1951) 264
Page, M. 10–11
Palace of Culture and Science 27, 135–137, *136*
Panzhihua 116, *116, 118,* 119–121, *121*
parasitic urbanisation 185–186
Partial Test Ban Treaty 232
Le partisans de la Paix (*La paix* from 1951) 264
Passalacqua 238, 243–246, *245,* 248, 250–252
Patpong 281, 290, 294, 297
peace flag 273
peace movement 219–221, 223–237, 229–232, 258, 260, 270
Peace Partisan movement 259
peace week 261, 262, *262*
Peakhurst peace group 226
Peleggi, M. 286
Penn-Lincoln Parkway 41, 43
Peter Snow Millers 191, 192, *193*
Pilsudski, Jozef 129, *129*
Pittsburgh 28, 37–38, 50–51; laboratory 43–47; Research and Development Center 47–50, *48*; research and suburbanization 38–42, *41*

Index 321

Pittsburgh Area Industrial Dispersion Committee (PAIDC) 46
Pittsburgh Press 44
Pittsburgh Regional Planning Association (PRPA) 39
Pittsburgh Renaissance 39
Pizzi, K. 4, 5, 28
Plac Wilsona metro station 143–144, *144*
Plutopia: Nuclear Families, Atomic Cities, and the Great Soviet and American Plutonium Disasters (Brown) 7
PM-2A 195, 197
Popov, Blagoj 175
Popovski, Živko 179
population 2, 7, 8, 11, 12, 14, 15, 18–20, *21*, 25, 26, 29, 56, 57, 60, 61, 65, 70–72, 84, 86, 89, 92–94, 136, 141, 142, 168, 170, 176, 197, 203, 204, 214, 220, 223, 225, 226, 230, 237, 238, 260, 263, 264, 271, 285, 287, 288, 292, 301
prefabrication process *176*, 176–177
Project Iceworm 196
propagandista 264
public protest 221, 222, 225–258, 230, 231
Pugh O'Mara, M.: *Cities of Knowledge: Cold War Science and the Search for the Next Silicon Valley* 6–7

Raadi airbase 148–149, *152, 161,* 164n5; military personnel 155; Raadi Manor 150–152, 158–160, 162; Soviet occupation 152–155, *153*; Tartu, Estonia 149–150, 155–157, *156*; urbanisation 157–160, 162
radial march 228–232
Radio-Activity (Warning Organisation) 73
RAND Corporation 5, 12
Red Army 83, 128, 131–133
Red Bologna 256, 273, 274; *see also* Bologna
"Red Emilia" 256, 260
Regional Commissioner 61, 70, 71
Regional Industrial Development Corporation 39
regional seats of government (RSG) 18
Reiko Tomita 106
Rescue Training Ground 59, *59, 60*
Research and Development Center 47–50, *48*
Rotor programme 66

Royal Observer Corps (ROC) 66, 67, *67,* 73
Rozanski, Stanislaw 130
Russell, Dora 249
Rybar, Ctibor 143

Salerno, Joseph P. 295
Sarnoff, David 249
Scalmer, S. 226
Schelling, Thomas 139
Schoharie Valley Townsite project 187, 188
Schregel, S. 220
Second World War 1, 3, 7, 14–16, 18–20, 24, 26, 78, 81–83, 86, 150–152, 158, 167, 170, 236, 265, 287, 290, 309
Secret State: Preparing For The Worst 1945–2010, The 6
Sector Operations Centre (SOC) 68
shinkenchiku 100–102
Shirai Sei'ichi 101, 102, 111n18, 111n19
Shute, Nevil: *On The Beach* 222
Siam Intercontinental 294–297, *295*
Sillamäe 77, 78, *79*, 81, *81*, 84–89, *85–88*, 91–94, 155
Siple, Paul 188
Skidmore, Owings and Merrill design 15, 48
SKM metro system 133, 134
Skopje 3, 27, 167–170; architecture and urban planning 170–172; East and the West 178–180; post-1963 172–175; solidarity 175–177
Slagelse *206*, *210*
Socialist Realism 78, 82, 91, 93, 141
Songkhram, Phibun 287, 290, 294
Sontag, Susan 104, 108
Sosnowski, Oskar 130, 131
Soubry, M. A. 136
Southern European Task Force (SETAF) 236–242, 244, 246–248, 251
South Head Peace Committee 226
Soviet Communist Party 82
Soviet occupation 81, 85, 93, 148, 150–155, *153*, 157, 160
Soviet Socialist Republic (SSR) 78, 80
Soviet Union 2, 5, 14, 23, 24, 55, 77, 78, 81, 83–87, 92–94, 98, 115, 116, 129, 139, 140, 143, 154, 155, 157, 167, 168, 177, 185, 189, 190, 196, 201, 224, 232, 257, 287
Sputnik launch 8, **8**, 24, 29, 140, 249
Stalingrad (Volgograd) 82–83

Index

Stalinist planning 77–81; Estonia 83–84, 91–93; Kohtla-Järve 89–91; principles of 81–83; Sillamäe 85–89
Staniszkis, Magdalena 131
Stanzani, D. 272
Stare Miasto (the Old City) *135*, 135–137
Starostin, I. 80
Starzynski, Stefan 130, 131, 144
Steinbach, J. 222
Stockholm Appeal 259–261, 263, 265, 266, 272
Strath Report 72–73
Student Dormitory 'Goce Delčev' (Konstantinovski) 179, *179*
Stupar, A. 17
sub-controls 70–72
subsurface urbanism 185, 188
Sultson, S. 81
Sutherland Shire District Peace Committee 227
Sydney 219–222, 231–232; civil defence 223–225; CND 226, 228, 229, 231, 232; nuclear attack 222–225; peace movement 219–221, 223–227, 229–232; public protest 225–228; radial march 228–230; youth protest and public dissent 230–231
Syrkus, Szymon 130
Szczepanski, Jan 126

Tange, Kenzō 99–107, 109, 111n20, 173–175, *174*
Tartu 25, 84, 148–160, *156*, 162, 163n1
telecommunications centre (Konstantinovski) 179
telephone exchange 301–303, *304–309*
Television Act of 1954 310
Temporary Office Blocks (TOB) 60, 61
Territorial Army 68
Thailand 281, 282, 284–289, 293, 296, 297; consumerism and urban development 290–293; and United States 287–290, 296–297; *see also* Bangkok
Third Front Construction 26, *115*, 116, 117, 120, *121*
Thule Air Base 185, 189, 191, 192
Tihomirov, V. 80
Tippel, V. 80, 90
Tito, Josip Broz 170, 171
Tito–Stalin split (Yugoslav–Soviet split) 170
Togliatti, Palmiro 249, 272

Tokyo 98, 101, 105, 106, 110n2
Tokyo Bay Plan 105–107
Trotsky, Leon 128, 142
Truman, Henry 259, 261, 265
Tsar Bomba 138–139

undercut trench concept 192, *193*
Union of Italian Women (UDI) 258, 259, 261, 263, 272, 273
Union of Soviet Socialist Republics (USSR) 8, 82, 83, 85, 86, 89–91, 125, 137, 139, 140, 142, 148–150, 152–156, 158, 162, 176, 177, 257, 261, 265
United Kingdom Warning and Monitoring Organisation 73
United States Information Agency (USIA) 285, 287, 297n2
United States Information Service (USIS) 281, 285, 288, 297n2
Urban History 4–5
urbanisation 157–160, 162
urbanism 4–7, 115–121
US Army 182, 185, 191, 192, *193*, 194–196, 243
U.S. Distant Early Warning (DEW) Line 24

Vale, L. J. 14
Vanderbilt, T. 188
Verona 236–237; DC party 237, 240–242, 248–250, *250*; dystopia 251–253; NATO and US 237–240; Passalacqua 238, 243–246, *245*, 246, 248, 250–252; population 237, 238; recreation and consumerism 246–251; schooling 240–246; SETAF 236–242, 244, 246–248, 251
Vistula river 129, 131, 133, *134*, 136
Volkov, L. 80
Vseviov, D. 81, 93

Wager, W. H. 196, 197
Walesa, Lech 143
wall newspaper 269
Warfare State (Edgerton) 6
Warsaw Pact and metro 125–126, 144, 201, 236; archives, reconstruction and the communist 131–133; armed camp 126–127; camps notion 140–141; hiatus 137–138; metro 133–135, 141–142; Old City and the Palace of Culture and Science *135*, 135–137, *136*; Plac Wilsona

metro station 143–144, *144*; postwar reconstruction, railways and connectivity 128–130; thaw and deterrence 139–140; trains, 'Homo Sovieticus' and pistols 128; tunnels and Tsar Bomba 138–139; tunnels and war 130–131; Warschau 131
Warschau 131
Weisman, Ernest 178
Weissmann, Ernest 172
Westad, O.A. 6
Westinghouse 27, 28, 37–39, 43–50
Wharton, A.J. 282–283
Wiener, N. 12

Williamson, G. 190
Wills, John 125
Wilson, Harold 310

Yatsuka, Hajime 100, 104, 106, 112n29
Yayoi 100–104, 111n20
youth protest and public dissent 230–231
Yuasa Noriaki 97, 108, 109
Yugoslavia 26–27, 167, 168, 170–172, 175–180

Zanotto, Giorgio 237
Zholtovskij (Zholtovsky), I. V. 82, 92
Zipp, S. 5

Taylor & Francis eBooks

www.taylorfrancis.com

A single destination for eBooks from Taylor & Francis with increased functionality and an improved user experience to meet the needs of our customers.

90,000+ eBooks of award-winning academic content in Humanities, Social Science, Science, Technology, Engineering, and Medical written by a global network of editors and authors.

TAYLOR & FRANCIS EBOOKS OFFERS:

A streamlined experience for our library customers

A single point of discovery for all of our eBook content

Improved search and discovery of content at both book and chapter level

REQUEST A FREE TRIAL
support@taylorfrancis.com